Energy Engineering and Management

Energy Engineering and Management

Edited by George Thomson

SYRAWOOD
PUBLISHING HOUSE
New York

Published by Syrawood Publishing House,
750 Third Avenue, 9th Floor,
New York, NY 10017, USA
www.syrawoodpublishinghouse.com

Energy Engineering and Management
Edited by George Thomson

International Standard Book Number: 978-1-68286-492-0 (Hardback)

Cataloging-in-Publication Data

Energy engineering and management / edited by George Thomson.
p. cm.
Includes bibliographical references and index.
ISBN 978-1-68286-492-0
1. Power (Mechanics). 2. Power resources. 3. Renewable energy sources. 4. Energy consumption.
5. Energy conservation. 6. Electrical engineering. I. Thomson, George.
TJ163.9.E54 2017
621.042--dc23

Printed in China.

TABLE OF CONTENTS

PREFACE

Energy engineering and energy management are two vast interrelated areas of study. While energy engineering deals with aspects such as alternative engineering, energy efficiency, etc.; energy management on the other hand is concerned with the consumption, conservation as well as production of energy. This book elucidates the concepts and innovative models around prospective developments with respect to energy engineering and management. It presents researches and studies performed by experts across the globe. In this book, using case studies and examples, constant effort has been made to make the understanding of the difficult concepts of these fields as informative as possible, for the readers.

This book is a comprehensive compilation of works of different researchers from varied parts of the world. It includes valuable experiences of the researchers with the sole objective of providing the readers (learners) with a proper knowledge of the concerned field. This book will be beneficial in evoking inspiration and enhancing the knowledge of the interested readers.

In the end, I would like to extend my heartiest thanks to the authors who worked with great determination on their chapters. I also appreciate the publisher's support in the course of the book. I would also like to deeply acknowledge my family who stood by me as a source of inspiration during the project.

Editor

1

Hydrotreating reactions of tall oils over commercial NiMo catalyst

Jinto M. Anthonykutty[1], Juha Linnekoski[1], Ali Harlin[1] & Juha Lehtonen[2]

[1]VTT Technical Research Centre of Finland, Biologinkuja 7, Espoo, FI-02044 VTT, Finland
[2]Department of Biotechnology and Chemical Technology, School of Chemical Technology, Aalto University, PO Box 16100, Aalto FI-00076, Finland

Keywords
Commercial NiMo catalyst, decarboxylation, hydrogenation, hydrotreating, tall oil

Correspondence
Jinto M. Anthonykutty, VTT Technical Research Centre of Finland, Biologinkuja 7, Espoo, FI-02044 VTT, Finland.

E-mail: jinto.manjalyanthonykutty@vtt.fi

Funding Information
Jinto Manjaly Anthonykutty acknowledges the financial support from VTT Graduate School. The authors also acknowledge Stora Enso for supporting this research and their vision on wood-based olefins.

Abstract

Catalytic hydrotreating is an attractive method for upgrading bio-derived oils into renewable feedstocks with less oxygen content, suitable for producing valuable hydrocarbons through various petro-refinery processes. This study evaluates the catalytic activity of a commercial alumina (Al_2O_3) supported NiMo catalyst for hydrotreating tall oil feeds such as crude tall oil (CTO), distilled tall oil (DTO), and tall oil fatty acid (TOFA). Catalytic experiments carried out in a bench-scale fixed bed reactor set-up at different process conditions [space velocity (1–3 h^{-1}), temperature (325–450°C), and H_2 pressure (5 MPa)] produced a wide-range of products from tall oil feeds. Hydrotreating of TOFA produced highest yield of *n*-alkanes (>80 wt%) compared to DTO and CTO hydrotreating. A high conversion of fatty acids and resin acids was obtained in DTO hydrotreating. In CTO hydrotreating, a drop in conversion of fatty acids and resin acids was observed especially at the lowest temperature tested (325°C). The study revealed that there are various deoxygenation pathways preferential at different hydrotreating temperatures. As an example for TOFA, the decarboxylation route is dominant over the hydrodeoxygenation route at high temperatures (>400°C).

Introduction

Catalytic hydrotreating is an essential petro-refinery method to reduce the content of heteroatoms (S, N, and O) from raw materials, which has been widely employed to upgrade refinery feeds prior to such processes as catalytic reforming, catalytic cracking, and steam cracking [1]. In petro-refineries that use sulfur-rich fossil-fuel-based feeds, the catalytic hydrotreating technology is fully developed specifically for the hydrodesulfurization (HDS) of petroleum products and naphtha streams [2–4]. However, the recent research developments prompted by the increasing demand of bio-based fuels and chemicals reinvent the importance of developing optimal hydrotreating (hydrodeoxygenation, HDO) methods for reducing the amount of oxygenates from complex bio-derived feeds before the thermochemical conversion process into fuels or chemicals.

The amount of oxygenates in bio-derived oils vary depending on the origin of raw materials. Bio-oils produced by pyrolysis of lignocellulosic biomass are rich in oxygenates (35–40%). Therefore, deep hydrotreatment conditions are required to achieve complete deoxygenation from such bio-oils, inducing high processing cost [5]. Plant-based oils or vegetable oils are another class of bio-derived feedstock. Vegetable oils have been extensively studied as a feedstock for the production of bio-chemicals and biofuels through a catalytic HDO route [6–9]. The patented bio-synfining process produces naphtha range hydrocarbons for conventional steam cracking process from vegetable oils and fats [10, 11]. However, most of the widely studied vegetable oils are edible and remain as less favorable feedstocks as their selection to a refinery may create several environmental, economic, and societal issues. Therefore, a proper selection of a sustainable feedstock is vital. Tall oil, a by-product of the Kraft-pulping process, is nonfood chain affecting, economically feasible and low oxygen content feedstock [12]. Tall oil upgrading has already been received substantial attention as a sustainable bio-refinery method for producing biofuels and refinery feeds [12–14]. Tall oil comprises fatty acids, resin acids, and sterols. Different tall oil feeds such as crude tall oil (CTO), distilled tall oil (DTO), and tall oil fatty acid (TOFA) are commercially available, and can be used as valuable raw materials for chemical production. CTO is produced by means of the acidulation of tall oil soap skimmings from the black liquor which is obtained from pulping. DTO and TOFA are obtained from CTO distillation along with other fractions. Among the aforementioned tall oil feeds, CTO is regarded as the most cost-competitive raw material [15].

Sulfided molybdenum catalysts on alumina supports with nickel or cobalt as promoter metals are the most widely employed industrial catalysts for hydrotreating [16]. Supported CoMo and NiMo catalysts are more active in sulfided forms. There have been many studies reported on the origin of catalytic synergy between two main group elements in hydrotreating catalysts [17]. Different theories on the origin of catalytic activity and the nature of active sites can be found in literature [17, 18]. Kubicka et al. [19] reports that during hydroprocessing of vegetable oils over a sulfided NiMo catalyst, nickel sulfide phases are more active for decarboxylation than molybdenum sulfide phases. Therefore, the extent of HDO (hydrogenation/dehydration) versus decarboxylation reactions is highly dependent on factors such as promotor metal (Ni) concentrations and the dispersion of sulfidic phases. It is already learnt from previous studies that in comparison with CoMo catalysts, NiMo catalysts are more active for the HDO of aliphatic oxygenates [20]. Moreover, NiMo catalysts are found to be effective for hydrodeoxygenating cyclic oxygenates to cycloalkanes and also for causing ring opening reactions from cyclics at high temperatures, which produces valuable hydrocarbons for further refinery processes [12, 21]. Based on these observations, it can be suggested that NiMo catalyst is more applicable than a CoMo catalyst for hydrodeoxygenating bio-derived oils such as tall oil which contains aliphatic oxygenates as well as cyclic oxygenates. During catalytic hydrotreating of tall oil, oxygenates are removed by various deoxygenation routes which produces a wide range of hydrocarbons mainly paraffin range hydrocarbons [22, 23].

Among tall oil fractions, catalytic deoxygenation of TOFA has been studied with considerable interest for producing a hydrocarbon fraction suitable as a diesel fuel [24]. Egeberg et al. [2] reports that hydroconversion of TOFA can be slightly different than that of triglycerides (vegetable oils) as the former produces more methane by means of methanation reaction, whereas the latter produce more propane by the scission of glycerol backbone. Catalytic deoxygenation of TOFA investigated by other researchers reveal that a significant yield of *n*-heptadecane (C_{17}) with high selectivity can be obtained from TOFA using a palladium mesoporous carbon catalyst [24]. Low temperatures are found to be more favorable for the formation of *n*-octadecane from fatty acids through a HDO route. Hydrotreating of CTO has been discussed in the literature through several patents which mainly focuses the application of hydrotreated products as a diesel fuel [25–27]. Previously, we successfully reported on the use of hydrotreated DTO and CTO as a steam cracker feed [22, 23]. A high degree of deoxygenation was obtained at low temperatures (325–400°C) and space velocity of weight hourly space velocity (WHSV) = 1 h^{-1}, yielding a maximum in paraffins. Our group has also studied the hydrotreating chemistry of sterols in CTO and proposed a reaction scheme for the HDO of sterols at low temperatures in line with studies on the HDO of phenolic compounds and saturated alicyclic alcohols over sulfided catalysts [23]. Importantly, in our study, a high conversion of sterols was obtained at the tested conditions irrespective of space time and temperature.

In this study, a comparative study is presented using different tall oil feedstocks (TOFA, DTO, and CTO) The effect of space time and temperature on the hydrotreating of DTO and CTO is already been reported [22, 23], therefore, they are not discussed further in this study. More specifically, this study presents the results of TOFA hydrotreating under the applied conditions, which are then compared with the results of DTO and CTO hydrotreating. This comparison has carried out on the basis of achieved product distribution, composition of organic phase samples and the conversion of acid fractions at the most favorable space time (WHSV = 1 h^{-1}). The influence of feedstock type and reaction severity (temperature) during hydrotreating of tall oils (6 h of time on stream) over a sulfided NiMo catalyst is thus studied in this article.

Materials and Methods

Commercially available TOFA (SYLFAT® 2) and DTO (SYLVATAL® 25/30 S); and CTO obtained from Stora Enso pulping facilities in Finland were used as feeds for hydrotreating experiments. Detailed chemical and elemental composition and total acid number of the employed feeds are represented in Tables 1 and 2. A commercial alumina (Al_2O_3) supported NiMo catalyst in sulfided form was employed for hydrotreating studies. The presulfidation was carried out by using a H_2S/H_2 mixture for 5 h at 400°C (H_2S/H_2 = 5 vol. %). The test runs (6 h) with each tall oil feed were performed in a continuous down flow fixed bed reactor (stainless steel tube, length: 450 mm long, internal diameter: 15 mm) at different process conditions; temperature = 325–450°C, H_2 partial pressure = 5 MPa, weight hourly space velocity (whsv) = 1–3 h^{-1} and H_2/feedstock molar ratio = 17.4. Duplicate experiments were carried out for certain test runs so as to check the reproducibility of yield. In a typical experiment, a known amount of the catalyst (6, 3, and 2 g corresponds to WHSV 1 h^{-1}, WHSV 2 h^{-1}, and WHSV 3 h^{-1}) was loaded in the reactor which was then placed in an oven (Oy Meyer Vastus, Monninkylä, Finland). The catalyst bed temperature was monitored by a temperature controller (TTM-339 series). A set amount of feedstock was fed to the reactor along with hydrogen (62 mol/kg of CTO). TOFA and DTO were used as received for hydrotreating experiments, whereas CTO was preheated in order to reduce the viscosity, and thus to enhance the feeding of the material to the reactor. Gas chromatographic tools such as such as GC-MS (HP-5 MS column), GC×GC-MS (ZB-5 HT inferno and ZB-35 HT inferno columns)and GC×GC-FID/(TOF-MS) (Rtx-1 PONA and BPX-50 columns) were used for quantitatively (using external calibration method) and qualitatively analyzing the organic phase samples obtained from hydrotreating of tall oils [22, 23]. GC-FID (DB-23 and Zebron ZB-1 columns) analyses were carried out to calculate the residual amounts of fatty acids, resin acids, and sterols from organic phase samples. The amount (wt%) of water in aqueous phase samples was determined by Karl Fischer (KF) titration (method: ASTM E 203). Analysis of gas phase components was carried out by FT-IR and micro-GC instruments. The residual oxygen composition of organic phase samples from hydrotreated tall oils was determined by elemental analysis using CHN equipment based on ASTM D 5291 method. Sulfur content was determined by means of ASTM D 4239 method.

Terminology and calculations

For the sake of clarity, the term paraffins in this study correspond to *n*-alkanes and *i*-alkanes. Nonaromatics denote cycloalkanes or naphthenes (-mono, -di, or -tri) and other cyclic oxygenates as well as fatty alcohols and unsaturated hydrocarbons. The reactions which mention in this study under hydrotreating conditions are hydrogenation of

Table 1. Feedstock composition.

Composition (wt.%)	TOFA	DTO	CTO	
Free and bonded Fatty Acids	**92.6**	**71.3**	**48.1**	
			Free	Bonded
(16:0) Palmitic acid	0.40	0.20	2.1	0.04
(17:0) Margaric acid	0.80	0.30	0.4	0.01
(18:0) Stearic acid	0.90	0.70	0.8	0.01
(18:1) Oleic acid	27.4	15.3	8.6	0.16
(18:1) 11-octadecenoic acid	0.70	0.50	0.3	0.01
(18:2) 5.9-octadecadienoic acid	0.50	0.30	0.2	0.02
(18:2) conj. octadecadienoic acid	6.00	8.30	5.5	0.36
(18:2) Linoleic acid	23.1	24.3	19.1	2.4
(18:3) Pinolenic acid	8.00	4.4	1.6	0.5
(18:3) Linolenic acid	4.30	0.60	0.6	0.02
(18:3) conj. octadecatrienoic acid	0.20	1.80	0.4	0
(20:0) Arachidic acid	0.80	0.40	0.7	0.01
(20:3) 5.11.14-eicosatrienoic acid	7.90	7.60	1.4	0.08
(20:3) 7.11.14-eicosatrienoic acid	1.30	0.60	0.2	0.2
Other fatty acids	10.3	6.00	1.8	0.28
Free and bonded Resin Acids	**1.3**	**23**	**28.5**	
			Free	Bonded
8,15-isopimaradien-18-oic acid	0.1	0.5		
Pimaric acid	0.7	4.8	2.6	0.03
Sandaracopimaric acid	–	0.3	0.8	0.03
Diabietic acid	–	0.5	–	–
Palustric acid	–	2.2	3.9	0.01
Isopimaric acid	–	1.1	1.3	0.01
7. 9 (11) -abietic acid	–	0.4	0.1	0
13-B-7.9(11)-abietic acid	–	0.3	0.2	0.02
Abietic acid	0.1	7.7	9.1	0.1
Dehydroabietic acid	–	3.6	2.4	0.04
Neoabietic acid	–	0.4	3.3	0.02
Other resin acids	0.4	1.3	2.8	0.7
Neutrals			**23**	
Campesterol	–	–	0.02	–
Campestanol	–	–	0.17	–
β-Sitosterol	–	–	2.5	–
β-Sitostanol	–	–	0.5	–
Lupeol	–	–	0.5	–
Cycloartenol	–	–	0.2	–
24-methylenecycloartenol	–	–	2.5	–
Prenol-7	–	–	0.07	–
α-Sitosterol	–	–	1.1	–
Methyl betulinate	–	–	0.02	–
Betulin	–	–	0.3	–
Betulinic acid	–	–	0.5	–
Prenol-8	–	–	0.9	–
Other	–	–	12.6	–

Table 2. Elemental composition total acid number (TAN).

Composition	TOFA	DTO	CTO
Elemental composition			
C (wt%)	76.8	77.4	78.6
H (wt%)	11.6	11.1	11.2
O (wt%)	11.6	11.4	9.9
N (ppm)	12	30	100
S (ppm)	53	161	1800
P (ppm)	–	–	36
Alkali metals and alkaline earth metals (ppm)	–	–	56
Total acid number (TAN) mg/kg	326	300	129.5

double bonds, decarboxylation, decarbonylation, HDO, isomerization, hydrocracking of alkane and cyclic structures, and thermal cracking reactions. HDO is represented as a reaction resulted from complete hydrogenation with saturation of double bonds as the first step. Selective deoxygenation denotes the catalytic deoxygenation reactions at low temperatures (Mainly HDO, removal of oxygen as water). Nonselective deoxygenation represents the deoxygenation reactions which occur at high temperatures (thermal or catalytic cracking, removal of oxygen as CO_2).

The mass based yield (%) of various product streams in this study is calculated as a relative yield based on the amount (g/h) of feedstock entering to the reactor. Conversion of fatty acids and resin acids was calculated by assuming that residual acid fractions are mostly present in the organic phase. The equation used for calculating the conversion is

$$\text{Conversion (\%)} = \frac{n_A,\text{Feed} - n_A,\text{O.P.}}{n_A,\text{Feed}} \times 100,$$

n_A, Feed, is the total mol of acids (fatty acid/resin acid) present in the feed, n_A, O.P., is the total mol of acids (fatty acid/resin acid) present in the organic phase.

Product yield of paraffins (*n*-alkanes + *i*-alkanes) was calculated by assuming that paraffins are solely formed from fatty acid fraction in tall oils. Product yield was calculated by using the following equation.

$$\text{Product yield (\%)} = \left[\frac{\text{Total mol of paraffins in O.P}}{\text{Total mol of fatty acids in Feed}}\right] \times 100$$

The reaction steps discussed in this study in association with the hydrogenation and dexoygenation of fatty acids (linoleic acid and oleic acid) in tall oils over a sulfided NiMo catalyst are shown below by Equations (1), (2), (3), (4), (5), and (6)

Hydrogenation:

$$C_{17}H_{31}COOH + H_2 \rightarrow C_{17}H_{33}COOH \quad (1)$$

$$C_{17}H_{33}COOH + H_2 \rightarrow C_{17}H_{35}COOH \quad (2)$$

$$C_{17}H_{31}COOH + 2H_2 \rightarrow C_{17}H_{35}COOH \quad (3)$$

Deoxygenation:

$$C_{17}H_{35}COOH + 3H_2 \rightarrow n\text{-}C_{18}H_{38} + 2H_2O \quad (4)$$

$$C_{17}H_{35}COOH + H_2 \rightarrow n\text{-}C_{17}H_{36} + CO + H_2O \quad (5)$$

$$C_{17}H_{35}COOH \rightarrow n\text{-}C_{17}H_{36} + CO_2 \quad (6)$$

Experimental Results

Hydrotreating of TOFA

Hydrotreating experiments with TOFA on a NiMo catalyst resulted in liquid product streams which differ in color and yield depending on the conditions in the reactor. Importantly, the samples from TOFA hydrotreating at low temperatures (<350°C) appeared as oily samples with the tendency of solidification at room temperature. It was also observed that the temperature, 300°C, tested with TOFA resulted the excessive deposition of wax; thus, reactor plugging. In line with earlier studies [22, 23], mass balance error (±5%) was higher in experiments at elevated temperatures (>400°C) in comparison with low-temperature experiments. Liquid products obtained from TOFA hydrotreating were separated into organic and aqueous phases. Table 3 shows the detailed product distribution in organic phases obtained from TOFA hydrotreating at various process conditions. As hydrotreating of TOFA produces more paraffins than other tall oil feeds, the product distribution of major paraffinic hydrocarbons such as *n*-octadecane and *n*-heptadecane obtained in TOFA hydrotreating at different process conditions is presented in this study as a separate section.

Product distribution of C_{17} and C_{18} hydrocarbons in TOFA hydrotreating

The amount of *n*-octadecane and *n*-heptadecane obtained in TOFA hydrotreating on a sulfided NiMo catalyst is shown in Table 3. It is clear from Table 3 that a significant drop in the production *n*-octadecane occurred with TOFA from 325 to 450°C especially at the longest space time tested (WHSV = 1 h^{-1}), whereas *n*-heptadecane production steadily increased until 400°C and then sharply decreased at 450°C. As already noted from earlier studies [22, 23], low temperatures (<400°C) are more favorable for the formation of *n*-octadecane through a HDO route. As the temperature increases, routes consuming less hydrogen such as decarboxylation and decarbonylation are

Table 3. Product distribution of the organic phase obtained from the hydrotreating of TOFA at different process conditions.

Temp (°C)	325	350	400	450	325	350	400	450	325	350	400	450
WHSV (h^{-1})	1	1	1	1	2	2	2	2	3	3	3	3
Composition (wt%)												
Paraffins												
nC_7-C_9	0.4	1.2	6.8	10.1	0.9	1.8	4.4	8.3	0.2	0.8	3.1	9.2
nC_{10}-C_{16}	1.4	2.2	4.3	8.6	4.2	2.9	3	14.1	2.8	1.9	4.4	8.3
nC_{17}	27.7	36	36	27	24	31	31	20	21.5	29	30	23
nC_{18}	53	44	32	22	49.8	49	35	29	43	41	33	33
nC_{19}	1.2	1	1.3	2.1	3.8	2.4	0.9	1.1	0.8	0.7	0.6	0.4
nC_{20}	2	0.8	1.4	1.5	2.4	2.1	0.2	0.9	2.9	1.5	1.4	0.2
*i*alkanes	2.1	2.5	4.1	6	1.4	1.7	3.4	3	1.4	2.1	2.7	2.9
Total	87.6	87.7	86	78.1	86.5	90.9	77.9	76.4	72.6	77	75.2	77
Nonaromatics												
Cycloalkanes	0.08	0.2	1.4	3	0.9	0.4	0.4	2.5	0	0.03	0.9	0.3
Olefins	0.07	0.4	0	0	0.08	0.03	0.01	0.6	0.1	0.2	0.5	0.6
Alcohols	2.3	3.5	2.7	0.3	0.4	0.4	1.8	1	0.3	1	1	0.9
Other	9.5	7.9	4.9	15.8	5.1	7	9	9.1	13.9	5.9	6.8	6
Total	11.9	12	13	19.1	6.4	7.83	11.2	13.2	14.3	7.13	9.2	7.8
Aromatics												
Monoaromatics	0	0	1.7	4.1	0.07	0.04	0.9	1.4	0	0.04	0.8	1.2
Overall wt%	99.5	99.7	96.7	91.3	93	98.7	90	90	86.9	84	85.2	86
Residual oxygen	0.8	0.8	0.46	0.4	1.3	0.9	0.7	0.9	1.6	1.9	1.1	1

prominent, producing a significant amount of *n*-heptadecane at these temperatures. Higher concentration of *n*-heptadecane over *n*-octadecane was observed beyond 350°C at the longest space time tested (WHSV = 1 h^{-1}). This observation is in agreement with the earlier report on the promotion of decarboxylation route over HDO as a function of increasing temperature on a sulfided NiMo catalyst [28]. It may also imply that decarboxylation + decarbonylation reactions predominate over HDO reaction solely at longer space times, that is, WHSV = 1 h^{-1} with TOFA. Importantly, in comparison with hydroprocessing of vegetable oils such as rapeseed oil and sunflower oil over a NiMo catalyst [29], a similar yield of *n*-alkanes from TOFA hydrotreating at lower temperature (350°C) was achieved, especially for the n-C_{17} + n-C_{18} fraction. It is noteworthy that the amount of oleic acid ($C_{18:1}$) and linoleic acid ($C_{18:2}$) is considerably lower in TOFA compared to the aforementioned vegetable oils. Interestingly, in TOFA hydrotreating, irrespective of space time a similar product distribution of *n*-heptadecane and *n*-octadecane was observed at 400°C.

Comparison of TOFA, DTO, and CTO hydrotreating

Overall product distribution

Figure 1 shows the yield of various product streams obtained as a function of temperature at WHSV = 1 h^{-1} in hydrotreating of tall oils. As noted from Figure 1, among the three tall oil feeds tested, TOFA was found to give the highest yield of liquid products and organic products at different temperatures. It was also observed with the employed tall oil feedstocks that the lowest temperature tested (T = 325°C) resulted to a maximum yield of liquid products. As expected, maximum gas product yield was obtained at 450°C with each tall oil feeds. At this temperature, the highest yield of gas products was obtained from CTO in comparison with DTO and TOFA. The FT-IR detected gas phase components were CO_2, CO, methane, ethane, ethylene, propane, propylene, and butane. Importantly, lighter alkanes and alkenes were obtained solely at high temperatures especially from CTO and DTO. As can be seen from Figure 1, among three tall oil feedstocks the highest yield of water was obtained from TOFA. The trend with the production of water from tall oils followed the decreasing order as TOFA–DTO–CTO. Water is obtained as a by-product of HDO (hydrogenation/dehydration) reaction during hydrotreating.

Composition of organic phase

Figure 2 shows a comparison of the product composition of organic phase samples as a function of temperature obtained in TOFA, DTO, and CTO hydrotreating at WHSV 1 h^{-1}. As can be seen from Figure 2, maximum amount (wt%) of paraffins was obtained from TOFA among three tall oil feeds. This observation appear as an

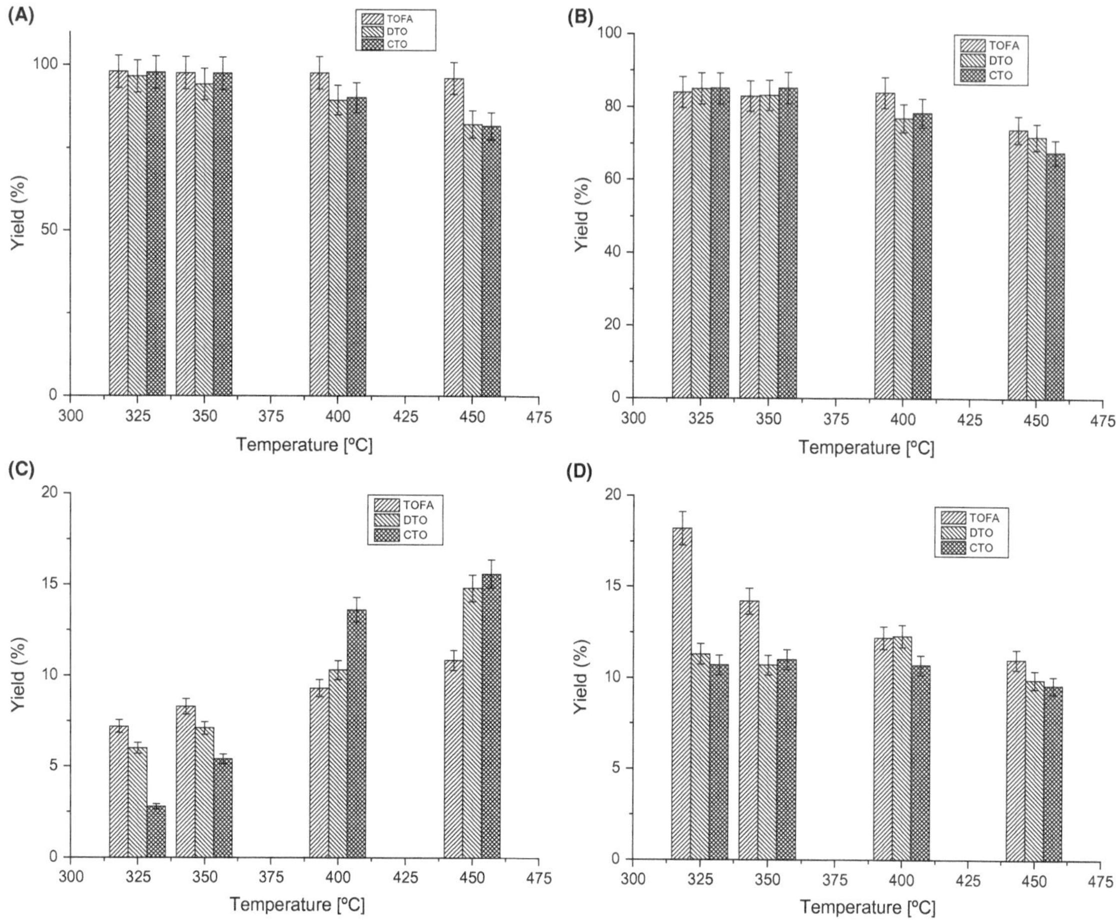

Figure 1. Yield (%) of various product streams obtained in hydrotreating of tall oils over a NiMo catalyst, WHSV = 1 h^{-1}, T = 325–450°C (A) liquid product yield, (B) organic phase product yield, (C) gas product yield, and (D) water yield.

obvious result in relation to the higher amount (wt%) of fatty acids in TOFA.

Product distribution of aromatic and nonaromatic hydrocarbons

As can be seen from Figure 2, nonaromatics such as cyclic structures (monocyclic and polycyclic hydrocarbons), alcohols, unsaturated hydrocarbons, and sterol derivatives were obtained in higher amount with CTO in comparison with other tall oil feeds. Among the three tall oil feeds, maximum production of aromatics was obtained from DTO. It has been reported previously that the aromatics from DTO hydrotreating comprised primarily of monoaromatics such as substituted benzenes in addition to polyaromatics [22]. Organic phases obtained from three feedstocks especially from CTO also comprised some sulfur (100–500 ppm) and nitrogen (10–20 ppm) [23].

Fatty acid and resin acid conversion

Fatty and resin acid conversions were evaluated based on GC-FID analysis. Figure 3 shows the fatty and resin acid conversions obtained with different tall oil feedstocks at the most favorable space time (WHSV = 1 h^{-1}) as a function of temperatures. Figure 3A shows that fatty acid conversion achieved in a high extent with TOFA and DTO. A similar conversion level achieved with fatty acids in TOFA and DTO in low-temperature hydrotreating experiments (325–400°C). In comparison with TOFA and DTO, a drop in conversion of fatty acids was obtained in CTO hydrotreating under the tested conditions. Figure 3B shows that elevated conversion level was reached for resin

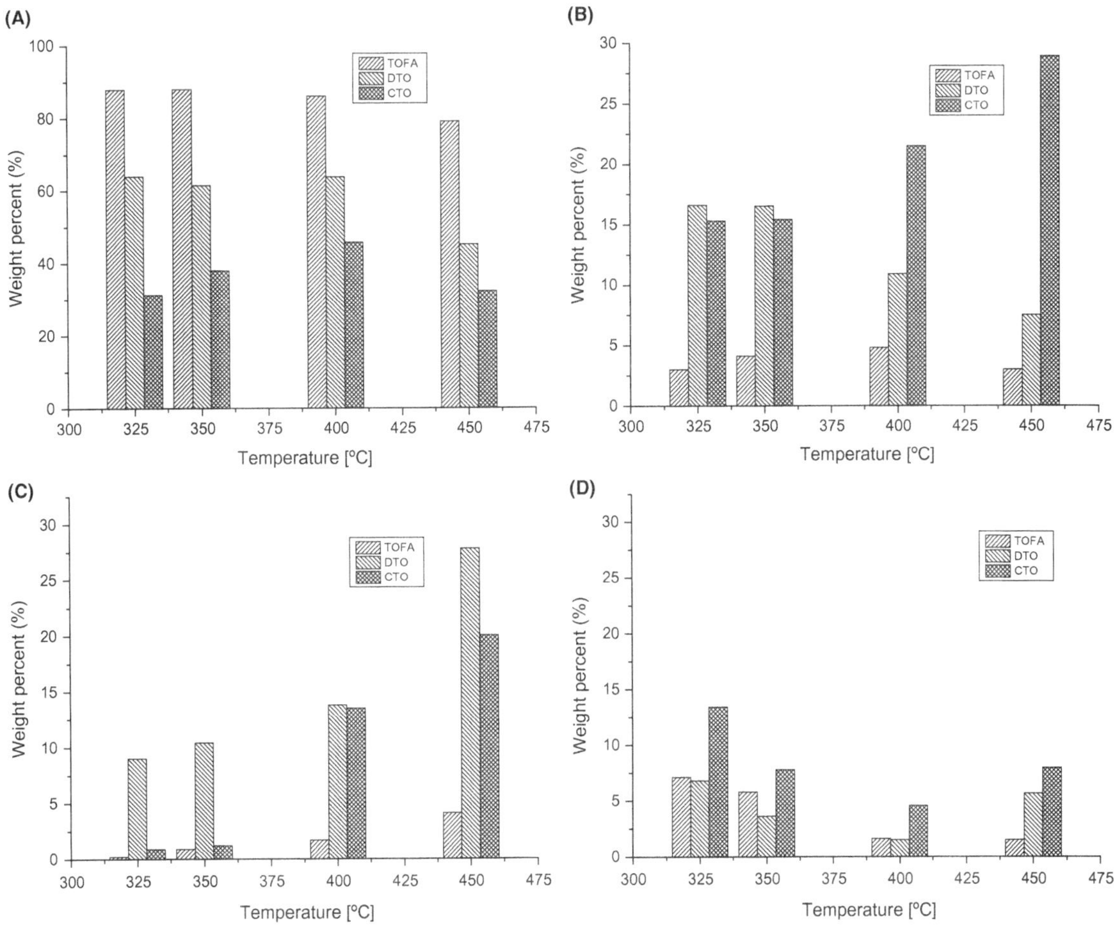

Figure 2. Product composition (wt%) of organic phase samples in tall oil hydrotreating (A) paraffins,(B) nonaromatics, (C) aromatics, and (D) residual fractions; WHSV =1 h^{-1}, T = 325–450°C.

acids as well in DTO hydrotreating. In CTO hydrotreating, a low conversion level of resin acids was obtained in line with the conversion of fatty acids from the same feedstock.

Furthermore, in order to get more insight into the behavior of fatty acids and resin acids during hydrotreating of tall oils, the conversion of individual fatty acids and resin acids in each tall oil feedstock was assessed. Figure 4A–I shows the conversion of linoleic acid, oleic acid, and dehydroabietic acid obtained during the hydrotreating of TOFA, DTO, and CTO at different space times (WHSV = 1–3 h^{-1}) as a function of temperature. It should be stated that all fatty acids and resin acids other than the aforementioned ones have achieved more or less full conversion levels during the hydrotreating of tall oils under the tested conditions; therefore, they are not included in the results.

Discussion

Based on distribution of gaseous products obtained from tall oil hydrotreating experiments, it can be assumed that in TOFA and DTO hydrotreating at high temperatures (>400°C), mainly nonselective deoxygenation occurs with a sulfided NiMo catalyst by means of cracking (thermal or catalytic) [12, 28] which requires no hydrogen and produces short-chain oxygenates and CO_2 as well as hydrocarbons (cycloalkanes and aromatics) [30]. With CTO, it is suggested that nonselective deoxygenation can be lower at high temperatures (>400°C) from sterols [23] although sterol hydrotreating chemistry at high temperature has not been elucidated yet. The gas product composition with high concentration (~60% yield of total gaseous products) of lighter alkanes and alkenes from CTO hydrotreating may imply that sterols in CTO are mainly converted by thermal

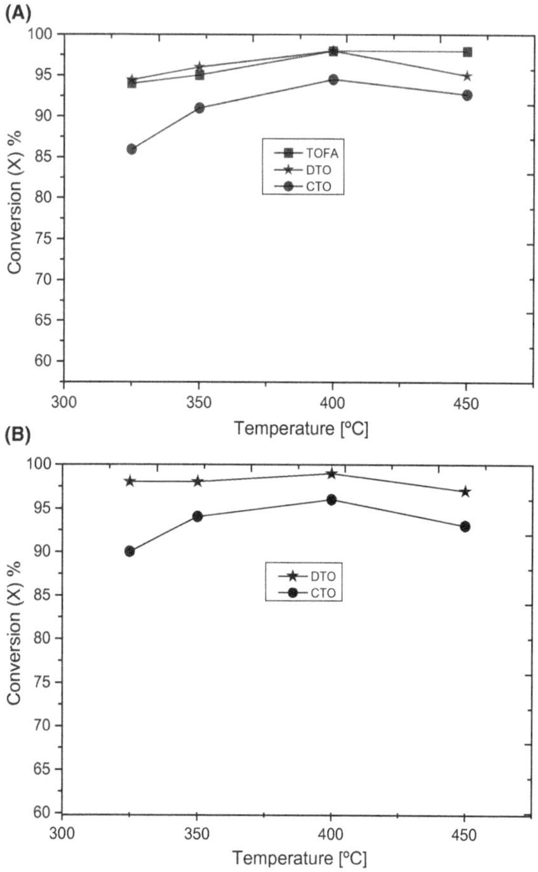

Figure 3. Conversion of fatty acids (A) and resin acids (B) obtained in hydrotreating of TOFA, DTO, and CTO at WHSV = 1 h^{-1}, T = 325–350°C.

decomposition (cracking) of side chains as well as the ring opening at high temperatures (>400°C). The thermal decomposition and ring opening reactions from sterols produce gas phase light hydrocarbons along with oxygenates and unsaturated cyclic structures [31]. A ring opening and thermal decomposition route is also expected from resin acids in CTO and DTO at high temperatures, producing lighter alkanes and alkenes [32] On the other hand, the increased yield of gaseous products in low temperature (<400°C) TOFA and DTO hydrotreating in comparison with the CTO hydrotreating at similar conditions is attributed to the increased formation of CO_2 (~40%), CO (~20%), and methane (~20%) through selective deoxygenation routes. Interestingly, in contrast with methane, propane was obtained only in minor yields in low temperature (<400°C) tall oil hydrotreating.

n-Octadecane and *n*-heptadecane were obtained as the major paraffinic hydrocarbons from tall oil hydrotreating through HDO, decarboxylation and decarbonylation reactions. In order to further evaluate the extent of decarboxylation + decarbonylation reactions versus HDO reaction in tall oil hydrotreating, the C_{17}/C_{18} ratio of the liquid products was calculated. Figure 5 shows the comparison of the C_{17}/C_{18} ratio obtained from the hydrotreating of TOFA, DTO, and CTO at different process conditions (space time and temperature). It is interesting to note from Figure 5 that in agreement with TOFA hydrotreating, DTO hydrotreating at different WHSV's also results an increase in C_{17}/C_{18} ratio as a function of increasing temperature until 400°C. Surprisingly with CTO, a decrease in C_{17}/C_{18} ratio was observed from 325 to 400°C in all tested cases. The higher C_{17}/C_{18} ratio obtained in low temperature (325°C and 350°C) experiments irrespective of space time is attributed to the decrease in formation of *n*-octadecane. This result suggests that the complex nature of CTO feedstock plays a significant role during its hydrotreating over a NiMo catalyst. The selective deoxygenation reactions may be reduced with CTO at lower temperatures (<400°C) due to a number of reasons, which is related to the reactivity of fatty acids in CTO and discussed in a separate section in this study. At 450°C, it appears that a comparison of C_{17}/C_{18} ratio as a function of WHSV is challenging for the tested tall oil feeds. Presumably, the formation *n*-octadecane and *n*-heptadecane can be significant at shorter space times (WHSV = 2 h^{-1} and WHSV = 3 h^{-1}), whereas cracking reactions which produce shorter chain alkanes from C_{17} and C_{18} hydrocarbons can also be significant as a function of increasing space time.

Regarding the other fractions in organic phase, monoaromatics are mainly formed from fatty acid fractions through intermediate *n*-alkanes by means of cyclization and aromatization reaction routes [30]. In addition to this, a direct route for the formation of cycloalkanes and aromatics prior to deoxygenation has also been proposed in the literature [33]. Monoaromatics can also be resulted from resin acids and sterols in tall oil by selective ring opening and dehydrogenation reactions which are valid on the employed catalyst at high temperatures. As fatty acids act as the major precursor for monoaromatics formation at high temperatures, it is obvious that lower amount of fatty acids in CTO in comparison with DTO causes a low yield of monoaromatics [30]. Interestingly, the ratio of polyaromatics to monoaromatics was higher with CTO compared to DTO, and it can be credited to the supplement formation of polyaromatics from the sterol fraction in CTO, either prior to or after deoxygenation.

Based on the obtained composition of reaction products in organic phase samples, a reactivity scale is proposed for the formation of major products from each tall oil feeds as a function of temperature (325–450–°C), and represented as Figure 6. It should be noted that the

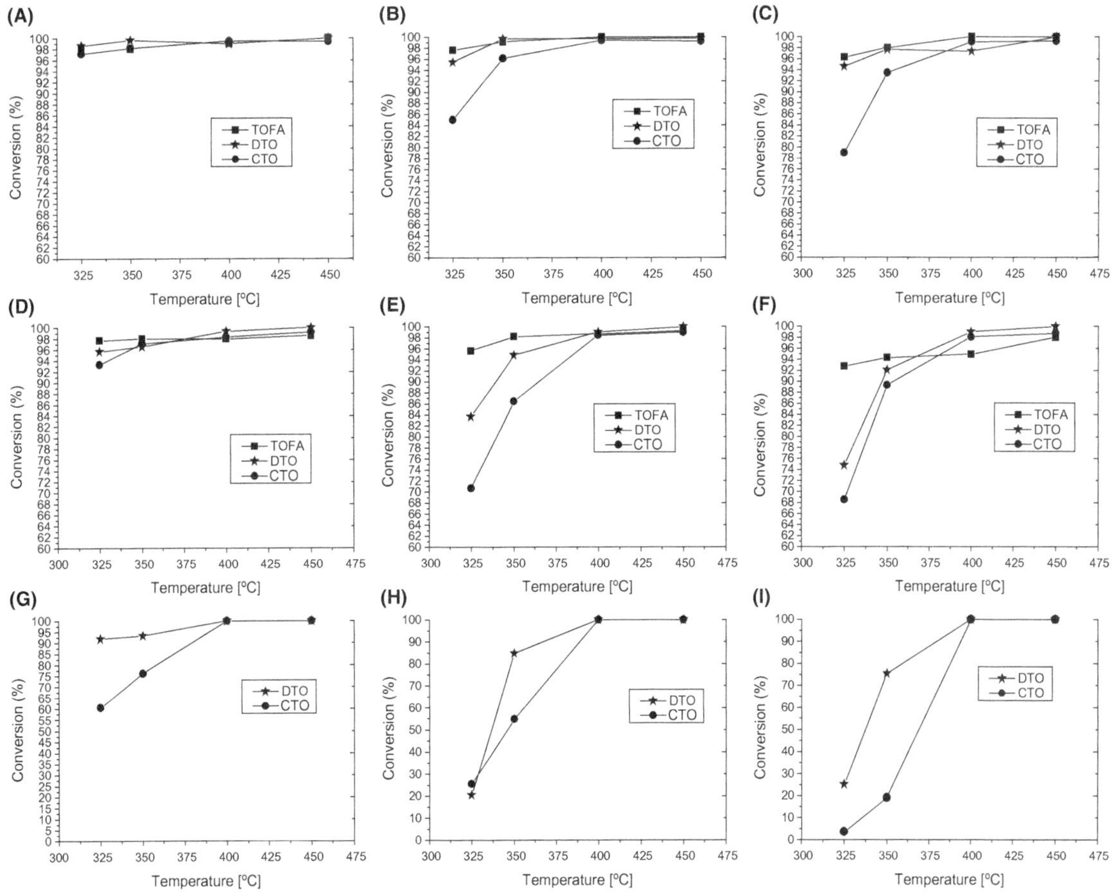

Figure 4. Conversion of individual fatty acids and resin acids; linoleic acid: (A) WHSV =1 h^{-1}, (B) WHSV =2 h^{-1}, (C) WHSV = 3 h^{-1}; oleic acid: (D) WHSV = 1 h^{-1}, (E) WHSV = 2 h^{-1}, (F) WHSV = 3 h^{-1};dehydroabietic acid: (G) WHSV = 1 h^{-1}, (H) WHSV = 2 h^{-1}, and (I) WHSV = 3 h^{-1}.

proposed reactivity scale excludes the possible thermal or catalytic cracking reactions which occur directly from tall oil fractions in high-temperature experimental runs.

Conversion and reactivity of acid fractions in tall oil

Fatty acids

As noted from Figure 3A, a decrease in conversion of fatty acids was obtained in CTO hydrotreating. Therefore, it is important to assess the effect of other fractions on the conversion of fatty acids in CTO. It is already learnt from DTO hydrotreating that resin acids are not inducing any significant effect on fatty acids in terms of inhibiting its reactivity. In addition to resin acids CTO also contains a considerable amount of neutrals (sterols). In view of the earlier study on the HDO of phenolic compounds and alicyclic alcohols [20], it can be proposed that the sulfur anion vacancy (CUS) site which is responsible for the HDO of acid fractions are also responsible for the HDO of sterols (direct C–O hydrogenolysis). Interestingly, in our previous study on CTO hydrotreating [23], a high conversion range of sterols occurred at the tested conditions irrespective of space time and temperature. Therefore, based on these observations we may only conclude that a high conversion of sterols is achievable in spite of the competitive adsorption on CUS between acid fractions and sterols in CTO.

The effect of sulfur on fatty acid conversion

It is widely accepted that metal impurities present in the feed influences the hydroprocessing reactions on a sulfided NiMo catalyst [3]. However, the effects of metal impurities on the conversion of fatty acids are ruled out in this research study as the test runs were carried out

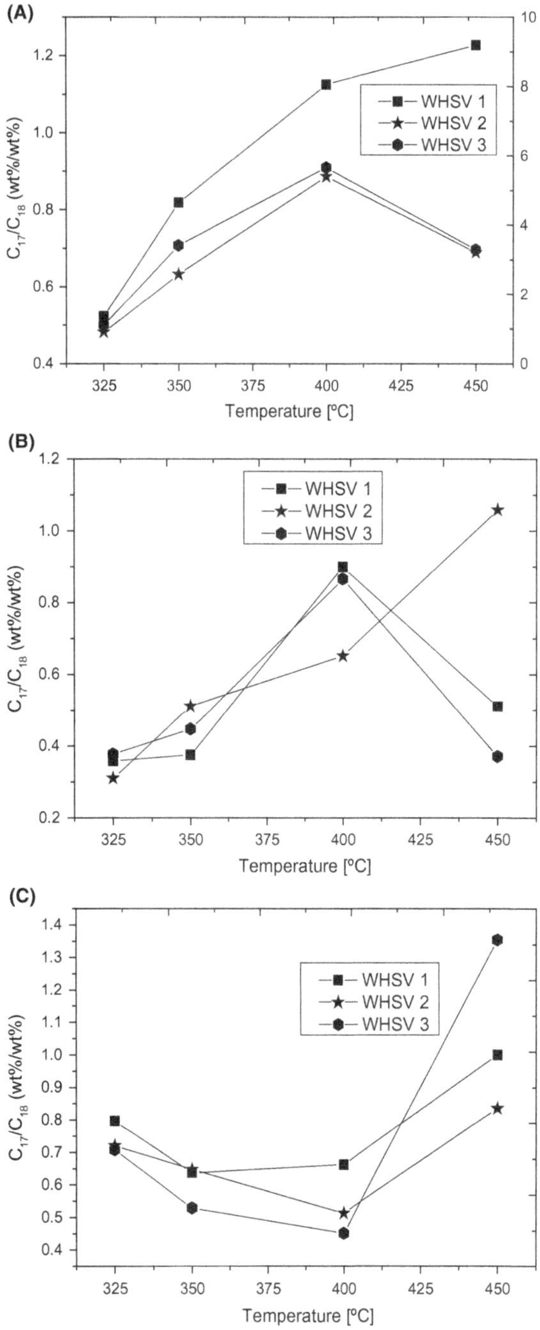

Figure 5. C_{17}/C_{18} (wt%/wt%) ratio obtained from tall oil hydrotreating (A) TOFA (B) DTO (C) CTO; WHSV = 1–3 h^{-1}, T = 325–450°C.

only for 6 h of time-on-stream. The drop in conversion of fatty acids in CTO hydrotreating is considered in this study mainly based on the sulfur content in the feedstock.

It has been reported that the sulfur content in the feed, for instance, elemental sulfur, preferably in a range 2000–5000 w ppm is able to enhance the extent of decarboxylation reactions over a sulfided NiMo catalyst [26]. This increase in the magnitude of decarboxylation reactions is attributed to the increased catalyst acidity by means of adsorption of sulfur species onto the catalyst surface. Senol et al. [34] and Ryymin et al. [35] studied the effects of sulfur additives (H_2S and dimethyl disulfide (DMDS)) in the HDO of aliphatic oxygenates and published that in the presence of sulfur additives, reductive reactions such as hydrogenation (saturation of double bonds), and HDO (addition of hydrogen adjacent to carbonyl carbon, which results the step-wise formation of alkanes and water through intermediate alcohol) are suppressed on a sulfided NiMo catalyst. On the basis of this aspect, we can conclude that the unexpected higher C_{17}/C_{18} ratio obtained in CTO hydrotreating at low temperature (<350°C) (Fig. 5) is presumably due to the higher concentration (1800 ppm) of sulfur in CTO compared to DTO and TOFA, and their influence on selective deoxygenation routes from fatty acids. However, a significant increase in the decarboxylation route (C_{17} formation) in comparison with HDO route (C_{18} formation) is not observed in CTO hydrotreating by virtue of the sulfur content in the feed, but rather it seemed like both HDO and decarboxylation are occurred in a similar extent in CTO hydrotreating. It may presume that in comparison with TOFA and DTO hydrotreating, decarboxylation route was more prominent in CTO hydrotreating in low-temperature experiments apparently due to the effects of sulfur content. The drop in HDO of fatty acids in CTO at low temperatures compared to acids in other feedstocks can be attributed to the decreased reductive reactions due to the effect of sulfur. However, no conclusive comments are drawn here associating the structure–activity relationship of the employed catalyst during hydrotreating of tall oils.

Reaction mechanism assessment of fatty acids

The reaction mechanism of fatty acids is discussed here based on the conversion of individual fatty acids presented in Figure 4. Figure 4A–C shows that linoleic acid, one of the main fatty acids present in tall oil (present in similar concentrations in TOFA, DTO, and CTO), was converted in similar extent with all feedstocks at longer space time (WHSV = 1 h^{-1}). However, at shorter space times (WHSV = 2 and 3 h^{-1}) a prominent drop in reactivity of linoleic acid especially at low temperatures (325–350°C) is observed with CTO compared to other feedstocks. It is also observed from Figure 4D–F that other major fatty acid, oleic acid, converted in a slightly lesser extent in CTO hydrotreating even at WHSV 1 h^{-1} compared to the hy-

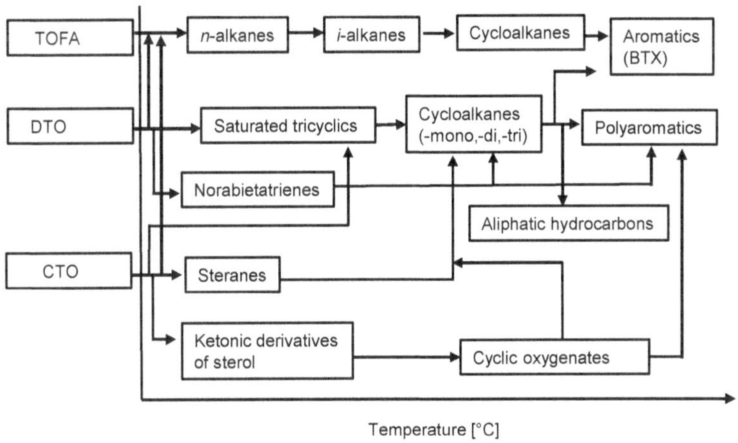

Figure 6. Proposed reactivity scale for the formation of major products from tall oil feeds in hydrotreating (T = 325–450°C).

drotreating of TOFA and DTO. At shorter space times (WHSV = 2 and 3 h^{-1}), a significant drop in the conversion of oleic acid was also observed in DTO hydrotreating. Oleic acid can be formed from linoleic acid by partial hydrogenation [36], therefore, we perceive the possibility of oleic acid formation from linoleic acid at least favorable space times such as WHSV = 2 and 3 h^{-1}. The formation of oleic acid at these conditions indirectly results a drop in the conversion of oleic acid. Furthermore, Figure 7 is plotted to show the molar amount (moles) of stearic acid (a saturated fatty acid (C_{18}) which is reported to form as an intermediate acid during the hydrotreating of C_{18} unsaturated fatty acids) in feedstock and hydrotreated products, respectively, at WHSV=1 h^{-1}. It is evident from Figure 7 that high concentrations of stearic acid is obtained in low temperature (<400°C) hydrotreating experiments. This observation confirms that fatty acids (linoleic acid and oleic acid) undergo deoxygenation through an initial fatty acid chain double-bond hydrogenation as proposed by other researchers [8, 36, 37]. The saturated fatty acid formed; stearic acid in this case, can then undergo deoxygenation via different routes.

It can be suggested that hydrogenation is prominent at elevated temperatures as well, and produce saturated fatty acids. However, as shown by reactions (4), (5), and (6), n-alkanes are formed (up to an optimum temperature level) at the expense of intermediate fatty acids, which in turn results a drop in the concentration of saturated fatty acids (stearic acid) in the product stream from high temperatures (>400°C); as evident from Figure 7. Interestingly for all feeds short-chain fatty acids as well as partially hydrogenated fatty acids were detected in product streams, where for CTO the concentration of partially hydrogenated fatty acids especially at low temperatures

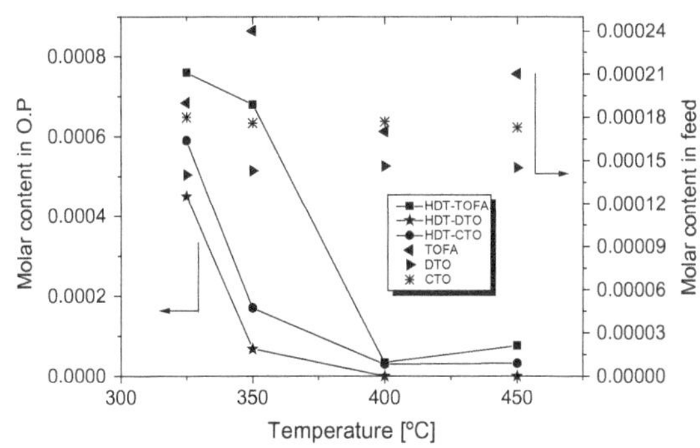

Figure 7. Concentration (molar content) of stearic acid in feeds and hydrotreated products at WHSV = 1 h^{-1}.

was markedly higher than for DTO and TOFA. With TOFA and DTO, a high yield (wt%) of short-chain fatty acids at high temperatures are resulted by means of cracking reactions.

Fatty acid conversion versus paraffin yield

The product yield (%) of paraffins was calculated in this study in order to correlate fatty acid conversion and paraffin formation. It is important to observe that product yield of paraffins corresponds to the amount of fatty acids in feedstocks especially in DTO and CTO hydrotreating experiments. Palanisamy et al. [38] proposes a route for the formation of n-alkanes and i-alkanes from resin acids (abietic acid) through a ring opening mechanism over a sulfided NiMo catalyst. This ring opening mechanism through C–C bond cleavage may be valid in our experimental conditions producing acyclic paraffinic hydrocar-

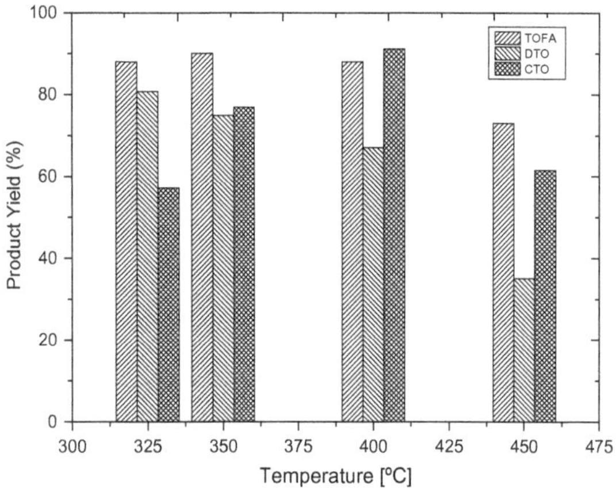

Figure 8. Product yield of paraffins (%) from TOFA, DTO, and CTO as a function of hydrotreating temperature (WHSV = 1 h^{-1}).

bons (*n*-alkanes and *i*-alkanes). Figure 8 shows the yield of paraffins as a function of temperature, obtained from TOFA, DTO, and CTO hydrotreating at WHSV = 1 h^{-1}. It should be taken into consideration that product yield of paraffins can be lower at a high temperature (>400°C) due to the successive formation of cycloalkanes and aromatics from paraffins. Considering this, it can be seen from Figure 8 that a high yield (%) of paraffins was obtained from TOFA hydrotreating, which is no surprise as fatty acids were converted in high extent in that case. A high yield of paraffins was obtained from low-temperature experimental runs in DTO hydrotreating. Remarkably, a high yield of paraffins was obtained from CTO hydrotreating at temperatures 350 and 400°C. As already discussed, fatty acids in CTO are converted in a lesser extent than fatty acids in TOFA and DTO. Furthermore, fatty alcohols and fatty acid esters were detected in significant concentrations in samples from low temperature (325–400°C) hydrotreating of CTO. These fatty alcohols and fatty acids are resulted from the incomplete HDO of fatty acids. This observation also signifies that even at low temperatures all converted fatty acids in CTO were not turned into paraffins. Therefore, a possibility of paraffin formation from fractions other than fatty acids is proposed in CTO hydrotreating. It is known that CTO feed contains a considerable fraction of fatty alcohols (6–8 wt %) along with other fractions; consequently, one route can be suggested as paraffin formation from fatty alcohols via hydrogenation. Ring opening activity of sulfided NiMo catalyst with resin acids as proposed by Palanisamy et al. [38] can also be extend to sterols in CTO. Nonetheless, it is suggested that detailed studies are needed in order to further evaluate the possible routes for the formation of paraffins from cyclic structures in tall oils.

Resin acids and their reaction mechanism assessment

From DTO hydrotreating, it is known that fatty acids did not impose any inhibiting effects on resin acid conversions over a sulfided NiMo catalyst, and vice versa [22]. Figure 3 shows that conversion of resin acids in DTO and CTO follows the same trend as of fatty acids. Therefore, it is proposed that resin acid conversion is also altered by the same effects which are related to fatty acid conversion. Figure 4G–I shows that the conversion of resin acid, dehydroabietic acid, was lower in the case of DTO and CTO hydrotreating at low temperatures (<400°C), which implies that a temperature above 350°C may be needed for achieving a complete conversion of dehydroabietic acid. Earlier, it has been reported by our group that 18-norabietane ($C_{19}H_{34}$) is formed from resin acids [e.g., abietic acid ($C_{19}H_{29}COOH$)] via complete hydrogenation and deoxygenation [22]. Coll et al. propose that deoxygenation from resin acids can occur by means of complete hydrogenation which produces intermediate aldehydes and alcohols and finally a hydrocarbon structure with same number of carbon atoms as the parent resin acids [12]. However, the formation of 18-norabietane as a major product from resin acids at low-temperature hydrotreating [22] implies that deoxygenation was mainly occurred through a hydrogenation and decarboxylation/decarbonylation step and not through an intermediate aldehyde–alcohol route as proposed by Coll et al. [12] The formation of a small fraction of abietane from resin acids also implies that hydrogenation/dehydration mechanistic route is still valid from resin acids. Furthermore, it was reported [22] that norabietatrienes and monoaromatic tricylic structures are formed even at 350°C from abietic-type resin acids presumably through a parallel dehydrogenation and decarboxylation reaction route as proposed by Dutta et al. [21] The decarboxylation step involved in this reaction route can be catalytic (in the presence of a HDO catalyst under H_2 pressure) or noncatalytic (thermal reaction in the absence of a HDO catalyst). Interestingly, as evidence to noncatalytic route, it has been reported that norabietatrienes are formed during the thermal treatment of abietic acid at 350°C [22, 39]. Therefore, different reaction routes from resin acids are proposed for low-temperature hydrotreating (<400°C) as in the case of fatty acids. Moreover, it may infer that during the hydrotreating of DTO at low temperatures (<400°C), most of the resin acids are consumed through a decarboxylation or decarbonylation as well as through the noncatalytic step (direct decarboxylation + dehydrogenation) without inducing much interaction with the active site responsible for HDO (hydrogenation/dehydration) reaction.

Conclusions

The activity of a commercial alumina supported NiMo catalyst was evaluated for hydrotreating of different tall oil feedstocks. In this study, sulfided NiMo catalyst was found to be active for producing wide-range of products from tall oil feedstocks through various reaction routes existing at different temperatures. At lower temperatures (<400°C), selective deoxygenation routes such as, HDO, decarboxylation, and decarbonylation routes were prominent. At higher temperatures reactions such as cracking (thermal or catalytic) and dehydrogenation were prominent, which produced cycloalkanes and aromatics as main products. Importantly, the trend obtained for the formation of *n*-octadecane and *n*-heptadecane in this study especially from TOFA is found to be in well agreement with the behavior of sulfided NiMo catalyst reported in the literature for hydrotreating of vegetable oils. Moreover, fatty acids were found to convert in high extent in TOFA and DTO hydrotreating. In CTO hydrotreating, a decrease in conversion of acid fractions was observed at low temperatures, which is attributed to the complex nature of CTO with significant amount of sulfur compounds in it. Sulfur compounds are proposed to alter the deoxygenation reactions particularly decarboxylation reactions from acid fractions at low temperatures in CTO hydrotreating. Furthermore, it is proposed based on the results in this study that deoxygenation reaction from resin acids at low temperatures mainly occurs through a hydrogenation and decarboxylation/decarbonylation step.

Acknowledgments

Jinto Manjaly Anthonykutty acknowledges the financial support from VTT graduate school. The authors also acknowledge Stora Enso for supporting this research and their vision on wood-based olefins.

Conflict of Interest

None declared.

References

1. Furimsky, E. 2000. Catalytic hydrodeoxygenation. Appl. Catal. A: Gen. 199:147–190.
2. Egeberg, R., N. Michaelsen, L. Skyum, and P. Zeuthen. 2010. Hydrotreating in the production of green diesel. PTQ Q2 Available at www.digitalrefining.com/article/1000156 (accessed 31 May 2014).
3. Furimsky, E., and F. E. Massoth. 1999. Deactivation of hydroprocessing catalysts. Catal. Today 52:381–495.
4. Furimsky, E., and F. E. Massoth. 1993. Regeneration of hydroprocessing catalysts. Catal. Today 17:537–659.
5. Elliott, D. C. 2007. Historical developments in hydroprocessing bio-oils. Energy Fuels 21:1792–1815.
6. Kubička, D., P. Šimáček, and N. Žilková. 2009. Transformation of vegetable oils into hydrocarbons over mesoporous-alumina-supported CoMo catalysts. Top. Catal. 52:161–168.
7. Kikhtyanin, O. V., A. E. Rubanov, A. B. Ayupov, and G. V. Echevsky. 2010. Hydroconversion of sunflower oil on Pd/SAPO-31 catalyst. Fuel 89:3085–3092.
8. Guzman, A., J. E. Torres, L. P. Prada, and M. L. Nunez. 2010. Hydroprocessing of crude palm oil at pilot plant scale. Catal. Today 156:38–43.
9. Kovács, S., T. Kaszaa, A. Thernesz, I. W. Horváthb, and J. Hancsóka. 2011. Fuel production by hydrotreating of triglycerides on NiMo/Al_2O_3/F catalyst. Chem. Eng. J. 176–177:237–243.
10. Abhari, R., and Havlik P. Z. 2009. Hydrodeoxygenation process. US patent publication 2009/0163744.
11. Abhari, R., H. L. Tomlinson, and E. G. Roth. 2009. Biorenewable naphtha. US patent publication 2009/0300971) A1.
12. Coll, R., S. Udas, and W. A. Jacoby. 2001. Conversion of the rosin acid fraction of crude tall oil into fuels and chemicals. Energy Fuels 15:1166–1172.
13. Sharma, R. K., and N. N. Bakhshi. 1991. Catalytic conversion of crude tall oil to fuels and chemicals over HZSM-5: effect of co-feeding steam. Fuel Process. Technol. 27:113–130.
14. Furrer, R. M., and N. N. Bakhshi. 1988. Catalytic conversion of tall oil to chemicals and gasoline range hydrocarbons. Res. Thermochem. Biomass Convers 956.
15. Kirshner, M. 2005. Chemical profile: tall oil. Chemical Market Reporter: 34.
16. Furimsky, E. 1998. Selection of catalysts and reactors for hydroprocessing. Appl. Catal. 17:177–206.
17. Furimsky, E. 1983. Chemistry of catalytic hydrodeoxygenation. Catal. Rev-Sci. Eng. 25:421–458.
18. Topsøe, H., R. G. Egeberg, and K. G. Knudsen. 2004. Future challenges of hydrotreating catalyst technology. Prepr. Pap-Am. Chem. Soc. Div. Fuel Chem. 49: 568–569.
19. Kubička, D., and L. Kaluza. 2010. Deoxygenation of vegetable oils over sulfided Ni, Mo, and NiMo catalysts. Appl. Catal. A: Gen. 372:199–208.
20. Şenol, O. I., E. M. Ryymin, T. R. Viljava, and A. O. I. Krause. 2007. Effect of hydrogen sulphide on the hydrodeoxygenation of aromatic and aliphatic oxygenates on sulphided catalysts. J. Mol. Catal. A: Chem. 277:107–112.
21. Dutta, R. P., and H. H. Schobert. 1993. Hydrogenation/dehydrogenation reactions of rosin. Fundam. Studies Coal Liquefaction 38:1140–1146.

22. Anthonykutty, J. M., K. M. Van Geem, R. D. Bruycker, J. Linnekoski, A. Laitinen, J. Räsänen, et al. 2013. Value added hydrocarbons from distilled tall oil via hydrotreating over a commercial NiMo catalyst. Ind. Eng. Chem. Res. 52:10114–10125.
23. Anthonykutty, J. M., J. Linnekoski, A. Harlin, A. Laitinen, and J. Lehtonen. Catalytic upgrading of crude tall oil into a paraffin-rich liquid. Biomass Conv. Bioref. doi: 10.1007/s13399-014-0132-8
24. Rozmysłowicz, B., P. Mäki-Arvela, S. Lestari, O. A. Simakova, K. Eränen, I. L. Simakova, et al. 2010. Catalytic deoxygenation of tall oil fatty acids over a palladium-mesoporous carbon catalyst: a new source of biofuels. Top. Catal. 53:1274–1277.
25. Knuuttila, P., P. Kukkonen, and Ulf Hotanen. 2010. Method and apparatus for preparing fuel components from crude tall oil. US patent publication WO 2010/097519 A2.
26. Stigsson, L., and V Naydenov. 2009. Conversion of crude tall oil to renewable feedstock for diesel range fuel compositions. European patent publication WO 2009/131510.
27. Diaz, M. A. F., R. A. Markovits, and S. A. Markovits. 2005. Process for refining a raw material comprising black liquor soap, crude tall oil or tall oil pitch. EP1568 760 A1.
28. Kubičková, I., and D. Kubička. 2010. Utilization of triglycerides and related feedstocks for production of clean hydrocarbon fuels and petrochemicals: a review. Waste Biomass Valorization 1:293–308.
29. Mikulec, J., J. Cvengros, L. Jorı'kova, M. Banic, and A. Kleinova. 2010. Second generation diesel fuel from renewable sources. J. Clean. Prod. 18:917–926.
30. da Rocha Filho, G. N., D. Brodzki, and G. Djéga-Mariadassou. 1993. Formation of alkanes, alkylcycloalkanes and alkylbenzenes during the catalytic hydrocracking of vegetable oils. Fuel 72:543–549.
31. Wagner, J. L., V. P. Ting, and C. J. Chuck. 2014. Catalytic cracking of sterol-rich yeast lipid. Fuel 130:315–323.
32. Severson, R. F., and W. H. Schuller. 1972. The thermal behavior of some resin acids at 400–500°C. Can. J. Chem. 50:2224.
33. Scharmann, H., W. R. Eckert, and A. Zeman. 1969. Elucidation of structure of the methyl esters of cyclic fatty acids V: mass-spectrometry of isomeric methyl esters of phenyl undecanoic acid. Fette Seifen Anstrichmiftel 71:118.
34. Senol, O. I., T.-R. Viljava, and A. O. I. Krause. 2007. Effect of sulfiding agents on the hydrodeoxygenation of aliphatic esters on sulphided catalysts. Appl. Catal. A: Gen. 326:236–244.
35. Ryymin, E.-M., M. L. Honkela, T.-R. Viljava, and A. O. I. Krause. 2010. Competitive reactions and mechanisms in the simultaneous HDO of phenol and methyl heptanoate over sulphided NiMo/γ-Al2O3. Appl. Catal. A: Gen. 389:114–121.
36. Monnier, J., H. Sulimma, A. Dalai, and G. Caravaggio. 2010. Hydrodeoxygenation of oleic acid and canola oil over alumina-supported metal nitrides. Appl. Catal. A 382:176–180.
37. Snåre, M., I. Kubicˇkova, P. Malki-Arvela, K. Eralnen, and D. Y. Murzin. 2006. Heterogeneous catalytic deoxygenation of stearic acid for production of biodiesel. Ind. Eng. Chem. Res. 45:5708–5715.
38. Palanisamy, S. 2013. Co-processing fat-rich material into diesel fuel. Doctoral thesis, Chalmers University
39. Bernas, A., T. Salmi, Y. D. Murzin, J. Mikkola, and M. Rintola. 2012. Catalytic transformation of abietic acid to hydrocarbons. Top. Catal. 55:673–679.

2

Reactivity of CaO with CO_2 in molten CaF_2–NaF: formation and decomposition of carbonates

Viktorija Tomkute[1], Asbjørn Solheim[2], Simas Sakirzanovas[3] & Espen Olsen[1]

[1]Department of Mathematical Sciences and Technology, Norwegian University of Life Sciences (NMBU), PO Box 5003, Drøbakveien 31, NO-1432 Ås, Norway
[2]SINTEF Materials and Chemistry, Post Office Box 4760, Sluppen, NO-7465 Trondheim, Norway
[3]Institute of Chemistry, Center for Physical Sciences and Technology, A. Gostauto 9, LT-01108 Vilnius, Lithuania

Keywords
Calcium fluoride, calcium oxide, carbon dioxide, carbonate looping, molten salt, sodium fluoride

Correspondence
Espen Olsen, Department of Mathematical Sciences and Technology, Norwegian University of Life Sciences (NMBU), Post Office Box 5003, Drøbakveien 31, NO-1432 Ås, Norway.
E-mail: espen.olsen@nmbu.no

Funding Information
No funding information provided.

Abstract

The characteristics of CO_2 reacting with CaO in a molten eutectic mixture of CaF_2 and NaF has been investigated. Calculations of the Gibbs free energy, temperature analysis of the decomposition of the formed carbonates, and XRD analyses of quenched samples taken during CO_2 absorption or desorption were employed to identify the phases present in the melt. Efficient CO_2 absorption from a simulated flue gas was observed, due to a combined reaction where CaO initially reacts with CO_2 and forms $CaCO_3$. Subsequently, Na_2CO_3 is formed by an ion exchange reaction between $CaCO_3$ and NaF. It was found that the CaO activity is highest in the temperature range 826–834°C. Increasing the CaO concentration from 5 to 20 wt% in the molten salt resulted in reduced CO_2 reactivity efficiency, probably because of precipitation and agglomeration of the sorbent. The total carbonation conversion was independent of the CO_2 concentration in the inlet gas, and the sorbent carrying capacity was in the range 0.722–0.743 g CO_2/g CaO corresponding to 0.037–0.144 g CO_2/g total liquid. Decarbonation was conducted by raising the temperature. 40% conversion back to CaO was recorded at 1160°C. The recorded curves for the CO_2 concentration in the outlet gas exhibited a rapid desorption step followed by a slow step.

Introduction

Sorption of CO_2 by CaO-based solid-state substances is being developed for the use in cyclic absorption–desorption processes for capture of CO_2 from waste gases emitted by a wide range of industrial processes [1, 2]. Examples are power generation [3], cement production [4, 5], and metallurgical industry [6]. CaO from naturally occurring limestone sources may be used directly with high absorption capacity. However, these materials exhibit degradation upon cycling with limited sustained capacity; after 10 cycles, around 10% of the initial capacity is maintained without H_2O present in the gas. With water present, around 25% sustained capacity is obtained after 10 cycles [7]. Solid CaO-based materials with enhanced properties both with regards to cycling and carrying capacity are being developed [8–10]. In this work, we explore a somewhat different approach. The dissolution and/or dispersion of sorbent particles in a molten salt matrix may enhance the reactivity of the sorbent with CO_2 because efficient gas-liquid-solid reactions can be employed [11–13]. Various alkali and alkaline-earth metal halide salts ($CaCl_2$, $CaCl_2$–CaF_2, CaF_2–NaF and CaF_2–LiF) have previously been studied for the use as solvents for dissolution and/or dispersion of CaO to enhance sorbent activities in the CO_2 capture process [11–13]. CaF_2–NaF and CaF_2–LiF eutectic mixtures have been used. The results indicate that the highest carrying capacity (0.722–0.743 g CO_2/g CaO) may be achieved with the CaO–CaF_2/NaF system due to thermodynamically stable Na_2CO_3 being formed through an ion exchange reaction [11, 14]. Studies of cyclic CO_2 absorption/desorption by 5.32 wt% CaO in molten $CaCl_2$ showed a stable carrying capacity of 0.504 g CO_2/g CaO [12]. In this case, the decomposition of the formed carbonates proceeded rapidly and completely, but when using alkali and alkaline-earth metal halide salts, full decomposition of the carbonates formed in the melt could not be achieved [11].

Notably, besides the technical design, the selection of appropriate molten salt composition for dissolution or partial dissolution of the active sorbent is an essential aspect in development of a technologically and economically efficient CO_2 capture process. Therefore, the aim of this work was to investigate systematically the efficiency of a process based on CaO dissolved/dispersed in a eutectic mixture of CaF_2 and NaF. Previously, the phase transitions of CaO, $CaCO_3$ and Na_2CO_3 in CaF_2/NaF were investigated experimentally and mapped using FactSage software [14]. The eutectic composition of CaF_2/NaF (41.85 wt% CaF_2 in NaF, 815°C) was chosen because of its suitable melting point for the carbonation reaction and low vapor pressure, which allows low-pressure operation [15, 16]. The CO_2 sorption/desorption behavior was studied using a Fourier transform infrared (FTIR) gas detector and gravimetric analysis. Particularly, the impacts of the CaO mass proportion in the melt, the temperatures of the system for the carbonation/decarbonation reactions, the CO_2 content and the simulated flue gas flow rate were examined to determine the efficiency of the CO_2 sorption process by the CaO–CaF_2–NaF chemical system.

Materials and Methods

Materials

CaO (Sigma-Aldrich reagent, 96–100% purity), CaF_2 (Sigma-Aldrich reagent, 99.0–100.0% purity), and NaF (Sigma-Aldrich reagent, 98.5–100.0% purity) were used for preparing the molten salt mixtures. The powders were dried at 200°C for 24 h. All sorbents were prepared by direct mixing of NaF–CaF_2 (41.85 wt% CaF_2 and 48.15 wt% NaF) and CaO (5, 10, 15, 20 wt%) powders to form a 10 cm high column in a nickel crucible (52 mm dia. × 350 mm height) in the liquid state. The weight varied from 630 to 650 g depending on the composition, unless otherwise mentioned. The mixtures were melted in argon atmosphere and kept at 970°C for 10 h before carbonation, unless otherwise mentioned. Pure gases provided by AGA were used for all experiments and the mixtures of carbon dioxide (purity of 99.99%) in nitrogen (purity of 99.999%, $H_2O \leq 3$ ppm, $O_2 \leq 3$ ppm, $C_nH_m \leq 1$ ppm) were controlled by mass flow controllers (MASS-STREAM, M+W Instruments GmbH, Allershausen, Germany).

Experimental set-up and methods

Absorption and desorption of CO_2 were performed in an automated one-chamber atmospheric pressure reactor. The nickel crucible mentioned above was placed inside a stainless steel outer chamber [12]. This assembly was fixed in a vertical ceramic tube furnace. The carbonation/decarbonation experiments were monitored using a FTIR gas analyzer (Nicolet 6700, Omnic 8 software; Thermo Scientific, Waltham, MA, USA) and an industrial weighing scale (MS8001S, accuracy of 0.1 g; Mettler Toledo, Columbus, OH, USA). Before each experiment the melts were stirred by bubbling N_2 (0.5 L/min) through the liquid at the chosen temperature (820–950°C) for 2 h before carbonation. The N_2 and CO_2-N_2 gas mixtures (total flow 0.6 L/min, 1.3–14.0 vol % CO_2) were bubbled into the CaO–CaF_2–NaF melt via a nickel pipe passed through the top of the sealed reactor assembly and immersed to a depth of 1 cm from the bottom of the crucible. A constant flow of argon (0.20 L/min) was applied as a purge flow over the outer stainless steel assembly to minimize corrosion of the chamber. The temperature measurements were performed using a type S (Pt–Pt10Rh, ±1°C) thermocouple immersed in the melt. The tightness of the system was controlled by a mass flow meter (Sierra 820 Series; Sierra Instruments, Inc., Monterey, CA, USA) supplied at the outlet of the FTIR gas cell. In each test, the carbonation process was considered to be complete when the inlet and outlet CO_2-N_2 stream compositions were nearly matching. Decomposition of the formed carbonates back to oxide was conducted at temperatures ranging from of 994 to 1163°C. The total weight of the assembly as well as the gas composition was continuously monitored. The experimental parameters (gas flow, temperature) were controlled and data collected by a computer integrated system (National Instruments, Austin, TX, USA).

Weight increase due to absorption of CO_2 (carbonation), and weight decrease due CO_2 liberation during thermal decomposition of the formed carbonates (decarbonation) were recorded during the experiments. Correction for the weight change due to oxidation of the stainless steel compartment was taken into the calculations. For this purpose, a baseline study of the oxidation of the reactor was performed at different temperatures. Weight loss due to evaporation of the liquids was deemed negligible as no condensate was found anywhere outside the reaction chamber.

Characterization

Powder X-ray diffraction (XRD) analysis was performed for phase identification. Samples were collected both during the carbonation and the decarbonation stages of 20 wt% CaO in NaF–CaF_2. The samples were collected by quenching the melts on a stainless steel cylinder. XRD data were collected from $5° \leq 2\theta \leq 110°$ (scan rate 0.01 degree per second) using Ni-filtered Cu K_α on a Bruker D8 Advance diffractometer working in Bragg-Brentano ($\theta/2\theta$) geometry.

All XRD measurements were performed at room temperature and ambient pressure in air.

Results and discussion

Evaluation of the CaO–CaF_2–NaF system for the CO_2 sorption processes

Figure 1 depicts measurements of the melt temperature, CO_2 concentration in N_2 leaving the sealed reactor, and the weight changes of the reactor assembly detected during the experiments. First, the melt consisting of 5 wt% CaO, 50.8 wt% NaF, and 44.2 wt% CaF_2 was kept under N_2 atmosphere at 824°C for 2 h before initiating the carbonation. The CO_2 flow was started at time $t = 0$. This caused temperature increase, which can be attributed to the exothermic reaction of CaO with CO_2,

$$CaO(s) + CO_2(g) \leftrightarrow CaCO_3(s) \quad (1)$$

The weight and CO_2 concentration curves reveal a period with fast carbonation followed by slow carbonation. The initial fast carbonation reaction represents a highly efficient CO_2 removal stage where close to 99.9% of the applied CO_2 is captured; this stage can be related to CO_2 reaction with dispersed CaO. The second carbonation stage may be assigned to the reaction with dissolved CaO, and possibly with deposited CaO (which may be covered by a layer of carbonate). This slow stage is considered to be completed when the CO_2 concentration recorded by the FTIR analyzer becomes similar to the inlet concentration. Decomposition of the formed carbonates was conducted by raising the temperature to 1156°C and bubbling pure N_2 through the melt, which also resulted in a rapid step and a slow step (Fig. 1). The total conversion efficiency of CaO to carbonates at 824°C was around 94 wt%, but complete decomposition of the formed carbonates was not reached (65 wt% of CO_2 released). The high carbonation efficiency may be enabled by the ion exchange reaction described by equations 2 and 3:[12]

$$CaCO_3 + 2NaF \leftrightarrow Na_2CO_3 + CaF_2 \quad (2)$$

$$CaO + CO_2 + 2NaF \leftrightarrow Na_2CO_3 + CaF_2 \quad (3)$$

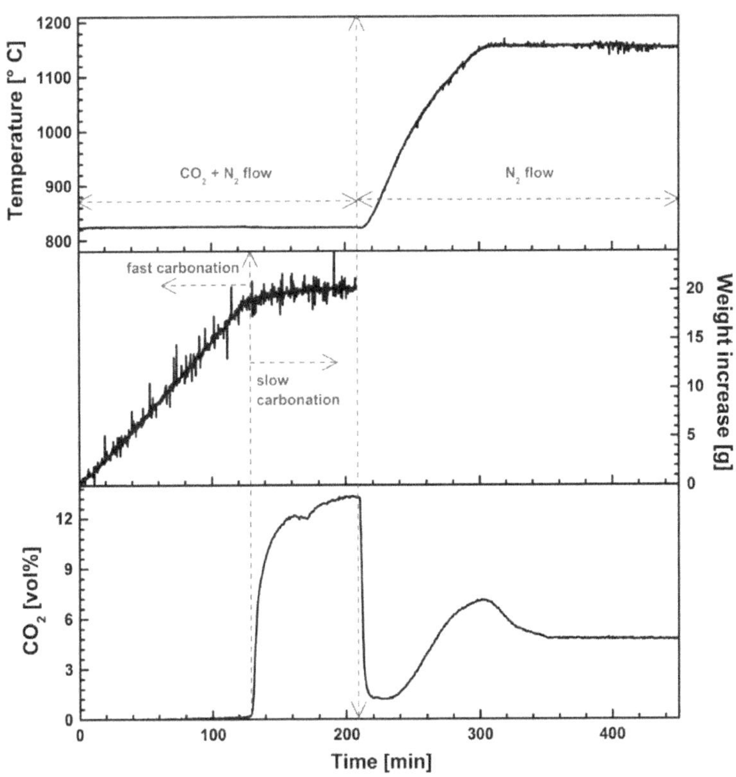

Figure 1. Separation of CO_2 from a simulated flue gas (N_2 + 13 vol% CO_2) by CaO/CaF_2/NaF (5/44.2/50.8 wt%) at 824°C. Decomposition of the formed carbonates was conducted at 1156°C under N_2 atmosphere.

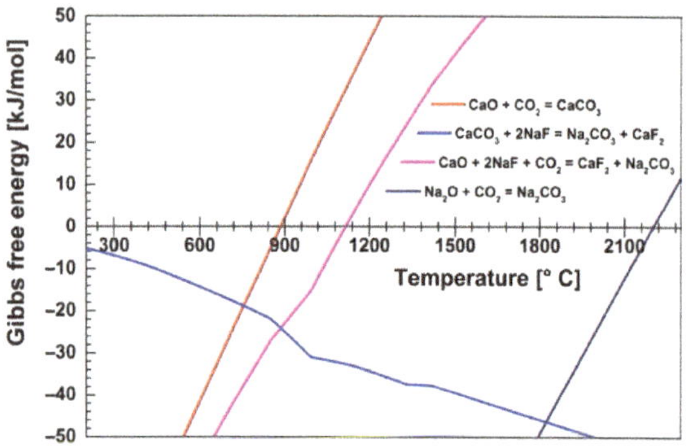

Figure 2. Gibbs free energy versus temperature for possible carbonation/decarbonation reactions in the system CaO–CaF_2–NaF and in solid Na_2O.

At temperatures lower than 335°C, formation of shortite ($Na_2Ca_2(CO_3)_3$) may take place (eq. 4), but above this temperature shortite decomposes to nyerereite ($Na_2Ca(CO_3)_2$) which melts at 817°C (eq. 5) [17].

$$3CaCO_3 + 2NaF \leftrightarrow Na_2Ca_2(CO_3)_3 + CaF_2 \quad (4)$$

$$Na_2Ca_2(CO_3)_3 \leftrightarrow Na_2Ca(CO_3)_2 + CaCO_3 \quad (5)$$

The Gibbs free energy for possible carbonation/decarbonation reactions, as described by equations 1, 2, and 3, was calculated using the HSC Chemistry program (version 6.1, Outotec, Espoo, Finland), and the results are provided in Figure 2. In addition, the Gibbs free energy for the reaction between solid sodium oxide and CO_2 was calculated. The results show that Na_2CO_3 is thermodynamically more stable than $CaCO_3$, because the Gibbs free energy for equation (2) is negative in the temperature range 400–2300°C. Decomposition of $CaCO_3$ occurs at temperatures above 900°C. The value of the Gibbs free energy for the combined, total reaction (eq. 3) is negative at low temperatures and becomes positive above 1120°C, indicating that the equilibrium will be shifted toward the left as the temperature increases. The occurrence of this reaction was supported experimentally by conducting a baseline study using the CaF_2–NaF system without addition of CaO (Fig. 3). In this study, 14 vol% CO_2 in N_2 was bubbled through the CaF_2–NaF melt at 820°C after approximately 2 min of pure N_2 flow. CO_2 removal from the gas mixture could not be observed.

Structural analysis

X-ray diffraction analysis was performed on quenched samples collected during the carbonation step using a melt consisting of CaO/CaF_2/NaF (20/37.2/42.8 wt%) and 14 vol% CO_2 in N_2 at 830°C. The results are shown in Figure 4. For the sample taken before CO_2 was passed through the melt, the XRD pattern peaks can be assigned to CaO, CaF_2, and NaF. The intensity of the peaks attributed to NaF and CaO decreased systematically as carbonation commenced. The X-ray diffraction pattern of the sample taken after 100 min of CO_2 absorption

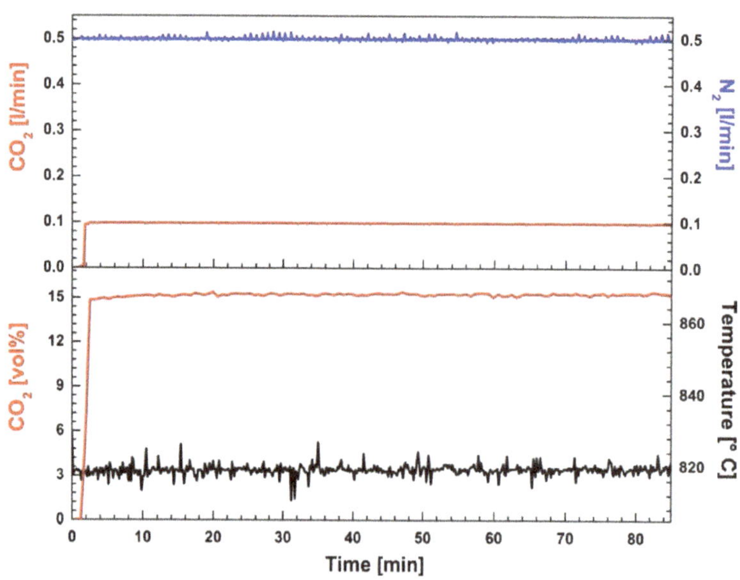

Figure 3. Baseline study of CO_2 capture at 820°C by bubbling 15 vol% CO_2 in N_2 through the melt consisting of 48.15 wt% NaF in CaF_2.

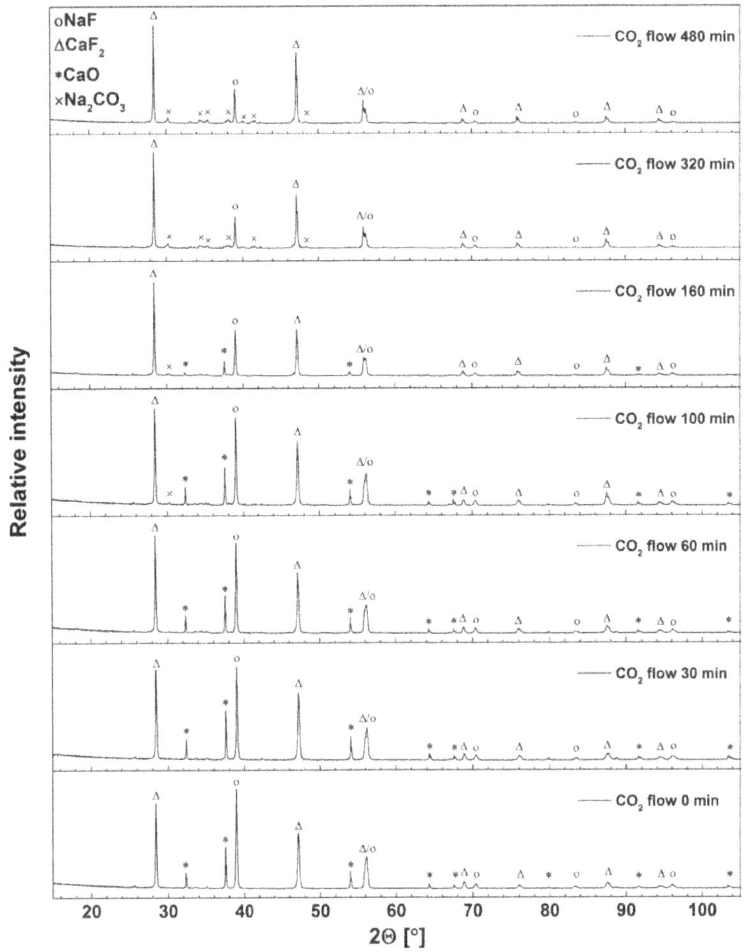

Figure 4. X-ray diffraction (XRD) patterns of the quenched samples collected during the carbonation step. Initial composition of the melt: 20 wt% CaO and 37.2 wt% CaF_2 in NaF. Carbonation of CaO–CaF_2/NaF was conducted by bubbling 14 vol% CO_2 in N_2 through the melt at 830°C.

indicates formation of Na_2CO_3 (space group P6(3)/m) [18]. We suggest that the ion exchange reaction forming Na_2CO_3 takes place immediately during the carbonation step (eq. 2) [14]. The total reaction occurring can be expressed as equation (3). After 480 min of CO_2 absorption most of the CaO was converted to CaF_2, and the NaF has been partly converted to Na_2CO_3 as shown in Figure 4. No peaks attributable to $CaCO_3$, $Na_2Ca(CO_3)_2$, or $Na_2Ca_2(CO_3)_3$ were found during carbonation. The formation of $Na_2Ca(CO_3)_2$ and $Na_2Ca_2(CO_3)_3$ may be expected at temperatures lower than 400 and 817°C, respectively [17]. We suggest that complete sodium cation substitution in $CaCO_3$ occurs, and that Na_2CO_3 and CaF_2 are formed as described by equation (3).

The formed carbonates were decomposed by increasing the temperature to 1170°C under pure N_2 atmosphere. Samples were removed from the melt as described above. The X-ray diffraction patterns of the samples are shown in Figure 5. The intensity of the Na_2CO_3 peaks decreased with increasing decarbonation time. In addition, the presence of CaO can clearly be seen after 120 min of CO_2 desorption. As expected, the intensity of peaks characteristic for CaO increased with time, indicating that the reversible reaction occurs (eq. 3). Finally, there were no intensive Na_2CO_3 peaks detected in the XRD pattern after 480 min of decarbonation, which indicates that full decomposition of the formed carbonates was accomplished.

Because of the complexity in the undergoing reactions between $CaCO_3$ and NaF, it is difficult to calculate the amount of CaO that reacts in the carbonation/decarbonation process based on the diffraction data. The formation of small amounts of complex carbonates in the

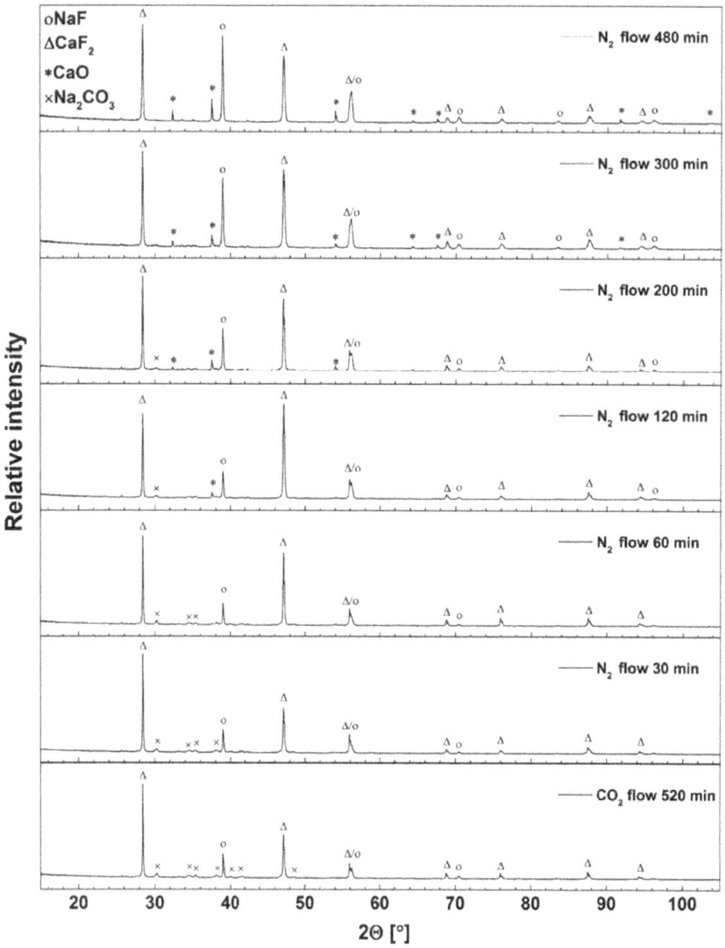

Figure 5. X-ray diffraction patterns of the quenched samples collected during the decarbonation step, after carbonation of the sample (CaO/CaF_2/NaF =20/37.2/42.8 wt%) for 520 min under 14 vol% CO_2 in N_2. Decarbonation of the samples was carried out at 1170°C under pure N_2 for the time indicated.

CaO–CaF_2–NaF system not detectable in the XRD pattern may influence the results. Therefore, the conversions of the sorbent to carbonates for all tests were calculated on the basis of the initially added amounts of CaO in the CaF_2–NaF eutectic mixture.

Effect of carbonation temperature

The effect of temperature on the carbonation reaction in CaO/CaF_2/NaF (5/44.2/50.8 wt%) was investigated to determine the most efficient conditions for the sorbent conversion (Fig. 6). A temperature increase of 4–6°C was detected upon start of carbonation. The slow carbonation reaction stage starts when a rapid increase in the recorded CO_2 concentration in the gas stream from the reactor is observed. It can be suggested that the fast reaction stage can be related to the heterogeneous reaction between CO_2 and dispersed CaO, whereas the slow stage is related to the homogeneous reaction between CO_2 and dissolved CaO, or between CO_2 and agglomerated CaO where there will be strong mass transfer limitations. The results calculated from the amount of CO_2 having been absorbed from the gas fed to the reactor indicate that CaO conversion in the initial fast carbonation reaction is close to 90 wt% of total reacted sorbent (Fig. 7). The highest carbonation efficiency was observed at 826–834°C, where the fast reaction was completed after 117 min and the final CaO conversion was 94–95 wt%. At temperatures higher than 834°C the total conversion efficiency decreased from 95 to 91 wt%. The carbonation conversion observed in the temperature range 821–950°C indicates formation of the thermodynamically stable carbonate (Na_2CO_3, see eqs 2 and 3).

Effect of decarbonation temperature

To investigate the decarbonation reaction mechanism, decomposition of the carbonates was carried out while flowing pure N_2 through the reactor at temperatures ranging from

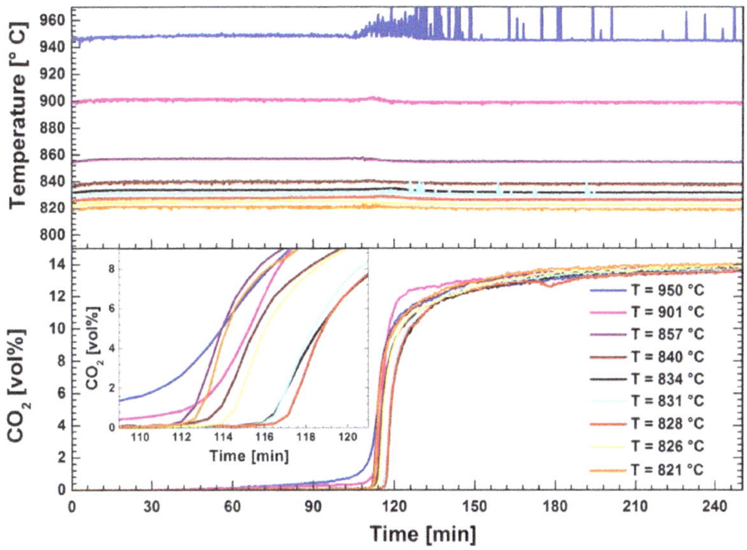

Figure 6. Effect of the temperature on the carbonation reaction in $CaO/CaF_2/NaF$ (5/44.2/50.8 wt%). The carbonation step was conducted by transporting 14 vol% CO_2 in N_2 through the melt. Inset: CO_2 concentration versus time (109–121 min).

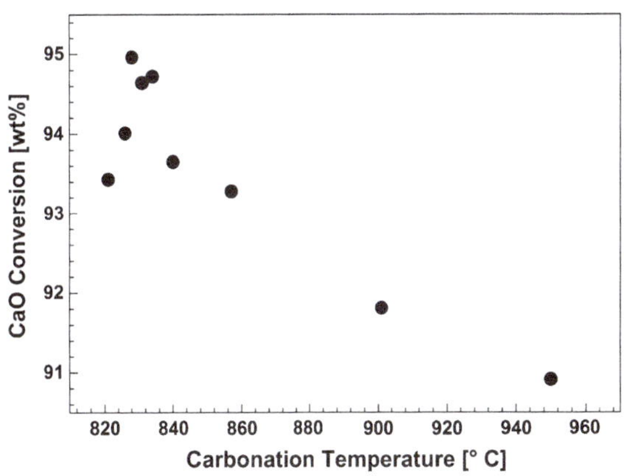

Figure 7. CaO conversion dependence on carbonation temperature. Effect of the temperature on the carbonation reaction in $CaO/CaF_2/NaF$ (5/44.2/50.8 wt%) was investigated by bubbling 14 vol% CO_2 in N_2 through the melt for 300 min at temperatures in the range 821–950°C.

994 to 1163°C (Fig. 8). The initial carbonation was conducted at 830°C by bubbling 14 vol% CO_2 in N_2 through the melt consisting of dissolved/dispersed 5 wt% CaO in CaF_2–NaF for 300 min. The decarbonation was performed by stopping the flow of CO_2 and increasing the temperature. It was observed that the CO_2 concentration curves recorded at 1048–1163°C exhibit two peaks, and the curves recorded at the lower temperature show one single peak (Fig. 8). As described above, the calculation of the standard Gibbs free energy for the reaction where CaO reacts with CO_2 and NaF to form CaF_2 and Na_2CO_3 (eq. 3) indicates that the equilibrium may be achieved at around 1120°C (Fig. 2). Therefore, the faster decomposition of the formed sodium carbonate is supposed to appear at temperatures higher than 1120°C (this temperature is not corrected for activities lower than unity for all substances other than CaO). We suggest that the first peak in the CO_2 concentration curve may be assigned to the decomposition of a small amount of $CaCO_3$, while the second peak is because of decomposition of Na_2CO_3. The results in Figure 9 demonstrate the effect of decarbonation temperature on CaO regeneration. The degree of conversion of the formed carbonates was calculated from the amount of CO_2 released from the reactor during 170 min. It can be observed that the release of CO_2 increases rapidly at higher temperatures. However, complete decomposition of the formed carbonates was not observed because of the reactor design; it was observed that the nickel tube which was used to transport the gases into the melt was blocked due to precipitation of CaO.

Effect of variation in CaO content in the fluoride melts

The influence of the CaO content in the CaF_2/NaF (46.5/53.5 wt%) eutectic mixture on the CO_2 uptake is shown in Figure 10. As noted above, the conversion of CaO in the molten fluoride based salt was found to be highest between 826 and 834°C (Figs. 6 and 7). Accordingly, the experiments with variable CaO concentration were

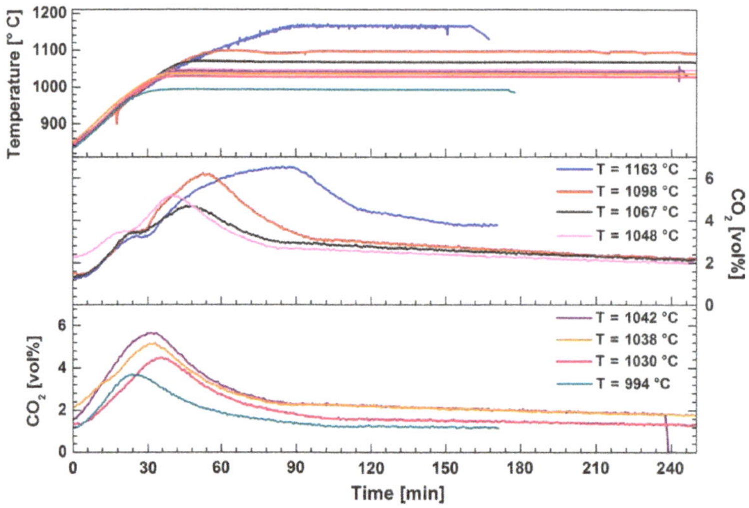

Figure 8. Effect of the temperature on decarbonation in the melt consisting of $CaO/CaF_2/NaF$ (5/44.2/50.8 wt%). Carbonation was conducted by bubbling 14 vol% CO_2 in N_2 through the melt at 830°C for 300 min. The decarbonation step was performed under pure N_2.

carried out within this temperature range. The gas composition in the carbonation step was 14 vol% CO_2 in N_2. It was found that 99.9% of the total applied CO_2 was removed from the inlet gas in the initial carbonation stage with 5 wt% CaO in CaF_2/NaF. The initial, fast CO_2 absorption efficiency decreased to 98.1 % by increasing the CaO concentration from 5 to 20 wt%. The total CaO conversion efficiency decreased rapidly from 94.6 to 75.2 wt% when increasing the CaO concentration from 10 to 20 wt% (Fig. 11). The loss in CaO activity above 10 wt% may be attributed to inactivation of CaO due to deposition and agglomeration of solid deposits of CaO (s) in stagnant parts of the crucible with low circulation with these highly supersaturated liquids.

Regeneration of the CaO was performed by raising the temperature to 1160–1170°C under pure N_2 (Fig. 10). Incomplete CO_2 desorption was recorded by gas analysis for CaO contents of 5, 10, and 15 wt%. This was due to clogging of the purge gas tube due to deposition of CaO in the outlet as previously mentioned. This is evidenced by the abrupt change in recorded value after the initial, rapid release of CO_2. In this case, the regeneration efficiency of CaO was calculated as an estimate from the measured amount of CO_2 released, for the total reacted CaO with CO_2 (Fig. 11). The highest CO_2 release rate from the carbonated sample (84.5% of total captured CO_2) was achieved with 20 wt% CaO in CaF_2–NaF. The formation of carbonates in the CaO–NaF–CaF_2 system may influence the liquidus temperature of the mixture, and also the solubility of CaO. The carbonates have higher solubility than CaO, so higher carbonate concentration would lead to lower melting points and maybe to higher

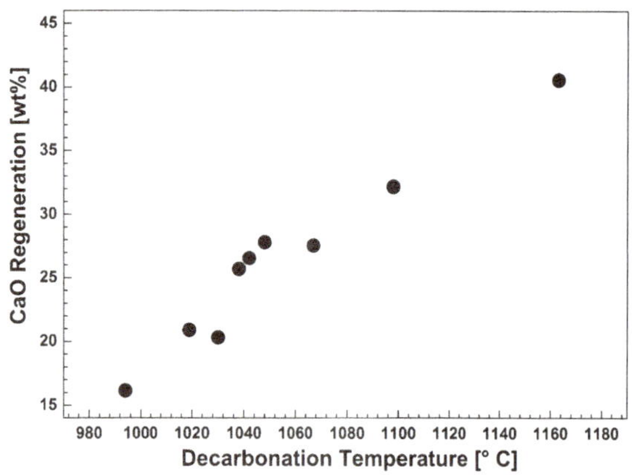

Figure 9. CaO regeneration as a function of the decarbonation temperature. The decarbonation efficiency was determined after 170 min. Carbonation was conducted by bubbling 14 vol% CO_2 in N_2 through the melt consisting of $CaO/CaF_2/NaF$ (5/44.2/50.8 wt%) at 830°C for 300 min. The decarbonation step was subsequently performed under pure N_2.

solubility of CaO or a more complex substance being formed. However, these observations are hard to explain and will be the subject for further studies.

Effect of CO_2 concentration in the simulated flue gas

The impact of the inlet gas composition on sorbent conversion efficiency was examined by varying the CO_2 concentration in N_2 between 1.3 and 14 vol% at 825–835°C. For this purpose, a melt containing 5 wt% CaO was used.

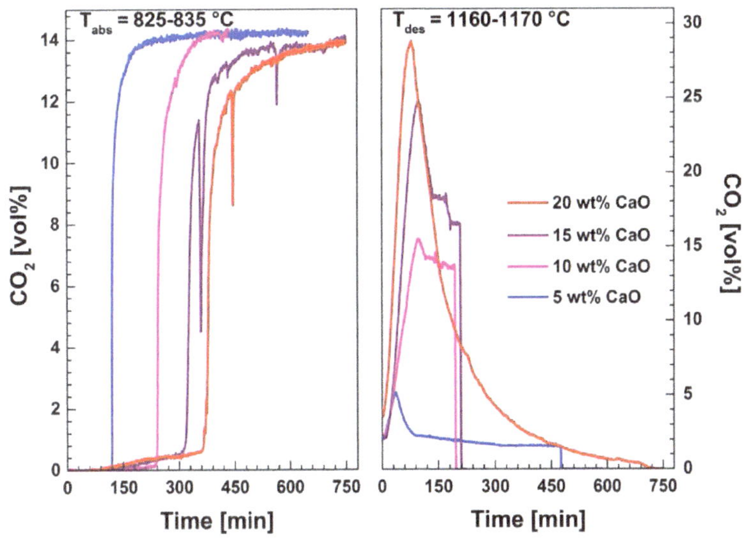

Figure 10. The influence of the CaO concentration on CO_2 capture in a CaF_2–NaF eutectic mixture The carbonation step was conducted by bubbling 14 vol% CO_2 in N_2 through the melt at 825–835°C. The decarbonation step was performed at 1160–1170°C under pure N_2.

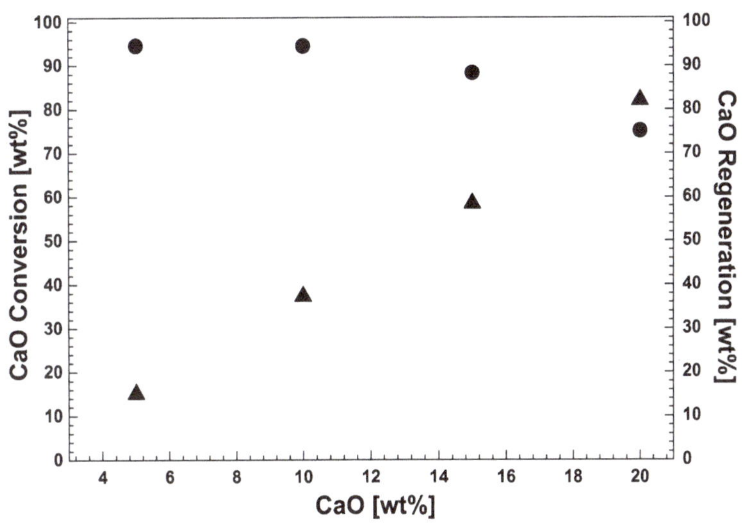

Figure 11. CaO conversion (●) and regeneration (of total reacted CaO, ▲) as a function of the CaO concentration in the CaF_2–NaF eutectic mixture.

Figure 12 shows the concentration of CO_2 versus time in the outlet gas as recorded by the FTIR gas analyzer. It was found that the total carbonation conversion is independent on the inlet CO_2 concentration in N_2, and it was around 93.3 wt% (0.733 g CO_2/g of CaO). However, a closer examination of these results reveals that the regions of fast and slow carbonation reaction are affected by the inlet gas composition. The conversion of CaO in the fast carbonation stage is reduced from 89.5 to 70.1% by decreasing the CO_2 concentration from 14 vol% to 1.3 vol%. In addition, the duration of the slow carbonation reaction was prolonged when the CO_2 concentration decreased. This may be related to phase transitions in the CaO–NaF–CaF_2 system because of the formation of carbonates which influence the solubility of unreacted CaO as well as the liquidus temperature of the melt. As mentioned above, the fast carbonation reaction may be attributed to reaction between CO_2 and dispersed, solid CaO. Therefore, the additional dissolution/dispersion of solid CaO could influence the fast as well as the slow step in the carbonation reaction.

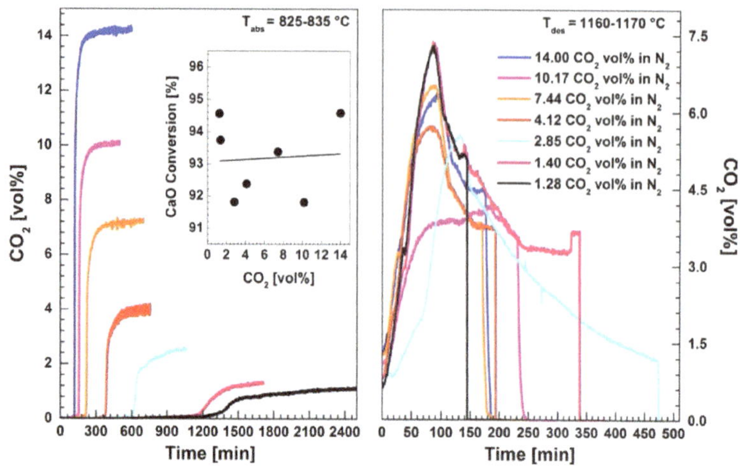

Figure 12. The influence of the CO_2 concentration in N_2 on carbonation/decarbonation reactions in CaO/CaF_2/NaF (5/44.2/50.8 wt%). The carbonation step was conducted by bubbling 1.3–14 vol% CO_2 in N_2 through the melt at 825–835°C. The decarbonation step was performed at 1160–1170°C under pure N_2. Inset; CaO conversion as a function of the CO_2 concentration.

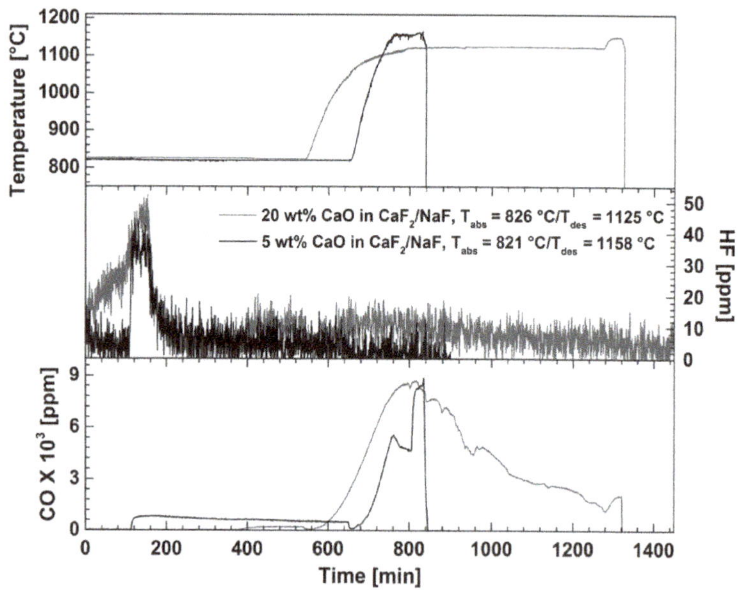

Figure 13. Hydrofluoric acid (HF) and carbon monoxide (CO) concentration in the outlet gas during CO_2 absorption/desorption with 5 and 20 wt% of CaO in a CaF_2–NaF eutectic mixture. The carbonation was carried out at 821 and 826°C, and the decarbonation temperatures were 1125 and 1158°C, respectively. 14 vol% CO_2 in N_2 was applied in the carbonation step and pure N_2 in the decarbonation step.

Effect of CO and HF formation

The concentrations of carbon monoxide (CO, from the reduction in CO_2) and hydrofluoric acid (HF, from the hydrolysis of fluoride salts) are shown in Figure 13. The data were recorded during carbonation experiments using 14 vol% CO_2 in N_2 and 5 or 20 wt% CaO in CaF_2–NaF. Desorption of CO_2 was conducted under pure N_2. The results indicate that the CaO concentration has little or no effect on the HF formation. The HF peak at 50 ppm was detected after 100 min of carbonation at 821 and 826°C, decreasing to 10 ppm. The same value was found during the decarbonation process at elevated temperature (1125 and 1158°C). This indicates that metal fluoride hydrolysis is independent on the temperature in the range 821–1158°C, and it may be related to minor contents of moisture in the inlet gas. The weight change because of the hydrolysis of the fluoride salts was considered negligible.

The carbon monoxide concentration curves indicate that the formation of CO was affected by the CaO concentration (Fig. 13). The CO emission from the reactor in the initial fast carbonation step is close to 0 ppm for both compositions (5 and 20 wt% of CaO in CaF_2–NaF), but during the slow carbonation step, a CO peak of 470 ppm is detected at low CaO concentration. However, during the decarbonation step CO contents up to 9000 ppm are observed, depending on the CO_2 release from the reactor. The carbon monoxide behavior may be explained by CO_2 reduction by iron at elevated temperatures, or by the Boudouard reaction.

Concluding Remarks

The purpose of the present research is to contribute in the ongoing development of CO_2 capture by the calcium- or carbonate looping process. The aim is to investigate if using CaO in a liquid as being dissolved or partly dissolved in molten CaF_2–NaF may be an alternative to solid CaO-based sorbents. By dissolving the active substances in a liquid, the total carrying capacity by total mass is reduced accordingly. However, this may be an alternative approach to modifying solid sorbents by adding secondary oxides or other substances in order to enhance the physical stability of the particles upon cycling. We regard the process as being promising; the laboratory data reported here indicate that carbonation was fast and efficient, and the total efficiency did not deteriorate with increasing number of cycles. Presently, the bottleneck seems to be the decomposition of the carbonates formed, as CO_2 desorption was incomplete because of sedimentation and deposition of CaO. However, this might be possible to solve with a better reactor design, since structure analysis of melt samples indicated that full decomposition of the formed carbonates should be possible. The formation of Na_2CO_3 seems to enhance the absorption, however, the desorption temperature is increased due to the higher stability of sodium carbonate. This will lead to higher energy consumption in a potentially larger scale process.

It should be kept in mind that the present research is still at its fundamental stage. In future investigations it will be necessary to analyze the effect of side reactions on the stability of the molten salt and the efficiency of the overall process. Particularly, reactions involving water vapor, which is always present in real flue gases, may be important. Also, there is still room for improvement of the salt mixture and optimization of the process conditions. Moreover, further research using different reactor designs is needed, and it can be foreseen that the choice of materials will be an important issue in the future as metallic construction materials may not show long-physical stability at these elevated temperatures.

Acknowledgments

The authors thank the Research Council of Norway for financial support through the CLIMIT research program.

Conflict of Interest

None declared.

References

1. MacDowell, N., N. Florin, A. Buchard, J. Hallett, A. Galindo, G. Jackson et al. 2010. An overview of CO2 capture technologies. Energy Environ. Sci. 3:1645–1669.
2. Wang, M., A. Lawal, P. Stephenson, J. Sidders, and C. Ramshaw. 2011. Post-combustion CO2 capture with chemical absorption: a state-of-the-art review. Chem. Eng. Res. Des. 89:1609–1624.
3. Zaman, M., and J. H. Lee. 2013. Carbon capture from stationary power generation sources: a review of the current status of the technologies. Korean J. Chem. Eng. 30:1497–1526.
4. Barker, D. J., S. A. Turner, P. A. Napier-Moore, M. Clark, and J. E. Davison. 2009. CO(2) capture in the cement industry. Energy Procedia 1:87–94.
5. Naranjo, M., D. T. Brownlow, and A. Garza. 2011. CO2 capture and sequestration in the cement industry. 10th Int. Conf. Greenhouse Gas Control Technol. 4:2716–2723.
6. Kuramochi, T., A. Ramirez, W. Turkenburg, and A. Faaij. 2012. Comparative assessment of CO2 capture technologies for carbon-intensive industrial processes. Prog. Energy Combust. 38:87–112.
7. Donat, F., N. H. Florin, E. J. Anthony, and P. S. Fennell. 2012. Influence of high-temperature steam on the reactivity of CaO sorbent for CO2 capture. Environ. Sci. Technol. 46:1262–1269.
8. Liu, W. Q., J. J. Yin, C. L. Qin, B. Feng, and M. H. Xu. 2012. Synthesis of CaO-based sorbents for CO2 capture by a spray-drying technique. Environ. Sci. Technol. 46:11267–11272.
9. Valverde, J. M. 2013. Ca-based synthetic materials with enhanced CO2 capture efficiency. J. Mater. Chem. A 1:447–468.
10. Gupta, H., and L. S. Fan. 2002. Carbonation-calcination cycle using high reactivity calcium oxide for carbon dioxide separation from flue gas. Ind. Eng. Chem. Res. 41:4035–4042.

11. Olsen, E., and V. Tomkute. 2013. Carbon capture in molten salts. Energy Sci. Eng. 1:144–150.
12. Tomkute, V., A. Solheim, and E. Olsen. 2013. Investigation of high-temperature CO2 capture by CaO in $CaCl_2$ molten salt. Energy Fuel 27:5373–5379.
13. Tomkute, V., A. Solheim, and E. Olsen. 2014. CO2 capture by CaO in molten CaF_2–$CaCl_2$: optimization of the process and cyclability of CO_2 capture. Energy Fuel 28:5345–5353.
14. Tomkute, V., A. Solheim, S. Sakirzanovas, B. Oye, and E. Olsent. 2014. Phase equilibria evaluation for CO_2 capture: CaO–CaF_2–NaF, $CaCO_3$–NaF–CaF_2, and Na_2CO_3–CaF_2–NaF. J. Chem. Eng. Data 59:1257–1263.
15. Forsberg, C. W., P. F. Peterson, and H. H. Zhao. 2007. High-temperature liquid-fluoride-salt closed-Brayton-cycle solar power towers. J. Solar Energy Eng. 129:141–146.
16. Beilmann, M., O. Benes, R. J. M. Konings, and T. Fanghanel. 2011. Thermodynamic investigation of the (LiF + NaF + CaF_2 + LaF_3) system. J. Chem. Thermodyn. 43:1515–1524.
17. Cooper, A. F., J. Gittins, and O. F. Tuttle. 1975. System Na_2CO_3–K_2CO_3–$CaCO_3$ at 1 Kilobar and Its Significance in Carbonatite Petrogenesis. Am. J. Sci. 275:534–560.
18. Swainson, I. P., M. T. Dove, and M. J. Harris. 1995. Neutron powder diffraction study of the ferroelastic phase-transition and lattice melting in sodium-carbonate, Na_2CO_3. J. Phys. Condens. Matter 7:4395–4417.

3

The progressive routes for carbon capture and sequestration

Sonil Nanda[1], Sivamohan N. Reddy[2], Sushanta K. Mitra[3] & Janusz A. Kozinski[1]

[1]Department of Earth and Space Science and Engineering, Lassonde School of Engineering, York University, Toronto, Ontario, Canada
[2]Department of Chemical Engineering, Indian Institute of Technology Roorkee, Roorkee, Uttarakhand, India
[3]Department of Mechanical Engineering, Lassonde School of Engineering, York University, Toronto, Ontario, Canada

Keywords
Adsorption, biofuels, carbon dioxide, carbon sequestration, flue gas, oceanic carbon storage

Correspondence
Janusz A. Kozinski, Lassonde School of Engineering, York University, Toronto, Ontario M3J 1P3, Canada.
E-mail: janusz.kozinski@lassonde.yorku.ca

Funding Information
The authors thank Natural Sciences and Engineering Research Council of Canada (NSERC) for the financial support toward this renewable energy research.

Abstract

The global warming is directly related to the increased greenhouse gas emissions from both natural and anthropogenic origins. There has been a drastic rise in the concentration of CO_2 and other greenhouse gases since the industrial revolution primarily due to the intensifying consumption of fossil fuels. With the need to reduce carbon emissions and mitigate global warming certain strategies relating to carbon capturing and sequestration are indispensable. This paper comprehensively describes several physicochemical, biological and geological routes for carbon capture and sequestration. The trend of the increase in greenhouse gases over the years is illustrated along with the global statistics for fossil fuels usage and biofuels production. The physicochemical carbon capturing technologies discussed include absorption, adsorption, membrane separation and cryogenic distillation. The algal and bacterial systems, dedicated energy crops and coalbed methanogenesis have been vividly explained as the biological routes for carbon sequestration. The geological carbon sequestering route centers on biochar application and oceanic carbon storage. A systematic survey has been made on the origin and impact of greenhouse gases along with the potential for sequestration based on some fast-track and long-term sequestration technologies.

Introduction

Today, climate change and global warming are two of the hot topics for discussion at the global environmental panorama. Climate change is a long-lasting and irrevocable shift in the weather conditions recognized by the variations in atmospheric temperature, precipitation, air quality, the wind, and other indicators. The climate change has led to the experiencing of several extreme weather events worldwide, which are mostly attributed to anthropogenic global warming. Some of these unusual and unseasonal weather events include, but are not limited to, heat and cold waves, melting of ice cover, a rise in the sea level, drought, floods, violent storms, and tropical cyclones. The Earth's climate is seasonal and naturally variable on all timescales. Since the yesteryears, the increased concentrations of GHG (greenhouse gases) have led to the induced greenhouse effect resulting in the warming of the planetary surface.

According to IPCC, climate change can be defined as "any change in the climate over time, whether due to natural variability or as a result of human activity." Likewise, the UNFCCC (United Nations Framework Convention on Climate Change) defines climate change as "change in the climate that is attributed directly or indirectly to human activity thereby altering the composition of the global atmosphere and natural climate variability observed over comparable periods of time" [1]. Since the time of industrial revolution, there have been dramatic changes in the global agriculture, material manufacturing, transportation and infrastructure. The rapid urbanization has led to an increase in the emissions of the popular GHGs viz. CO_2 (carbon dioxide), CH_4 (methane) and N_2O (nitrous oxide). In addition to CO_2, CH_4, and N_2O, the GHGs also include SF_6 (sulfur hexafluoride), O_3 (ozone), water vapor, hydrofluorocarbon, and perfluorocarbon groups of gases.

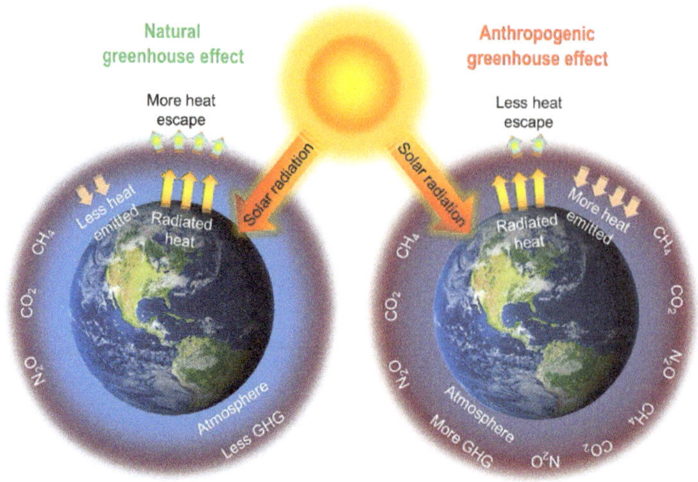

Figure 1. Graphical illustration of natural and anthropogenic greenhouse effects.

A major proportion of the GHGs in the atmosphere is due to anthropogenic reasons. The increased consumption of fossil fuels, use of chlorofluorocarbons in refrigerants, solvents, foam blowing agents and spray propellants are accountable not only for increased GHGs but also in the depletion of ozone layer. The industrial practices such as processing of minerals, metals, chemicals, solvents together with the production and utilization of halocarbons, and SF_6 also contribute to this effect. The GHGs aid in the greenhouse effect by trapping the outgoing infrared radiation from the Earth's surface and adding the heat to the net energy input of lower atmosphere [2]. Figure 1 is a graphical illustration of the natural and anthropogenic greenhouse effects. The higher levels of CO_2, CH_4 and N_2O cause depletion of the ozone layer along with the thickening of the layer of GHGs that trap the heat within the atmosphere making the Earth's surface warmer. The anthropogenic greenhouse effect has been recognized to adversely impact the global climate and functioning of the oceanic and terrestrial ecosystems by altering the temperature and rainfall patterns.

The attributions of CO_2, CH_4, and N_2O towards global warming can be considered to be 60%, 15%, and 5%, respectively [3]. While the concentration of CO_2 and CH_4 is increasing at the rate of 0.4–3% per year, N_2O is rising by 0.2% annually [4]. The Intergovernmental Panel on Climate Change (IPCC) has estimated a global rise in Earth's temperatures by 0.6°C over the last century. However, it can be predicted that the temperature will rise from 1.4°C to 5.8°C in the next two centuries if the anthropogenic emissions of GHGs keep enduring. The month of April 2015 was by far the warmest month on record since 1880s. The global average land surface temperature in April 2015 was 1.11°C above the twentieth century average [5].

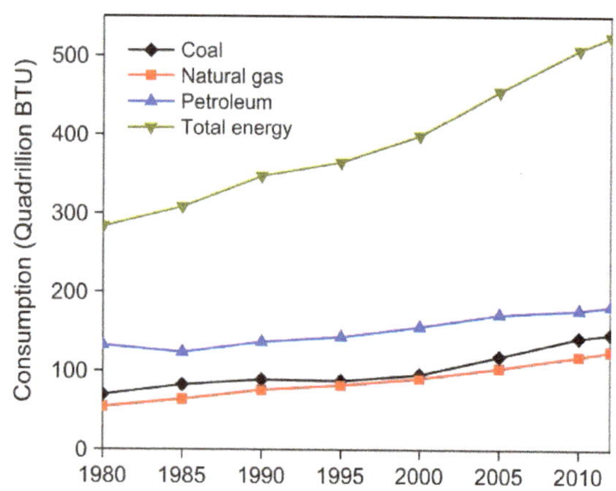

Figure 2. Worldwide consumption of fossil fuels from 1980 to 2012 (Data source: [8]).

The rapid global industrialization has led to the unprecedented consumption of fossil fuels including coal, petroleum and natural gas that releases surplus amounts of CO_2 into the atmosphere. The world human population is expected to grow from 7.3 billion today [6] to 9.2 billion by 2050 [7]. The per capita consumption of energy also increases with the gradual increase in population. The total energy use in 2012 was 524 Quadrillion Btu (quad); however, it is expected to rise by 60% by 2030 [8]. The use of petroleum and other liquid fossil fuels was 85.7 million barrels per day in 2008 with projections for an increase up to 112.2 million barrels per day by 2035 [9]. Figure 2 shows the trend of fossil fuels consumption since 1980. The use of petroleum increased from 156 quads in 2000 to 181 quads in 2012 [8]. Similarly, the demands for coal and dry natural gas also rose to 147 and 124 quads, respectively in 2012. It entreats for a smarter and sustainable way to manage the energy demand for the growing world population. In such a scenario, energy efficiency and conservation along with decarbonizing our energy sources are essential [10].

Climate change is already having significant impacts on the ecosystems, communities and global economy irrespective of any particular geographical region. The increased GHG emissions also threaten the human health due to freshwater shortages, smog, acid rain and other ecological disturbances. The government, policy makers and individuals should act in synergy for mitigating GHG emissions and subsequently lowering their impacts, risks, and associated vulnerabilities. Although the climate change is irreversible, yet it can be alleviated by curbing the GHG emissions (especially CO_2 and CH_4) with strategies to capture and sequester the carbon.

Many papers have reported the potential of CCS (carbon capture and sequestration) to mitigate global warming. Among these many reports, the information about the routes of carbon capturing is scattered, which makes it difficult to evaluate the efficiency of one technology over the other. This paper attempts for a comprehensive review of the underlying principles and current trends in the field of CCS to deter the global warming caused by GHGs. The review is focused on careful integration of technologies for sequestering CO_2 through physicochemical, biological and geological processes. However, it is important to note that the CCS technologies discussed in this paper are either under developmental stage or at a precommercial scale. The CCS technologies have not yet been fully integrated in full-scale commercial operation due to restrained CO_2 capturing efficiency, high costs and lack of regulatory framework.

Current Trend of Greenhouse Gas Emissions

The current top ten CO_2-emitting countries are China, USA, India, Russia, Japan, Africa, Germany, South Korea, Iran, and Saudi Arabia (Fig. 3). China and India rank as the first and third largest CO_2-emitting countries because of their escalating demands for fossil fuels, which are increasing at the rates of 3.5% and 3.9% per year, respectively [11]. As some developed nations such as USA, Russia, Japan, and Germany are in the list of high CO_2 emitters, the Kyoto Protocol places a heavier burden on the principle of "common but differentiated

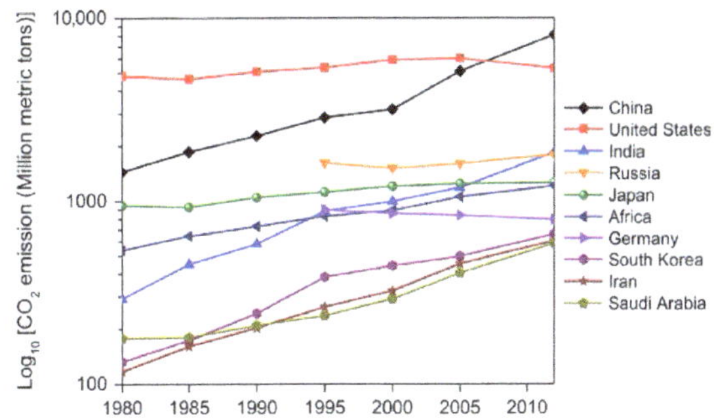

Figure 3. Top ten CO_2-emitting countries in 2012 (Data source: [8]).

responsibilities." The Kyoto Protocol is an international pact linked to UNFCCC that sets international GHG emission reduction targets. According to the Kyoto Protocol (an international environmental treaty proposed by UNFCCC on December 11, 1997 in Kyoto, Japan), it is mandatory for industrialized nations to reduce their anthropogenic GHG emissions by 5.2% from their 1990 levels within the commitment period of 2008–2012.

The "Doha Amendment to the Kyoto Protocol" was adopted in Doha, Qatar, in 2012. During its first commitment period (2008–2012), 37 industrialized countries including the European Union committed to reducing GHG emissions by 5% from their 1990 levels. In the second commitment period (2013–2020), the parties agreed to reduce the GHG emissions by 18% below their 1990 levels by 2020. The arrangement of parties in the second commitment period is different from the first, with Canada, Japan, and Russia withdrawing their commitments from the Protocol in 2011. The Canadian government withdrew from the Kyoto Protocol after ratification in December 2011. Although, Canada was committed to restraint its GHG emissions to 6% below 1990 levels, it showed 17% higher emissions in 2012. The penalties of $13.6 billion for not achieving the targets led to its withdrawal from the treaty [12]. The GHG emissions are debatable for the continual increase as two of the largest CO_2 emitters, namely, Russia and Japan also walked out of the Kyoto agreement.

Recently (November–December 2015) in Paris, France, the United Nations Climate Change Conference, that is, COP21 or CMP11 was held as the 21st annual session of the Conference of the Parties (COP) to the 1992 UNFCCC and the 11th session of Meeting of the Parties to the 1997 Kyoto Protocol. In this global meeting, 200 nations agreed to cut their carbon emissions to set a goal of curbing global warming to <2°C compared to the pre-industrial levels [13]. The GHG emissions have increased by 80% since 1970 and 30% since 1990, totaling 49 gigatonnes (Gt) of CO_2 equivalent in 2010 [14]. According

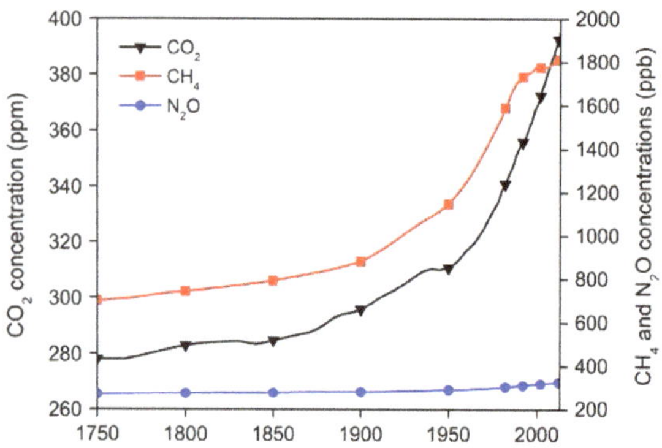

Figure 4. Worldwide greenhouse gas emissions from 1750 to 2013 (Data source: [16]).

to COP21, the parties will purse efforts to reduce the global GHG emissions by 40–70% by 2050 compared to 2010 levels and drop to zero emissions by 2100.

During the past one and half century, there has been a consistent increase in CO_2 concentration by 33% with the rate expected to rise by 4% per year [15]. Figure 4 illustrates the trend of GHG emissions since 1750. The current atmospheric concentration of CO_2 (404 ppm) compared to its 1750 levels (278 ppm) shows a dramatic increase in 265 years [16]. The CO_2 emission from fossil fuels combustion in 2014 was 36 Gt [17]. CO_2 accounted for 77% of the total anthropogenic GHG emissions in 2004 as its annual emissions have grown between 1970 and 2004 by 80%, that is, from 24 to 38 Gt, respectively [18]. CH_4 and N_2O are 20 and 300 times as potent as CO_2, respectively in retaining the atmospheric heat. The levels of CH_4 increased from 700 ppb in 1750 to 1814 ppb in 2013 indicating a 61% rise. Similarly, N_2O concentration grew by 17%, that is, from 270 ppb in 1750 to 326 ppb in 2013.

Despite the anthropogenic sources, the sources of natural GHG emissions are also of parallel consideration. Soil can be considered both as a chief source of atmospheric CO_2 and also a storehouse for carbon storage. Soil contributes a fraction of the total emission of CO_2 through soil respiration (from the plant root and heterotrophic respiration) [19]. Carbon released through soil respiration accounts for around 10% of the total atmospheric carbon pool [20]. Agricultural methods of manure management, field burning of agricultural residues, and rice cultivation are also classified as the sources of GHG emissions [21]. About 80% of CH_4 is produced biologically from rice cultivation, wetlands and sediments, landfills, enteric microbial fermentation and organic waste composting under conditions of low redox potential [22]. About 35% of the annual N_2O emission is from agriculture and change in land use [23]. N_2O is biologically generated through denitrification by heterotrophic microorganisms in oxygen deficient environments, nitrification by autotrophic and heterotrophic nitrifying microorganisms, and dissimilation of nitrate to ammonium by heterotrophic microorganisms in aerobic conditions [24].

The natural catastrophic events such as volcanic eruptions, forest fires and hydrothermal vents also release significant amount of GHGs into the atmosphere. The CO_2 released from volcanoes can be from the erupting magma and degassing of unerupted magma (i.e., recycled sub-ducted crustal materials and decarbonation of shallow crustal materials). They can restore the lost CO_2 from the atmosphere and oceans through weathering of silicates, carbonate deposition and burial of organic carbon. The CO_2 typically makes up to 10 mol% of the total volcanic gas emissions [25]. A hydrothermal vent is a fissure in Earth's surface from which geothermally heated water and hot gas bubbles explode. The dissolution of the gases causes local increases in water density and acidity resulting in sequestration of CO_2 [26]. The emissions from forest fires also have a direct impact on the regional and global carbon cycles by increasing CO_2 levels and affecting carbon sequestration by forests. However, compared to the land use changes (3.4 Gt CO_2 per year) and vehicular emissions (3 Gt CO_2 per year), volcanoes emit less CO_2 (up to 0.44 Gt) [27].

Climate change is predicted to influence biodiversity and agriculture to a large extent. The changes in climate system is unambiguous as evident from: (1) the increase in global average temperatures of soil, air, and water bodies; (2) widespread melting of snow, ice and glacier; (3) shorter freezing seasons of lake and river ice; (4) decrease in permafrost extent; and (5) rise in the average sea level [18]. The climate change and unseasonal weather conditions have also led to habitat shifting and alterations for several ecological species, for example, desertification, coral bleaching, tundra thawing, etc. The rising temperatures cause the meltdown of mountain ice caps, snow, glacier and permafrost. This results in a shift or loss of many species from their natural habitat. The behaviors of adaptation, endangering and extinction are the anticipated in several plant and animal species. The bird migration, egg-laying, leaf-unfolding, as well as the poleward and upward shift in plant and animal species are most common examples [18]. Community shift between fishes, coral reefs, aquatic plants and animals, plankton, algae, and marine bacteria are also associated with rising water temperatures, changes in ice cover, salinity, and oxygen availability.

The emission trading is an economically sensitive strategy at the national and international level for reducing the concentrations of GHGs, particularly CO_2. Carbon credits, carbon trading and carbon markets are some of the CO_2 emission trading approach. A carbon credit is a tradable permit or certificate representing the right to emit one ton of CO_2 or the mass of another GHG with one ton CO_2 equivalent. Carbon trading allows the industries (that cannot practicably reduce CO_2 emissions) to buy credits from those industries that have already reduced their emissions more than the committed level [28]. However, each carbon credit is worth one metric ton of CO_2. The carbon trading market is an economic approach as the cost of emission reduction is higher than the cost of credits. Alternatively, the industries can invest in reforestation projects for removing CO_2 from the atmosphere biologically via photosynthesis and carbon fixation [29]. Although this approach is typically called reduction rather than credit, the process is monitored over time and units are measured in tons of CO_2. Upon the reduction in other GHGs, carbon equivalents can be earned and traded.

Carbon Capturing and Storage Technologies

Carbon capture and storage (or sequestration), often abbreviated as CCS, is the process of capturing CO_2 from large-scale emitters such as fossil fuel refineries, power plants and product manufacturing industries, and transporting it to a storage site thus preventing its re-entry into the atmosphere. The cost of CO_2 separation and compression (to 11 MPa) is estimated to be \$30–50 per ton CO_2, whereas transportation (for every 100 km) and sequestration cost is about \$1–3 per ton CO_2 [30]. There are several routes available for CO_2 removal from the atmosphere or industrial flue gas stream. These routes can be categorized into physicochemical, biological and geological as shown in Figure 5.

The physicochemical route involves the application of absorption, adsorption, gas separation membranes and cryogenic distillation. This route is most popular for capturing CO_2 from industrial flue gas streams. The CO_2 can be isolated from the absorbent material for compression and transportation. The biological route includes algal systems and energy crops that can utilize CO_2 from the atmosphere for photosynthesis and fix carbon in the form of carbohydrates (in lignocellulosic plants) and polysaccharides (in algae). The carbohydrates and polysaccharides can be converted to hydrocarbon fuels via thermochemical (e.g., pyrolysis, torrefaction, liquefaction and gasification) and biochemical (e.g., fermentation and anaerobic digestion) conversion technologies [31]. Coalbed methanogenesis is also included in biological route owing to the involvement of methanogenic bacteria that lead to the biodegradation of coalbed and generating biogenic CH_4 instead to CO_2 that would otherwise be released upon coal combustion. The geological route involves storing carbon in the soil in the form of biochar or through underwater storage.

Carbon dioxide capture can be done in three ways, such as, postcombustion, precombustion, and oxy-fuel combustion [32–34]. The fossil fuels (e.g., coal and natural gas) are subjected to pyrolysis, gasification or reforming reactions in precombustion method to produce clean gas fuels such as H_2 that on combustion produces water [35]. The carbon present in the fuel gets converted to CO_2, and the effluent gas from reforming or gasification majorly comprises of H_2 and CO. The complete conversion of carbon content into CO_2 leads to its higher concentration with high pressure of the CO_2-rich effluent stream that implies lower operating costs for the separation process. High-pressure steam or pure oxygen is a prerequisite for reforming and partial oxidation to generate H_2 from fossil fuels.

The postcombustion process can capture CO_2 on-site from the flue gases emitted as a result of fossil fuel combustion. As the CO_2 concentrations in the flue gas are usually ≤15 vol%, indirect mode of operation (introducing a new phase to capture the required component) is employed to separate and sequester CO_2 [36]. The postcombustion process includes absorption, adsorption, membrane separation, and cryogenic distillation used in the capturing CO_2 from the flue gases. Each of these technologies has been vividly discussed in the following section. Table 1 summarizes different aspects of the pre-, post- and oxy-fuel combustion processes.

The oxy-fuel combustion process is the combustion of fuels with pure oxygen to generate huge quantities of energy at high temperatures [37]. The demand of pure O_2 for the combustion requires its separation from air, which is quite expensive and requires cryogenic distillation. Water and CO_2 are the typical products of the oxy-fuel combustion process. The exhaust gases obtained at

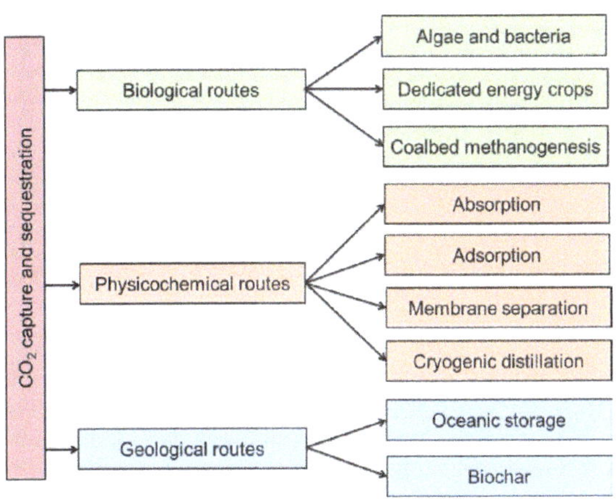

Figure 5. Routes of carbon capture and sequestration.

Table 1. Summary of CO_2 capturing technologies.

Parameter	Precombustion	Postcombustion	Oxy-fuel combustion
CO_2 concentration	15–40 vol%	4–14 vol%	75–80 vol%
Acid gases	Sulfur compounds need to be removed.	Contains NO_x, SO_x, COS and H_2S.	NO_x absent but gas desulfurization is required.
Combustion medium	Steam/air is required for gasification to generate CO_2.	Air is used.	High purity oxygen for combustion.
Equipment size	Medium size equipment.	Large size equipment required with high investment.	Low size equipment.
Temperature and pressure	Low temperature and high pressure (depends on the process employed).	Flue gas need to be cooled and pressure depends on CO_2 capture process.	Cryogenic temperature for separation of O_2. High temperatures are obtained for oxy-fuel combustion hence the flue gas is recycled to reduce the temperature.
Potential	Integrated gasification combined cycle and turbines which can effectively use H_2-rich syngas.	Can be applied to the existing coal combustion plants.	Novel cycles have been employed in synergy with integrated gasification combined cycle.
Pros	Low energy penalty than postcombustion processes. Regeneration can be achieved by altering pressure and temperature.	Small concentrations of CO_2 can be captured. Retrofit technology.	Efficiency of CO_2 capture reaches 100%. No presence of harmful NO_x.
Cons	Drying of syngas and its treatment prior to CO_2 capture. High investment and capital costs.	High operating and regeneration costs. High solvent losses.	High capital and operating cost. Annexing to the existing plants is difficult.
State of the art	Integrated gasification combined cycle and ammonia production plants are running currently.	Amine scrubbing plants (reaction with monoethanolamine) are in practice. Power plants currently employ this technology	Efficient CO_2 separation.

high temperatures are cooled and further recycled back to reduce the temperature of the flue gases. As a result, the exit stream contains higher concentrations of CO_2 for capture and storage. The economy of the process depends on separating O_2 from N_2, which is very expensive [38]. The typical techniques employed to separate CO_2 from the flue gas includes absorption, adsorption, membrane separations, and cryogenic distillation [34, 35].

Chemical looping combustion, first developed by Richter and Knoche [39], is similar to oxy-fuel combustion where O_2 for combustion is supplied by using a metal oxide. By using a metal oxide, the exit flue gas is free of N_2 resulting in high concentrations of CO_2 making it viable for easy capturing. Chemical looping combustion is usually operated in two fluidized-bed reactors where the fuel comes in contact with metal oxide carrier (that supplies O_2 for combustion). Further, the metal oxide is transferred to a regenerative bed where it is reoxidized. The supply of O_2 from the metal oxide carrier depends on its oxidation and reduction cycles. The potential metal oxide carriers are copper, manganese, iron and nickel [40]. However, some technical issues encountered during this process are deactivation of metals, inefficient metal oxide–fuel contact and sulfur poisoning.

Physicochemical Routes for CO_2 Capture and Separation

Absorption

Sorption is a composite process involving absorption and adsorption. Sorption studies can evaluate the surface areas and pore structures of solid materials providing information on the sorption abilities of adsorbates by porous or nonporous solids [41]. Absorption is a gas-liquid operation where the gas components are absorbed into the new liquid phase. The solute is recovered from the solution either by reversing the process conditions or by other mass-transfer operation. The solubility requirements along with constraints such as nonvolatility, low viscosity, nontoxicity, nonflammability, and chemical stability restrict the choice of absorbent solvent to capture gases from effluents.

The capturing of gas molecules from the exhaust gases can be achieved by physical or chemical means. In the physical process, higher concentrations of flue gases at high pressures to capture CO_2 in a suitable solvent by dissolution process. The gas molecules dissolve into the chosen solvent at the given operating conditions to capture harmful gas effluents including CO_2, NO_x, SO_x, and H_2S. Henry's law comes into picture for the absorption of gases into the liquid solvents (partial pressure α concentration of the component in the liquid phase). High pressures and low temperatures are recommended for greater dissolution of gases in liquid solvents while the solvents are regenerated by altering the process temperature and pressure. An additional advantage of physical absorption process is that the solvents are capable of absorbing H_2S, carbonyl sulfide (COS) and hydrocarbons along with CO_2 [42].

The physical absorption of CO_2 by the solvent mixture of dimethyl ethers and polyethylene glycol solution is known as the Selexol process [43]. The typical formula for Selexol solvent is $CH_3(CH_2CH_2O)_{(3-9)}CH_3$ with pressures of 1.5–14 MPa and temperatures of 1–25°C [42, 44]. In the Rectisol process, methanol is employed as the absorbing solvent at −10 to −70°C and 3–8.1 MPa pressure [45]. The solubility of CO_2 in methanol is five times that of water at ambient temperature, although it can be enhanced by 8–15 times at temperatures lower than −0.15°C [46]. In addition to the methanol and Selexol solvents, other solvents used to capture CO_2 are N-methyl-2-pyrrolidine, morpholine, and propylene carbonate. Purisol process involves N-methyl-2-pyrrolidine as the solvent at −20 to −40°C and 1 MPa pressure. Similarly, Fluor process uses propylene carbonate as the solvent at 3–7 MPa and temperatures below 25°C [42]. Since the physical absorption processes require low temperatures (i.e. <25°C) and high pressures, the exit flue gas should be cooled and pressurized to the desired operating conditions. The solubility of CO_2 is usually high in Fluor process and appropriate for the gas streams where the partial pressure of CO_2 is greater than 0.4 MPa [42, 47].

Chemical absorption is one of the traditional techniques in which CO_2 from flue gases reacts with the solvent in an absorption column. The exhaust gas from power plants is obtained at high temperatures and cooled to temperatures below 50°C before feeding it to an absorption column. The cooled gas stream is allowed to pass through the absorption column in which nearly 85–90 vol% of CO_2 is absorbed. The CO_2-rich stream is fed to a regenerator or stripping column. The absorption of CO_2 operates at moderate conditions, that is, 0.1 MPa and 40–50°C while the regeneration of absorbent solvents takes place at 0.2 MPa pressure and high temperature (100–120°C) [47]. The solvents react with CO_2 present in the effluent gas and form a stable product in the absorber that is further passed to a stripping tower where the solvent can be regenerated by altering the temperature and pressure. The CO_2 lean stream solvent is recycled back to the absorption unit, and the energy intensity for recovering the solvent is high.

Table 2 summarizes the current worldwide status of research and development on several physicochemical CCS technologies. It is noteworthy that the technologies mentioned in this table are not fully implemented on a commercial scale; hence limited practical data is accessible on the fate of captured CO_2. The first choice of a feasible solvent is aqueous alkanolamines. The widely used alkanolamine to capture CO_2 from the gaseous stream is monoethanolamine. The reactivity of amines with CO_2 decreases with the increase in alkanol group (e.g., $RNH_2 < R_2NH_2 < R_3NH_2$) [43]. The reactions between amines and CO_2 can be considered similar to the reactions between a weak acid and a weak base. The zwitterions are formed as a result of the reactions between primary and secondary amines reaction. The zwitterions further form carbamates that can be reversed at rising temperatures. The carbamates thus formed are unstable in the presence of tertiary amine solvent, and bicarbonate ions become the main product. The carbamates undergo hydrolysis to form free amine molecules and bicarbonates from which the captured CO_2 is recovered in the regenerator.

The application of these solvents has constraints to capture CO_2 from the flue gas streams. The acidic gases such as SO_x and NO_x pose serious risks to these amine solutions due to their capability to form heat-stable salts that can be corrosive to the reactor in the power plants. The inert gases such as nitrogen and argon do not account for any safety or operational issues during the pipeline transport process or final storage. If not removed, 1 mol% of nitrogen can increase the energy requirement as well as CAPEX (capital expenditure) and OPEX (operational expenditure) in the transport chain with by 1% [48]. The presence of inert gases tends to reduce the volume of actual CO_2 in the transport pipeline and require the same compression energy. In addition, NO_x tends to react with some amine-based solvents and increase the overall reclaiming cost [49].

Solvent losses usually occur at high temperatures due thermal degradation and carbamate polymerization. The absorption usually happens at low temperatures while the regeneration of solvent involves high temperatures, and the solvent loss occurs in the stripping or regeneration tower. Furthermore, in the presence of oxygen, these amines degrade into undesired compounds demanding oxygen-free gas stream. To overcome these issues associated with amine solutions, mixed aqueous amines are employed to enhance their chemical characteristics to capture CO_2.

The economics of the process depends on solvent regeneration and compression of the captured CO_2 for transporting. The energy required for solvent regeneration is 3.2–4.2 GJ/ton of CO_2 and comprises of nearly 60%

Table 2. Current status of different physicochemical CO_2 capturing technologies.

CO_2 capture process	Separation technique	Status of research and development
Physical absorption	Rectisol, Selexol, etc. Mostly integrated gasification combined cycle.	1. Summit Power Group, LLC, Seattle, USA (Texas Clean Energy Project) 2. Don valley, Yorkshire, UK (Don Valley Power Project) 3. Nuon Power, Buggenum, The Netherlands (Integrated gasification combined cycle plant) 4. Elcogas, Puertollano, Spain (Integrated gasification combined cycle plant)
Chemical absorption	Amine, chilled ammonia, and amino acid salt solvent.	1. SaskPower, Saskatchewan, Canada (Boundary Dam Carbon Capture Project) 2. TransAlta Corporation, Alberta, Canada (Project Pioneer Keephills 3 Power Plant) 3. American Electric Power, Ohio, USA (Mountaineer Power Plant) 4. Dow Chemicals, West Virginia, USA (South Charleston plant) 5. PGE Bełchatów Power Station, Łódź Voivodeship, Poland 6. Hazelwood Power Station, Victoria, Australia (Hazelwood Carbon Capture Project) 7. Abu Dhabi Future Energy Company Masdar, Abu Dhabi, UAE (Siemens PostCap™ technology)
Adsorption	Pressure swing adsorption and Pressure-temperature swing adsorption	1. Under developmental stage.
Cryogenics	Cryogenic distillation	1. Air Products and Chemicals, Inc., Pennsylvania, USA
Membrane separation	Polymeric, inorganic and mixed membranes	1. Schwarze Pumpe power station, Spremberg, Germany (Oxy-fuel technology) 2. CS Energy: Callide Power Plant A, Queensland, Australia (Callide Oxy-fuel Project) 3. OxyCoal, UK (Oxy-fuel technology)
Chemical looping combustion	FeO, CuO, MnO, and NiO	1. Less large-scale demonstration plants.

of energy consumption [50]. Additionally, to capture a ton of CO_2, nearly 1.4 kg of monoethanolamine is required. In a study by Knudsen et al. [51], different process upgrades were employed to a 1 ton/h of CO_2 capture test facility operating on flue gas slipstream from a coal-fired power plant. Nearly 25% saving in regeneration energy was achieved with monoethanolamine (3.7 GJ/ton CO_2), process improvements and novel solvents. The improved process configurations and better solvents can reduce the overwhelming power demand of CO_2 removal by amine scrubbing from 0.37–0.51 MWh/ton CO_2 to 0.19–0.28 MWh/ton CO_2, equivalent to 20–30% of a power plant supply [52].

Recently, a novel solvent from amine family namely piperazine (cyclic diamine-$C_4H_{10}N_2$) has been found to enhance the performance both in capturing of CO_2 and maintaining its stability at higher temperatures and aerobic environment with minor solvent losses [53]. The piperazine solvent has a limited solubility in water; hence high temperatures are usually employed for absorption of CO_2. Aqueous ammonia solution has been found to be an alternative to the conventional amine solutions to capture CO_2 due to its inexpensiveness and low reaction heat. Due to the high volatility of ammonia, there is a high probability for its escape into the gas stream creating a technical impediment. The temperature employed in chilled ammonia process ranges from 0 to 20°C where the formed reaction products precipitate in the absorber. An advantage of ammonia solution is its ability to capture the acid gases such as SO_x, NO_x and Hg. The stripping tower operates at 50–200°C in the pressure range of 0.2–13.7 MPa [54].

Amino acid salts are also found to be beneficial as CO_2-absorbing solvents because they are nonvolatile (eliminating solvent loss via gas phase), nontoxic, nonexplosive, odorless, and biodegradable [55]. Recently, Siemens in partnership with Abu Dhabi Future Energy Company (Masdar) in UAE developed a proprietary postcombustion CCS technology called PostCap™. The Siemens PostCap™ technology involves selective absorption of CO_2 from flue gas using amino acid salt solvent and subsequent desorption to gain high purity CO_2. The technology was successfully validated with more than 9000 operational hours in a CO_2-capturing pilot plant adapted to a coal-fired and gas-fired power plant. The PostCap™ technology, capturing 1.8 million tonnes of CO_2 per year, is planned for use in enhanced oil recovery in the local depleted fields [55]. By capturing and transporting the CO_2 emitted from a steel-making process, the enhanced oil recovery project is targeted to benefit from approximately 5 million tonnes of CO_2 captured per year [49].

Another postcombustion process of CO_2 capture is the combination of carbonation and calcination. In carbonation, CaO reacts with CO_2 to form $CaCO_3$, whereas in the regenerator $CaCO_3$ decomposes to CaO and CO_2. After regeneration, CO_2 is stored while CaO is recycled saving the operating cost. While carbonation takes place at 600–700°C, calcination occurs at temperatures greater than 900°C [56]. The carbonation and calcination occur simultaneously in absorption and regeneration fluidized-bed towers. Although it has been proved to be economical than other CO_2 capture processes (i.e., amine-based process), a primary concern for this calcium looping is the decay in the activity of the absorbent.

Recently, ionic liquids have also gained interest in CO_2 capture. Ionic liquids are salts in the liquid state made of ions and short-lived ion pairs. The thermophysical properties such as low vapor pressure, high polarity, and thermal stability with nontoxic nature make ionic liquids as promising solvents in postcombustion CO_2 capture [57]. Depending on the ionic liquid, CO_2 can be both physically and chemical absorbed. The anionic ionic liquids exhibit higher absorbing capacity over the cationic ionic liquids. Amine-based ionic liquids can be synthesized to react with CO_2 to form complexes. The absorption capacity can be enhanced by loading the synthesized ionic liquids onto stable supports (e.g., silica gel) for chemical absorption [58].

Adsorption

The shortcomings of absorption process such as lower gas-liquid contact area, low CO_2 loading, and absorbent losses have shifted the attention toward adsorption. Adsorption is both a pre- and postcombustion CO_2 capture technology that involves a fluid-solid interface for effective capturing of CO_2. It is a gas-liquid operation, in which the molecules from the gas stream are adsorbed on the surface of the solid adsorbent. The component that is adsorbed is termed as the adsorbate while the solid that adsorbs is known as the adsorbent. The fluid molecules adhere to the surface of adsorbent either by physical forces or chemical reactions. While physical adsorption is called physisorption, the chemical adsorption is known as chemisorption [41]. Since adsorption is a surface phenomenon, the properties of solid adsorbents such as surface area, polarity, and porosity along with surface-reactive species play key roles in determining an ideal adsorbent to capture CO_2 from flue gases. The criteria for selecting effective adsorbents are: (1) low cost; (2) high availability; (3) fast reaction kinetics (i.e., adsorption and desorption); (4) low heat capacity; (5) high CO_2 loading and selectivity; and (6) chemical and thermal stability both during adsorption and regeneration. As the adsorption process is an exothermic process, it favors low temperatures, whereas desorption process (i.e., regeneration) is favored at high temperatures.

Physical adsorption is based on the interactions between the molecules of CO_2 and the solid adsorbent surface. The gas solute molecules of CO_2 preferentially interact and adhere to the adsorbent surface if the intermolecular forces between the adsorbent and gas molecules dominate the existing interactive forces between the gases. The heat of adsorption for physical adsorption process ranges from −20 to −50 kJ/mol [59]. The typical operating conditions are with partial pressures of CO_2 close to 0.1 MPa at normal temperatures (i.e., 25°C) due to the exothermic nature of the adsorption process.

Carbonaceous materials, microporous and mesoporous zeolites, chemically surface-modified polymeric materials along with metal-organic frameworks are found to be the potential adsorbents [47, 59]. Carbonaceous materials such as activated carbon, carbon nanotubes, and carbon molecular sieves can capture CO_2 by physical adsorption. Although with high thermal stability, activated carbons have low adsorption capacities. For improved adsorption characteristics by the carbon-based materials, a high surface area is desired along with chemical modification of adsorbent surface with basic groups to enhance the acidic nature CO_2. Activated carbons exhibit fast kinetics rendering less time for adsorption and desorption cycles. The major drawback of activated carbons is lower selectivity toward CO_2. Microporous carbon molecular sieves with narrow pore size distribution and high pore volume increase both selectivity and loading capacity. Carbon nanotubes (both single- and multi-walled) are growing in demand due to their selectivity toward CO_2 compared to other adsorbents.

Both natural and synthetic zeolites have been applied to capture CO_2 from flue gas streams. Among natural zeolites, mordenite, ferrierite, clinoptilolite, and chabazite exhibit an improved performance in the separation of CO_2 from N_2. The crystalline structure of zeolites results in narrow pore size distribution (i.e., 0.5–1.2 nm), which acts as molecular sieves for separation of gas mixtures [60]. The performance of natural zeolites with the high concentration of sodium and large surface area exhibit greater adsorption capacities with enhanced adsorption kinetics [61]. Conventional zeolites are synthesized by altering the silica-to-aluminum ratio and substitution of other metals (e.g., alkali and alkaline earth metals) which lead to rendering charge inside the pores [47, 59]. The structure of adsorbents can be modified by incorporating the reactive basic sites with amine or alkali/alkaline earth metals and enhance the loading capacities. The strong interaction between the basic groups with acidic CO_2 not only improves the adsorption but also increases the selectivity of CO_2 over other gases.

The adsorption capacities of zeolites depend on the pore size and characteristic ions within the pores. The charge induced by the cations in zeolite enhances the selective adsorption of oppositely charged molecules such as CO_2 (quadrupole moment 14.3×10^{-40} C.m^2)[62]. The adsorption kinetics on zeolites has been found to be quick in attaining the equilibrium loadings within a few minutes. The poor selectivity of zeolites toward CO_2 has directed to explore the incorporation of various cations into the adsorbent structures. The performance of synthetic adsorbents depends on the optimization of operating conditions (i.e., low temperature and high partial pressures), basicity (i.e., type of cations that are incorporated into the pore structure of zeolites), pore size, and pore volume.

Another class of materials named as metal-organic frameworks has received considerable attention to capture CO_2 due to high surface area and flexibility in altering the pore structures and surface properties [63]. The solid networks with the metal ion or cluster vertices coupled with organic spacers are usually known are metal-organic frameworks. The choice of organic linkers in the cluster of metal networks provides an opportunity to tune pore structure and its shape to alter CO_2 selectivity, kinetics and carbon loading capacities. Although metal-organic frameworks demonstrate high loading capacities of pure CO_2, yet in the presence of other gases the CO_2 adsorption is reduced dramatically. A considerable amount of research is being focused for improving the selectivity, stability, and recycling of metal-organic frameworks for multiple cycles. The physical adsorbents have poor selectivity and small capacities relatively at low pressures.

Unlike physical adsorption, chemisorption involves the reaction of CO_2 with the reactive groups on the surface of adsorbent, and the heat of adsorption is in the range of −60 to −90 kJ/mol [59]. Chemisorbents are classified into two classes, primarily amine-based adsorbents and alkali metal-based adsorbents [47, 59]. Amine-based adsorbents are synthesized with the help of impregnation and grafting methods. In impregnated adsorbents, weak interactions exist between the support and amine, whereas grafted amine exhibit strong covalent binding. Based on the interactions, the impregnated adsorbents demonstrate low thermal stability compared to the grafted amine-based adsorbents. Among alkali metal-based adsorbents, the carbonates of calcium, potassium, sodium and lithium have demonstrated better performance toward CO_2 capture. Although lithium-based adsorbents show high adsorption capacities, their high diffusional resistance hampers the commercial application [64, 65]. Sodium and potassium-based carbonated chemisorbents have showed high CO_2 adsorption capacities of 9.43 and 7.23 mmol/g, respectively [59]. The moderate adsorption and desorption cycle temperatures (60–200°C) make them potential adsorbents for carbon capture in the flue gases in the postcombustion processes. Microporous materials are competent with other adsorbents due to their high surface area. The polyphenylene material (PPN-4) showed adsorption capacities close to saturation due to its high surface area of 6460 m^2/g [66].

The regeneration of the applied material after the capture of CO_2 is a critical challenge to overcome the energy constraints imposed on the overall process economics. Different regeneration processes such as pressure swing, temperature swing, vacuum swing, hybrid (temperature and pressure), and electric swing adsorption are in practice [56]. The pressure swing adsorption operates at high pressures while desorption occurs at atmospheric pressures. In contrast to pressure swing adsorption, vacuum swing adsorption involves the adsorption and desorption processes at atmospheric pressure and vacuum, respectively [56]. High-temperature flue gas from the power plants is expected to reduce the temperatures for the implication of adsorbents to capture CO_2. Moreover, the flue gas often needs to be pretreated for removing any gases that can poison the adsorbent or reduce the selectivity by undesirable binding to the active sites.

Temperature swing and electric swing adsorptions are other regeneration processes where the adsorption capacities are altered with the change in temperature. Steam or hot air is used as the regeneration medium. In electric swing adsorption, the temperature of adsorbent is increased with the help of electricity (Joule effect) to liberate the adsorbates from adsorbents [56]. In situ heating of adsorbents in electric swing adsorption results in lower energy demand, fast kinetics, and dynamics without constraints on flow rate or heat transfer unlike temperature swing adsorption and vacuum swing adsorption [67]. The demand for longer regeneration time (i.e., in hours) in the case of temperature swing adsorption process compared to seconds in pressure swing adsorption makes the latter an economical option. A combination of pressure and temperature swing adsorption has also been attempted for CO_2 capture [47]. Adsorption also finds application in precombustion processes especially in steam reforming of hydrocarbons and water-gas shift reaction [68]. With the capture of CO_2, the equilibrium shifts toward the right and thereby enhances the productivity of H_2.

Membrane separation

Selective membranes have found versatile applications in the separation of gas mixtures such as in CO_2 capture from natural gas, separation of H_2 from CO_2 in synthesis gas and fuel cells. The two key parameters that decide the performance of the membranes are permeability and selectivity. The volume of gas passing through the membrane per unit area in unit time is termed as permeability. On the other hand, selectivity is defined as the ratio of permeability of the chief component to the other component in mixture. In the postcombustion processes, the flue gas stream comprises of NO_x and SO_x that have adverse impacts on the membranes. Prior to the application of membranes for separation of CO_2 from N_2, it is mandatory for the flue gas to be free of impurities. The flue gas needs to be cooled and compressed to create sufficient pressure differential for CO_2 transport.

The performance of the membranes depends on the driving force, that is, pressure differential (ΔP) for the transfer of the solute from feed to permeate side. The flux of a component (J_i) across the membrane of thickness (δ) can be obtained by Fick's law as shown in equation 1 [63].

$$J_i = \left(\frac{P_i}{\delta}\right) \Delta P \quad (1)$$

There is always a constraint on the pressure differential in terms of economics and selectivity since the increase in pressure decreases the purity of the CO_2 stream on the permeate side. The mechanism for transfer of the solute from feed to permeate depends on the type of membranes. For example, in porous membranes the solute passes through the membrane by pore diffusion. In contrast, in nonporous and facilitated transport membranes, the solute follows the solution-diffusion mechanism and reacts with carrier molecules [69].

A variety of organic (i.e., polymeric), inorganic, mixed membrane, and hybrid systems can be applied to capture CO_2 in postcombustion processes [63, 69]. Different polymeric organic membranes are synthesized for separation of CO_2 from N_2. The organic membranes such as polyimides, polysulfones, polyarylates, polyacetylenes, polycarbonates, polyaniline, etc. exhibit satisfactory permeability (85–450 Barrer) and selectivity (5–55) [63]. The polymeric membranes with reactive carrier molecules (either amines or carbonates) are flexible to fabricate providing high selectivity due to charging. In a recent investigation, researchers have developed a polyvinylamine membrane with CO_2 permeance of 1000 GPU (Gas Permeance Unit) and selectivity of 200 [70]. Moreover, with materials having a selectivity of 50 and permeance of 1000 GPU, the CCS cost has been estimated to be as low as \$23/ton [71].

Inorganic porous membranes such as zeolites, microporous silica, and carbon follow the pore diffusion mechanism for CO_2 transport [69]. Inorganic membranes have low CO_2 permeance with high selectivity. Microporous silica membranes exhibit comparable permeance and selectivity but are restricted for application due to the pore blocking with water and poisoning by SO_x [63, 69]. Membranes with inorganic particles dispersed in the continuous matrix of polymers, known as mixed matrix membranes, have also gained importance. The inorganic materials enhance the permeance and selectivity with good mechanical and thermal stability. In addition, the polymeric materials provide defect-free films at low processing cost. Research efforts are being invested to synthesize thin mixed matrix membranes with high inorganic loading to improve both permeance and selectivity. A hybrid membrane system is the combination of membrane and absorption process to enhance the selective capturing of CO_2. Hydrophobic membranes are usually employed to avoid the interactions between gas and absorption liquid. CO_2 from the gas stream diffuses through the membrane and gets absorbed into the liquid.

Similar to postcombustion processes, membranes are also applied to precombustion processes to selectively separate CO_2 from H_2. High purity CO_2 can be attained with CO_2-selective membranes. Polymeric rubber materials with ether groups, polyethylene glycol polymeric blends and polyethylene oxide block copolymers are considered for high selectivity of CO_2 in the precombustion process [69]. Cross-linked hydrophilic membrane matrix composed of stationary and mobile carriers facilitates high CO_2/H_2 selectivity showing excellent mechanical stability and processing flexibility.

The performance of the facilitated membranes depends on the carrier molecules as well as the highly reactive and stable carriers. Mixed matrix membranes for precombustion processes are targeted to improve CO_2 selectivity and enhance the stability characteristics. The incorporation of nanofillers in polymeric matrix offers high sorption of CO_2 and increases free volume leading to high selectivity and permeability. Carbon nanotubes are found to be potential fillers with high separation and excellent mechanical stabilities [72]. Although mechanical stability can be easily achieved with mixed matrix membranes, yet their low CO_2 separation efficiency is still a challenge to overcome.

The membrane separation processes usually have high energy (i.e., electricity) requirement. For instance, in the case of a target energy requirement of 2 GJ/ton for heat regenerated process, nearly 0.5–0.7 GJ/ton of heat is allowed for membranes [73]. Ho et al. [74] attempted to pressurize the flue gas stream from a coal-fired power plant up to 0.15 MPa while maintaining the permeate stream pressure at 0.008 MPa. About 35% reduction in the cost was achieved in this trail study. Compared to U.S. \$82/ton CO_2, the capture cost was U.S. \$54/ton CO_2 using membrane separation at the pressurized feed conditions.

Cryogenic distillation

Cryogenic separation of flue gases is based on the freezing points of the components to condense the desired product by reducing the temperature. From the flue gas mixture, CO_2 is separated by cooling the gas mixture below −73.3°C at atmospheric pressure to obtain it in liquid form [75]. The required cryogenic temperatures can be obtained by employing compressor, multi-stage heat exchangers, Joule–Thomson valve and cold traps. The composition of CO_2 in flue gas plays a key role on the operating temperature since the greater purity of CO_2 is attained at lower de-sublimation temperatures. For example, 2% CO_2 is obtained at de-sublimation temperature of −116°C, whereas 15% CO_2 is obtained at −99.9°C [63].

Currently, cryogenic separation has been performed in two ways for postcombustion processes. In the separation method proposed by Clodic and Younes [76], CO_2 is de-sublimated to solid CO_2 on the fins of heat exchangers, which is further heated and pressurized to obtain liquid CO_2 in the recovery stage. In another method by Tuinier et al. [77], packed beds are used for de-sublimation of CO_2. CO_2 is recovered from the packing material by feeding fresh gas stream to increase the temperature and enhance the concentration of CO_2 recovered from the packed bed. Though the cryogenic process is energy intensive, it does not involve any additional chemicals in the separation process.

Cryogenic distillation finds its application for the separation of O_2 from the air. The air separation unit employs cryogenic distillation for separation of O_2 from N_2 and other inert gases (mainly Ar). The energy requirement is proportional to the required purity of O_2. The condensation temperatures less than −182°C are applied to air for obtaining a pure stream of O_2 for oxy-fuel combustion [34]. Due to the oxy-fuel combustion, CO_2 concentration in flue gas stream reaches up to 89 vol% that is very high compared to other combustion processes. The efficiency of oxy-fuel CO_2 capture technology majorly depends on the method to obtain O_2. Typically, O_2 purity of 99.5 vol% is in demand, and an excess of 15% is supplied to the power plants for complete combustion [78]. The unreacted excess O_2 is recycled back, and the product stream is condensed to remove the product water and obtain pure CO_2 for compression and storage.

Some new routes to obtain the cold duty required for cryogenic separation of gases have been deployed. The cold duty from liquefied natural gas at its sites has been found to be an option to decrease the economics of cryogenic processes [56]. Cryogenic distillation is also applied in the separation of acid gases from natural gas. Controlled freeze zone process separates CH_4 and acid gases in a single unit [79]. CO_2 and other sulfur containing compounds along with heavier hydrocarbons result in the liquid product providing an economical route for transport and sequestration.

Hart and Gnanendran [80] have reported a cryogenic separation technology called CryoCell® that uses the distinctive solidification property of CO_2 as the basis of its separation from natural gas. The CryoCell® technology reduces water and chemical consumption as well as any corrosion-related issues. The cost savings in a CryoCell® plant are associated with gas treatment, CO_2 disposal, reduced electricity consumption, elimination of solvent pumping, and abridged heat requirement for amine reboiler duty.

Biological Routes for CO_2 Capture and Sequestration

Algal and bacterial systems

Microalgae, macroalgae, and cyanobacteria can be used to utilize the flue gas CO_2 as their carbon source along with sunlight and other nutrients to produce biofuels, value-added chemicals and byproducts. For the flue gas with 4% CO_2 and flow rate of 0.3 L/min, the carbon fixation rate of 15 g carbon/m^2 per day by microalgae has been attained [81]. The biofuels produced from algae are considered to be carbon-neutral as the CO_2 emitted from their combustion is consumed by fresh algae and plants during photosynthesis. Algae are simple photosynthetic aquatic organisms with estimated 300,000 species [82], that is, a diversity much greater than that of terrestrial plants. Microalgae photosynthesis can also result in the precipitation of $CaCO_3$ that is a potential long-lasting carbon sink [83].

The algal cells are made up of polysaccharides, especially hemicellulose (xylose, rhamnose, arabinose, mannose, and glucose); glucosamine; vitamin; proteins; fatty acids; lipids; and amino acids [84, 85]. Some typical green algae include *Botryococcus braunii*, *Chlorella* spp., *Chlamydomonas reinhardtii*, and *Dunaliella salina*. Algae also include diatoms such as *Phaeodactylum tricornutum* and *Thalassiosira pseudonana* as well as heterokonts such as *Nannochloropsis* and *Isochrysis* spp. [82]. Algae are used in BTL (biomass-to-liquid) and BTG (biomass-to-gas) conversion systems. They are attractive feedstock for the production of biofuels such as biodiesel, bioethanol, biobutanol, biogasoline, methane, hydrogen, and jet fuels. The lipids or the oily portion of algae are extracted and converted to biodiesel. *Botryococcus braunii*, although slow growing, is a rich source of hydrocarbons and ether lipids as it contains 60 wt% lipids within its cell wall [86]. The annual yield of oil from algae per unit area is estimated to be between 4.6 and 18.4 L/m^2 (5000 and 20,000 gallons per acre), which is about 30 times greater than the best oil-producing crop [87, 88].

The algal biomass is harvested and processed to release triacylglycerides for transesterification to produce biodiesel. However, extensive research is being done in recovering the lipids from the intracellular location of algae by energy-efficient and economical ways [89]. In addition, the primary consideration should be to convert most of the carbon from the algal biomass to biofuel with potential high-value byproducts. Biodiesel is one of the primary fuels of interest from algae because of: (1) its higher productivity than other plants; (2) higher accumulation of large amounts of triacylglycerides; and (3) lower cultivation cost with no arable land requirement. Today the top five biodiesel-producing countries in the world are USA, Germany, Argentina, Brazil, and France (Fig. 6).

After lipid extraction from algal cells, the *green* waste residue obtained is carbohydrate content of algae that can be fermented into bioethanol or biobutanol. Potts et al. [90] have demonstrated the production of butanol from macroalgae (*Ulva lactuca*) through fermentation by *Clostridium beijerinckii* and *C. saccharoperbutylacetonicum*. From the reducing sugar content of 15.2 g/L in the algal hydrolyzates, about 4 g/L of butanol was recovered as a result of fermentation. Furthermore, Ellis et al. [91] have also demonstrated the production of butanol from wastewater algae via fermentation using *C. saccharoperbutylacetonicum*. Algae such as *Gelidium amansii*, *Laminaria japonica*, *Sargassum fulvellum*, and *Ulva lactuca* have been used as raw materials for ethanol production by Kim et al. [92]. *Escherichia coli* KO11 employed for the fermentation of algal hydrolyzates resulted in ethanol yield of about 0.4 g/g of carbohydrate. The top five bioethanol-producing countries in 2012 were USA, Brazil, China, Canada, and France (Fig. 7).

Although algae utilize CO_2 during the growth cycle, their anaerobic digestion can result in the production of CH_4 as a gaseous fuel. The solar energy and CO_2 stored in the algal biomass in the form of carbohydrates could be released as CH_4 through anaerobic digestion. Yen and Brune [93] performed anaerobic co-digestion of *Scenedesmus* spp. and *Chlorella* spp. along with waste paper to generate CH_4. It is also suggested that the conversion of lipid extracted-algal biomass into CH_4 can recover more energy than the energy from the lipids alone [94]. Thermochemical liquefaction of microalga *Spirulina* using subcritical and supercritical ethanol has shown to produce high-quality bio-oil due to esterification of organic acids, ethanol, fatty acid methyl/dimethyl, and ethyl esters [95].

Although the benefits of algal cultivation are binary, particularly for CO_2 capturing and biofuel/biochemical production, yet there are some challenges to be addressed. Some disadvantages of microalgal cultivation in carbon sequestration are the high operating cost that makes the algal biofuels more expensive than fossil fuels. Sunlight is another limiting factor for algal growth. The artificial light source in the photobioreactors can be an alternative to sunlight, but their installation consequently adds to the overall production cost. Nevertheless, as the microalgal cultures also result in commercially viable products such as natural pigments, dyes, food additives, health

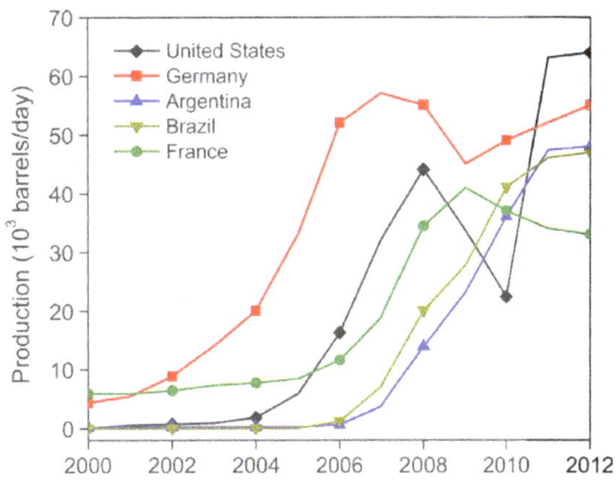

Figure 6. Top five biodiesel-producing countries in 2012 (Data source: [8]).

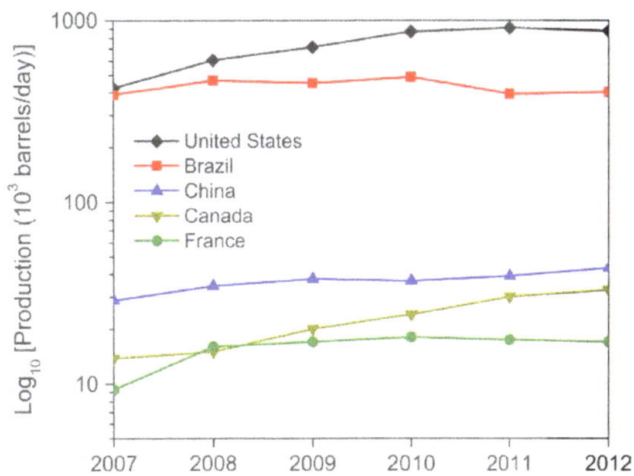

Figure 7. Top five bioethanol-producing countries in 2012 (Data source: [8]).

supplements, antioxidants, and bioactive compounds [96], their high market value may offset the high capital investment. Conversely, macroalgae have gained more attention than microalgae due to their higher biomass yield and utilization of flue gas as a direct source of CO_2, which significantly reduces the production cost [97]. The macroalgae such as *Gracilaria chilensis*, *Hizikia fusiforme*, and *Porphyra yezoensis* demonstrated nearly 2–3 times increased growth at elevated levels of CO_2 compared with the atmospheric CO_2 levels [98]. The use of flue gases containing 12–15% CO_2 is also found to maintain the desired pH range optimal for the macroalgal growth [99].

Cyanobacteria, also known as blue-green algae, are photoautotrophic bacteria that use CO_2 as the carbon source along with light and water as energy and electron sources, respectively. Cyanobacteria are photosynthetic in nature and require anaerobic growth conditions and sunlight. In addition to CO_2 fixation, cyanobacteria can also fix N_2 with the aid of nitrogenase enzymes. A few cyanobacteria that have been used for carbon fixation include *Anabaena* [100], *Aphanothece microscopica Nägeli* [101], *Chlorogleopsis* [102], *Fischerella* [103], and *Synechococcus elongates* [104]. *Synechococcus* is a promising cyanobacterium for CO_2 mitigation because of its high CO_2 uptake rate (i.e., 0.025 g/L/h or 0.6 g/L/day) at a cell concentration of 0.286 g/L [97]. Equation 2 gives the overall reaction of photosynthesis by cyanobacteria.

$$CO_2 + H_2O + \text{(sun)light} \rightarrow (CH_2O)_n + O_2 \qquad (2)$$

Cyanobacteria are also referred to as microbial fuel cells due to the ability to generate H_2 as a result of their metabolic pathway. A few cyanobacteria that can produce H_2 belong to the genus *Anabaena, Aphanocapsa, Calothrix, Chlamydomonas, Chroococcidiopsis, Cyanothece, Gloebacter, Gloeocapsa, Microcoleus, Microcystis, Microcystis, Nostoc, Oscillatoria, Rhodovulum, Synechococcus,* and *Synechocystis* [97, 105]. Both nitrogenase and hydrogenase enzymes in cyanobacteria can aid in biological H_2 production using CO_2 as the carbon source.

Dedicated energy crops

A dedicated energy crop is a non-food plant variety grown for the primary purpose of harvesting energy or biofuel production. The energy crops are mostly lignocellulosic in nature, and their cultivation is relatively less expensive with low-maintenance farming. The energy crops are used to generate biofuels such as bioethanol, biobutanol, bio-oil, biodiesel, and jet fuels, or combusted to generate electricity and heat. The energy crops can be woody (e.g., willow and poplar) or herbaceous in origin (e.g., *Miscanthus*, elephant grass, switchgrass, and timothy grass). These crops can be genetically modified for faster growth, greater biomass yields as well as less water and nutrient requirements. In addition to biofuel production, energy crops have tremendous potentials in carbon sequestration. In a study, *Miscanthus* and switchgrass have shown high yields with low production cost and less environmental impacts compared to conventional agricultural crops [106].

The pyrolysis of an energy crop can result in the production of combustible bio-oil and gases as well as biochar. Nanda et al. [107] and Mohanty et al. [108] have performed the carbon and energy (calorific value) balance of biochar and bio-oil from the pyrolysis of an energy crop – timothy grass. Timothy grass contains 43.4 wt% carbon (C), 6.1 wt% hydrogen (H), 1.3 wt% nitrogen (N), 0.1 wt% sulfur (S), and 45.4 wt% oxygen (O) [109]. The biochar from fast pyrolysis of timothy grass contained 63.7 wt% C, 3.6 wt% H, 1.9 wt% N, 0.04 wt% S and 30.8 wt% O [108], whereas bio-oil from the same process contained 49.2 wt% C, 9.3 wt% H, 2.2 wt% N, 0.9 wt% S, and 38.4 wt% O [107]. However, slow pyrolysis biochar from timothy grass contained 67.5 wt% C, 2.3 wt% H, 1.9 wt% N, 0.1 wt% S and 28.2 wt% O [108], whereas slow pyrolysis bio-oil contained 44.9 wt% C, 8.4 wt% H, 1.8 wt% N, 0.5 wt% S and 44.4 wt% O [107]. As the biochar had more carbon than the biomass, it can make the biorefinery process carbon-negative upon integration with soil amendment for carbon sequestration [110]. The thermochemical conversion of energy crops can also result in energy dense bio-oil and synthesis gas. The calorific value of timothy grass biomass was determined to be 15.9 MJ/kg [109]. While fast pyrolysis bio-oil from timothy grass had high yields, it also demonstrated higher calorific value (23.8 MJ/kg) compared to low yielding slow pyrolysis bio-oil (20.2 MJ/kg) [107].

The soil contains over 2000 Gt of carbon, and the vegetation cover helps in sequestering up to 3 Gt of carbon per annum making it a significant natural carbon pool [97]. Since, the carbon captured in living vegetation is estimated to be around 550 Gt, the trees work as a good carbon sink. The energy crops are lignocellulosic in composition as they contain cellulose, hemicellulose and lignin. The biofuels produced from energy crops are considered carbon-neutral (or near-zero CO_2 accumulation) as the CO_2 emitted from their combustion is recycled by new plants for photosynthesis. In addition, energy crops do not compete with cash crops in terms of arable lands as they can grow in low-value soil without the need for intensive agricultural practices [11].

The carbon-neutral biofuels generated from energy crops can also offset some amount of fossil fuel-derived CO_2. In addition, the carbon sequestered into the soil represents supplementary benefit for CO_2 mitigation that no alternative energy source could provide [111]. Figure 8 gives the trend for the worldwide production of biofuels, especially biodiesel and bioethanol. The continual proliferation in the generation of biofuels is a good indication for the alleviation of CO_2 emissions from fossil fuels [112–114].

The physiological stimulus of plants to CO_2 has received considerable attention because CO_2 is a substrate for photosynthesis. The chief indications of elevated CO_2 on plants include: (1) stimulated photosynthesis; (2) accumulation of non-structural carbohydrates; (3) reduced tissue N_2 concentration; and (4) increased root-to-shoot ratio [115]. In plants, a substantial portion of photosynthate (i.e., 12–54% of carbon fixed by photosynthesis) is allocated to stimulate the root growth [116]. Much of this additional photosynthate is made available to soil microorganisms as root exudates. Hence, elevated atmospheric CO_2 levels could have a significant impact on the energy flow through microbial food webs in the soil. Nevertheless, studies have shown that the rhizosphere of

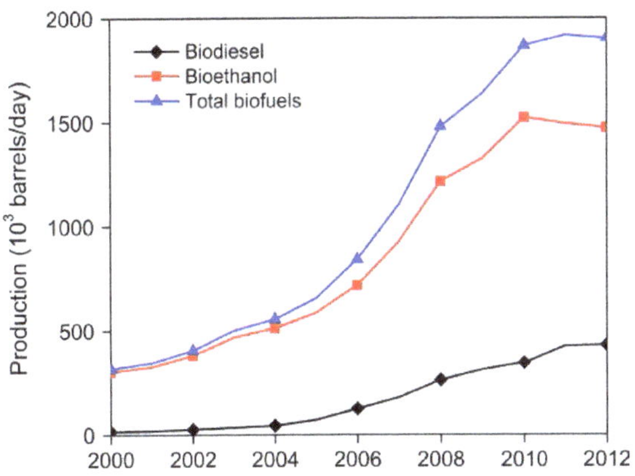

Figure 8. Worldwide production of biofuels from 2000 to 2012 (Data source: [8]).

the plants grown under elevated CO_2 contained 60% of soluble carbon [117]. In most terrestrial energy crop systems, much of the total carbon is below the ground because the soil contains two-third of the total terrestrial carbon in the biosphere [118].

As a response to elevated CO_2, the carbon assimilation in plants can increase through enhanced carbon fixation. This can also lead to higher carbon storage in the soil which is believed to be the largest and most stable terrestrial carbon pool. The high CO_2 levels lead to greater CO_2 fixation in energy crops through photosynthesis and higher carbon storage in the soil through soil-microbial-plant interactions. In addition, soil stores 2–3 times more carbon compared to the atmosphere. The temperature sensitivity of soil organic matter also determines the amount of soil carbon storage and emission over time. The long-term rise in atmospheric CO_2 may stimulate biomass production from energy crops through enhanced root growth resulting in greater carbon input in the soil through rhizodeposition and root exudates [119].

It should also be noted that increased soil microbial activity can lead to a loss of carbon from the soil by respiration and decomposition. However, increased CO_2 concentration can suppress soil organic matter degradation as microorganisms would primarily decompose plant root exudates as readily available carbon source before gearing up to utilize other organic matter from the soil via the priming effect [120]. The soil microorganisms are the chief regulators of soil organic matter and any alteration in its availability impedes their priming effect. In a study with an increase in CO_2 concentration, there was about 85% increase in rhizosphere bacterial population and 170% rise in respiring rhizosphere bacteria [121].

The natural decomposition and mineralization processes transforming soil organic matter into inorganic forms are affected by soil temperature and moisture content [122]. Soil temperature and moisture affect the soil microbial activity at microscale levels, which eventually impacts the cycling of soil organic carbon at local, regional and even global scenarios. The soil organic carbon decomposition and mineralization should be minimized to make an energy crop management system efficient for carbon sequestration. Soil disturbances such as change in the soil temperature, moisture, nutrients, and mineral matter maintain an optimal level of soil organic carbon and prevent its premature decomposition. The farming practices such as tillage is another soil disturbance factor that can disrupt soil aggregates that otherwise occludes soil organic matter [123]. Tillage can increase aeration and land surface area for triggered bacterial activity leading to the decomposition of soil organic matter.

Coalbed methanogenesis

The world coal reserve has dominated as one of the primary sources for fossil fuels. As mentioned earlier, there are significant consequences related to the direct combustion of coal concerning the GHG emissions. By reducing or eliminating the need for coal combustion, carbon emissions can be reduced dramatically, considering that a 500-megawatt coal-fired power plant produces 3 tonnes of CO_2 per year. This could potentially result in a net CO_2 emission reduction of 25% or more in the developed countries. A small but growing body of work related to coalbed methanogenesis exists today [124] including studies on methane-producing microbial communities from different coal seams in USA [125–127], Canada [128], China [129], and India [130].

Biogenic CH_4 is generated through anaerobic microbial process of methanogenesis and constitutes a significant fraction of the natural gas reserves. It is generally accepted that biogenic coalbed CH_4 is a product of coal biodegradation by methanogenic archaea and syntrophic bacteria inhabiting coal beds [129]. These microorganisms artificially stimulate the regeneration of biogenic coalbed CH_4. There have also been reports of field trials conducted in USA where stimulants have been added to coal seams to enhance biogenic CH_4 production [131]. Although the understanding of microbial coal conversion remains rudimentary, it is thought that both acetoclastic and hydrogenotrophic methanogens produce CH_4 by consuming small molecules generated from in situ biodegradation of coal components by other microorganisms [132]. It is reported that acetotrophic, hydrogenotrophic, and methylotrophic methanogenesis are the pathways for production of biogenic CH_4 in coalbed reservoirs [133, 134].

A study on the bioconversion of coal using core-flooding experiments revealed that coal can be economically converted into CH_4 and other value-added products [135]. Along with CH_4 generation, the methanogens produce CO_2, which can be pumped into a depleted oil reservoir for enhanced oil recovery [136], thereby enabling an efficient mechanism for geological carbon sequestration. However, anaerobic subsurface coal bioconversion science is in its infancy. While methanogenesis seems to be a feasible approach, other possibilities for biofuel generation are currently largely untapped. The understanding of microbial life in such coal seams, often refer to as the "microbial dark matter", is still being pursued vigorously due to the advancements of DNA sequencing and genomic analysis. New pathways for in situ bioconversion of coal to CH_4 would shed more light toward the overall reduction in GHGs.

Geological Routes for Carbon Sequestration

Biochar amendment

Increasing the soil carbon content is one of the best strategies to sequester carbon for longer timescales because more than 80% of organic carbon is preserved in the soil [137]. Biochar is one of the promising bioresources rich in recalcitrant carbon, thus making it a long-standing carbon pool. Biochar, a byproduct of biomass pyrolysis and gasification, is obtained along with crude bio-oil and gas components (e.g., H_2, CO, CO_2 and CH_4) [108, 138]. It comprises mostly of the recalcitrant aromatic forms of carbon that cannot be readily oxidized and released into the atmosphere as CO_2 [139]. The occurrence of biochar in the Amazonian *terra preta* soil is a historical indication that biochar can preserve carbon in the soil for hundreds to thousands of years. This makes biochar application a carbon-negative approach for CO_2 sequestration.

Biochar can also capture CO_2 through carbon sequestration and exclude NO_x and SO_x from flue gas through air purification. Biochar applied to soil can also help decrease emissions of N_2O and CH_4 whose respective potencies are 300 and 23 times higher than CO_2 [140]. The carbon sequestration by biochar is different to that of afforestation, conversion into trees, and no-tillage agriculture. While agricultural areas switched to no-tillage farming may lose carbon capturing efficiency within two decades, forestlands start maturing over the years and release CO_2 [141]. These divergences make biochar a long-term carbon sink for reducing CO_2 emissions. The non-charred carbonaceous materials such as dead plant or animal wastes decompose in the soil to release carbon slowly over time that might take a few decades. However, the complete loss of carbon from biochar is very slow and difficult to estimate suggesting it to preserve carbon from centuries to millennia.

Integrating pyrolysis with carbon sequestration can help a biorefinery earn carbon credits. In other words, pyrolysis of biomass converts about 50% of the biomass carbon to bio-oil and gases, while the remaining 50% of carbon is stabilized in biochar [142]. Although burning of bio-oil and combustible gases (H_2 and CH_4) for energy can release CO_2, it is also consumed by plants for photosynthesis making the process carbon-neutral. Consequently, amending biochar to the soil can capture the stable biomass carbon for longer durations making the entire refinery process carbon-negative [11]. This helps in CO_2 abatement and earning carbon credits. In Brazil, carbonization of sugarcane bagasse has shown to optimize the pyrolysis process by producing biochar for use in household and industrial applications [143].

Carbon credits are based in the ratio of 1:1 in relation to the tons of CO_2 stored or removed through carbon sequestration. Nearly 8 Gt of CO_2 is being accumulated per year through burning of fossil fuels [144]. This represents around 8–10 billion carbon credits being created every year. At present, CO_2 trading price by Chicago Climate Exchange is set at U.S. $4/ton [141]. In such a scenario, the carbon credit economy would be comparable in size to the current fossil fuel economy. Hence, the Earth's capacity for storing biochar would be limitless. For evaluating the carbon budget and carbon credit of an integrated bioenergy system, a few factors should be taken into consideration such as: (1) GHG emissions associated with the production of biomass; (2) CO_2 emissions and energy input during biomass harvest, transportation, pretreatment, and conversion; and (3) energy input in product processing and upgrading, flue gas purification, and biochar handling. These factors could address the amount of GHG emissions and carbon balance in each step of the overall biomass conversion chain with carbon sequestration potential by biochar.

Biochar also has several advantages to the soil upon its amendment. Biochar enhances the available water holding capacity, plants' root development, and soil faunal population [145]. The micropores in biochar allow the diffusion of air and water, thereby reducing the overall bulk density of the soil. Biochar, when mixed with the soil, has the increased internal surface area for better adsorption of soil nutrients and humus. Biochar retains substantial quantities of mineral matter in the soil to make them available to plants through their roots. Lehmann [139] suggested two main routes through which biochar can reduce soil or groundwater pollution, that are: (1) by retaining nutrients and minerals in the soil, thereby lowering their chances of leaching into groundwater; and

(2) by improving nutrient availability in the soil thus reducing the use of chemical fertilizers.

Oceanic storage and mineralization

The oceanic CO_2 storage has an expected retention rate of several hundreds of years [146]. The formations of underground carbonate minerals prolong the residence time reducing the chances of CO_2 escape into the atmosphere [147]. The ocean contains approximately 40,000 Gt of carbon compared to 750 Gt of carbon in the atmosphere and 2200 Gt of carbon in the terrestrial ecosystem [56]. The carbon sequestration in oceans is expected at the rate of 5 Gt of carbon per year by 2100 [148]. In the scenario of projected climate change where high CO_2 saturation in oceans may enhance the sequestration rate due to greater carbon fixation by marine photosynthetic organisms, higher temperature might also decrease the solubility of CO_2 in ocean water [97].

Fung et al. [149] performed a series of experiments with the National Center for Atmospheric Research (NCAR) – Climate System Model (CSM1.4) to understand the influence of land and ocean as repositories and sinks of CO_2. The CSM1.4 model included the modified terrestrial biogeochemistry model (i.e., Carnegie–Ames–Stanford Approach or CASA) and the modified Ocean Carbon Intercomparison Project 2 (i.e., OCMIP-2) oceanic biogeochemistry model. The modeling results suggested that carbon sink strengths vary with CO_2 emission rates. In simple words, the terrestrial and oceanic carbon storage capacity can decrease with rapid emissions of CO_2 and other GHGs. According to the model, the magnitude of oceanic carbon storage will depend on the oceanic circulation and sensitivity of marine ecosystem processes at changing climatic conditions (i.e., high temperature and elevated CO_2 levels).

The CO_2 after being separated from the flue gas mixture is compressed to liquid or supercritical fluid state for storage deep under the oceans via pipeline [56]. The CO_2 can be transported and injected into the ocean as dry ice or through vertical injection, inclined pipe transfer, or pipe towed by ship [150]. The longevity of CO_2 captured and injecting in deep reservoirs under the ocean is around thousands of years. Under supercritical conditions, the CO_2 is less dense than water and tends to migrate up the storage formation. Water above its critical temperature (T_c > 374°C) and critical pressure (P_c > 22.1 MPa) is called supercritical water. Likewise, CO_2 above its critical temperature (T_c > 31°C) and critical pressure (P_c > 7.4 MPa) is called as supercritical CO_2. In the event of increasing pressure or weakening of the geological reservoir, the highly pressured CO_2 inclines to escape causing uncertainties on its deep water storability [151]. Recently, saline aquifers were found to be attractive for geological CO_2 storage under water [152, 153]. However, the geological storage of CO_2 under the high-pressure system is under the developmental stage.

The inorganic carbon content in the oceans worldwide is estimated to be around 38,000 Gt with up to 2 Gt of carbon being sequestered annually [97]. The carbon sequestration in oceans can also occur naturally through several ways such as: (1) photosynthesis by deep water aquatic plants as well as photosynthetic bacteria and algae; (2) decomposition of dead aquatic plants and animals; and (3) carbonates formation. The carbon fixed by marine plants and microorganisms is also sequestered upon their decomposition and vertical flux for settlement in the ocean bed.

The photosynthesis by marine plants and microorganisms in the sunlit region and the turbulent diffusion of ocean water ensure the constant flux of carbon from the oceanic surfaces [154]. The biological carbon fixation and subsequent sequestration are enhanced through phytoplankton in the ocean by supplementing additional nutrients such as nitrogen and iron. Such a scheme can also help to earn carbon credits in response to the increased CO_2 emissions. The carbon fixed by the top surface photosynthetic organisms (e.g., aquatic plants, phytoplankton, algae, and photosynthetic bacteria) is eventually sequestered upon their death and descending to the ocean floor. However, any ecological disturbances and mutations in the plankton and photosynthetic microorganisms as a result of the additional nutrients cannot be ignored. The tumbling of surplus quantities of biomass formation from photosynthetic organisms and plankton can also lead to anaerobic digestion and CH_4 formation counteracting carbon sequestration [146].

The solubility pump is a technology used to dissolve CO_2 in ocean water for carbonate formation. This involves the conversion of CO_2 to inorganic carbonates through mineralization. The mineralization process offers an opportunity for the durable storage of CO_2, although it is subject to natural weathering process over longer durations. After the carbonates are generated, their complexes are carried away via water current towards the benthic ocean regions where they linger for longer periods [155]. In the oceans, the carbon is stored in the form of carbonates, especially $CaCO_3$ and $MgCO_3$, and their formation is summarized in equations 3 and 4.

$$CO_2 + CaSiO_3 \leftrightarrow CaCO_3 + SiO_2 \tag{3}$$

$$CO_2 + MgSi \leftrightarrow MgCO_3 + SiO_2 \tag{4}$$

The carbonate formation acts as a long-term source of carbon that is driven by the metabolic activities by photosynthetic organisms including cyanobacteria and algae [156]. The formation of carbonates especially $CaCO_3$ in the form of shells is also found in coral reefs by many aquatic organisms such as benthic molluscs, echinoderms, corals, and plankton belonging to the group coccolithophore. However, the particulate organic matter and dissolved organic matter in oceans have a tendency to undergo microbial mineralization releasing most of the organic matter to dissolved inorganic carbon [156]. However, a small proportion of particulate organic matter escapes mineralization as it reaches the sediment where organic carbon remains buried for hundreds of years [157].

The geological CO_2 storage can also have some limitations in the case of unforeseen CO_2 leakage. Due to asphyxiation, CO_2 poses life-threatening risks for aquatic organisms as it can accumulate locally due to its density. Higher CO_2 levels may also change the water pH causing acidification of groundwater, thereby dissolving many toxic heavy metals [158]. The ocean acidification by CO_2 leakage has a direct impact on the growth of corals. It also alters the chemical speciation and biogeochemical cycling of many elements and compounds. Ocean acidification also lowers the saturation states of $CaCO_3$, which has adverse impacts on the shell-forming marine organisms such as plankton, corals, mussels, snails, clams, etc. [159]. The broader implications and adaptation of marine organisms against increasing CO_2 is not well understood for ocean ecosystems; hence future research is required to advance the innocuous geological CO_2 storage approach.

Conclusions

Carbon sequestration is considered as a feasible approach to mitigate the increased greenhouse gas emissions that lead to global warming and climate change. The currently available technologies for carbon capture and sequestration can be classified under physicochemical, biological and geological routes. The physicochemical route includes some fast-track techniques for the separation of CO_2 from flue gas mixtures with the use of absorbents, adsorbents, gas separation membranes and cryogenic distillation. In contrast, the biological route involves a long-term approach such as carbon fixation through the cultivation of microalgae, macroalgae, cyanobacteria, and energy crops. The biofuels derived from energy crops (e.g., temperate grasses and short-rotation woody biomass) are accounted as carbon-neutral as the plants recycle the CO_2 released from biofuel combustion. Coal, a geological fossil fuel resource, is also subject to biogenic degradation by methanogenic bacteria resulting in CH_4 generation. This biogenic CH_4 could be considered as a clean fuel gas compared to the CO_2 that would otherwise be released upon coal combustion. Finally, the geological storage of carbon includes amendment of biochar into the soil, pumping of CO_2 deep into the oceans, and mineralization of CO_2 in the form of carbonates in seawater. Although these routes seem to be promising, yet their commercial applications are required for implementation toward carbon credits and carbon trading. The current challenge is to develop an international regulatory framework that would help decide the viability of these carbon sequestering technologies based on a long-term lifecycle assessment.

Acknowledgments

The authors thank Natural Sciences and Engineering Research Council of Canada (NSERC) for the financial support toward this renewable energy research.

Conflict of Interest

None declared.

References

1. Metz, B., O. R. Davidson, P. R. Bosch, R. Dave, and L. A. Meyer. 2007. Climate change 2007: mitigation of climate change. Cambridge Univ. Press, New York, USA.
2. Rastogi, M., S. Singh, and H. Pathak. 2002. Emission of carbon dioxide from soil. Curr. Sci. 82:510–517.
3. Watson, R. T., M. C. Zinyowera, and R. H. Moss. 1996. Climate change 1995: impacts, adaptations and mitigation of climate change. Cambridge Univ. Press, New York, USA.
4. Battle, M., M. Bender, T. Sowers, P. P. Tans, J. H. Butler, J. W. Elkins, et al. 1996. Atmospheric gas concentrations over the past century measured in air from fin at the South Pole. Nature 383:231–235.
5. NOAA (National Centers for Environmental Information). 2015. Climate Monitoring – Global Summary Information April 2015. Asheville, North Carolina. Available at https://www.ncdc.noaa.gov/sotc/summary-info/global/201504 (accessed 30 June 2015).
6. U.S. Census Bureau. 2016. U.S. and World Population Clock. Available at http://www.census.gov/popclock (accessed 21 March 2016).
7. United Nations. 2007. World Population Prospects: The 2006 Revision, Population database. Economics and Social Affairs. United Nations, New York.
8. USEIA (U.S. Energy Information Administration). 2015. Independent Statistics & Analysis. Available at http://www.eia.gov (accessed 17 October 2015).
9. IEO (International Energy Outlook). 2011. U.S. Energy Information Administration, Washington, DC. Available at http://www.eia.gov/ieo/pdf/0484(2011).pdf. (accessed 01 March 2011).
10. Chu, S., and A. Majumdar. 2012. Opportunities and challenges for a sustainable energy future. Nature 488:294–303.

11. Nanda, S., R. Azargohar, A. K. Dalai, and J. A. Kozinski. 2015. An assessment on the sustainability of lignocellulosic biomass for biorefining. Renew. Sustain. Energ. Rev. 50:925–941.
12. The Guardian. 2011. Canada pulls out of Kyoto protocol. Available at http://www.theguardian.com/environment/2011/dec/13/canada-pulls-out-kyoto-protocol (accessed 30 June 2015).
13. Sutter, J. D., J. Berlinger, and R. Ellis. 2015. Obama: climate agreement 'best chance we have' to save the planet. Cable News Network (CNN). Available at http://www.cnn.com/2015/12/12/world/global-climate-change-conference-vote (accessed 17 January 2016).
14. UNCCC. 2016. United Nations Climate Change Conference COP21/CMP11. Available at http://www.cop21.gouv.fr/en/learn/what-is-cop21/the-phenomenon-of-climate-disruption (accessed 17 January 2016).
15. Solomon, S., D. Qin, M. Manning, M. Marquis, K. Avery, M. M. B. Tignor, et al. 2007. Climate change 2007: the physical science basis. Cambridge Univ. Press, New York, USA.
16. EEA (European Environment Agency). 2015. Atmospheric concentration of carbon dioxide, methane and nitrous oxide. Copenhagen, Denmark. Available at http://www.eea.europa.eu
17. Earth CO_2 Home Page. 2016. CO_2 Earth: Are we stabilizing yet?. Available at www.co2.earth (accessed 21 March 2016).
18. Pachauri, R. K., and A. Reisinger. 2007. Climate change 2007: synthesis report. Intergovernmental Panel on Climate Change (IPCC). IPCC, Geneva, Switzerland.
19. Smith, W. N., P. Rochette, C. Monreal, R. L. Desjardins, E. Pattey, and A. Jaques. 1997. The rate of C change in agricultural soils in Canada at the landscape level. Can. J. Soil Sci. 77:219–229.
20. Raich, J. W., and C. S. Potter. 1995. Global patterns of carbon dioxide emissions from soils. Glob. Biogeochem. Cyc. 9:23–36.
21. UNFCCC (United Nations Framework Convention on Climate Change). 2008. Kyoto protocol reference manual: on accounting of emissions and assigned amount. UNFCCC, Bonn, Germany.
22. Yang, S. S., C. M. Lui, C. M. Lai, and Y. L. Lui. 2003. Estimation of methane and nitrous oxide emission from paddy fields and uplands during 1990–2000 in Taiwan. Chemosphere 52:1295–1305.
23. Isermann, K. 1994. Agriculture's share in the emission of trace gases affecting the climate and some cause-oriented proposals for sufficiently reducing this share. Environ. Pollut. 83:95–111.
24. Yu, K. W., Z. P. Wang, A. Vermoesen, W. H. Jr Patrick, and O. Van Cleemput. 2001. Nitrous oxide and methane emissions from different soil suspensions: effect of soil redox status. Biol. Fertil. Soils 34:25–30.
25. Burton, M. R., and G. M. Sawyer. 2013. Deep carbon emissions from volcanoes. Rev. Mineral. Geochem. 75:323–354.
26. Carey, S., P. Nomikou, K. C. Bell, M. Lilley, J. Lupton, C. Roman, et al. 2013. CO_2 degassing from hydrothermal vents at Kolumbo submarine volcano, Greece, and the accumulation of acidic crater water. Geology 41:1035–1038.
27. Gerlach, T.. 2011. Volcanic versus anthropogenic carbon dioxide. Eos 92:201–208.
28. McHale, M. R., E. G. McPherson, and I. C. Burke. 2007. The potential of urban tree plantings to be cost effective in carbon credit markets. Urban For. Urban Green. 6:49–60.
29. Cairns, R. D., and P. Lasserre. 2006. Implementing carbon credits for forests based on green accounting. Ecol. Econ. 56:610–621.
30. Wang, B., Y. Li, N. Wu, and C. Q. Lan. 2008. CO_2 bio-mitigation using microalgae. Appl. Microbiol. Biotechnol. 79:707–718.
31. Nanda, S., J. Mohammad, S. N. Reddy, J. A. Kozinski, and A. K. Dalai. 2014. Pathways of lignocellulosic biomass conversion to renewable fuels. Biomass Conv. Bioref. 4:157–191.
32. Gibbins, J., and H. Chalmers. 2008. Carbon capture and storage. Energ. Policy 36:4317–4322.
33. Thiruvenkatachari, R., S. Su, H. An, and X. X. Yu. 2009. Post combustion CO_2 capture by carbon fibre monolithic adsorbents. Prog. Energ. Combust. Sci. 35:438–455.
34. Markewitz, P., W. Kuckshinrichs, W. Leitner, J. Linssen, P. Zapp, R. Bongartz, et al. 2012. Worldwide innovations in the development of carbon capture technologies and the utilization of CO_2. Energ. Environ. Sci. 5:7281–7305.
35. Rubin, E. S., H. Mantripragada, A. Marks, P. Versteeg, and J. Kitchin. 2012. The outlook for improved carbon capture technology. Prog. Energ. Combust. Sci. 38:630–671.
36. IPCC (Intergovernmental Panel on Climate Change). 2005. Special report on carbon dioxide capture and storage. Cambridge Univ. Press, New York, USA.
37. Kather, A., and G. Scheffknecht. 2009. The oxycoal process with cryogenic oxygen supply. Naturwissenschaften 96:993–1010.
38. Burdyny, T., and H. Struchtrup. 2010. Hybrid membrane/cryogenic separation of oxygen from air for use in the oxy-fuel process. Energy 35:1884–1897.
39. Richter, H. J., and K. F. Knoche. 1983. Reversibility of combustion processes. Pp. 71–85 *in* R. A. Gaggioli, ed. Efficiency and costing ACS symposium series 235. Oxford Univ. Press, Washington, D.C., USA.
40. Lyngfelt, A., B. Kronberger, J. Adanez, J. X. Morin, and P. Hurst. 2004. The GRACE project. Development

of oxygen carrier particles for chemical-looping combustion. Design and operation of a 10 kW chemical-looping combustor. Proceedings of the 7th International Conference on Greenhouse Gas Control Technologies, Vancouver, Canada, September 2004.
41. Kwon, J. H., and L. D. Wilson. 2010. Surface-modified activated carbon with β-cyclodextrin—Part II. Adsorption properties. J. Environ. Sci. Health A Tox. Hazard. Subst. Environ. Eng. 45:1793–1803.
42. Chabanon, E., B. Belaissaoui, and E. Favre. 2014. Gas–liquid separation processes based on physical solvents: opportunities for membranes. J. Membr. Sci. 459:52–61.
43. Padurean, A., C. C. Cormos, A. M. Cormos, and P. S. Agachi. 2011. Multicriterial analysis of post-combustion carbon dioxide capture using alkanolamines. Int. J. Greenh. Gas Control 5:676–685.
44. Padurean, A., C. C. Cormos, and P. S. Agachi. 2012. Pre-combustion carbon dioxide capture by gas–liquid absorption for Integrated Gasification Combined Cycle power plants. Int. J. Greenh. Gas Control 7:1–11.
45. Burr, B., and L. Lyddon. 2008. A comparison of physical solvents for acid gas removal. 87th Annu. Conv. Proc. 2–5.
46. Kaneco, S., H. Katsumata, T. Suzuki, and K. Ohta. 2006. Electrochemical reduction of carbon dioxide to ethylene at a copper electrode in methanol using potassium hydroxide and rubidium hydroxide supporting electrolytes. Electrochim. Acta 51:3316–3321.
47. Yu, C. H., C. H. Huang, and C. S. Tan. 2012. A review of CO_2 capture by absorption and adsorption. Aerosol Air Qual. Res. 12:745–769.
48. Aspelund, A., and K. Jordal. 2007. Gas conditioning—The interface between CO_2 capture and transport. Int. J. Greenhouse Gas Control 1:343–354.
49. Birley, R., A. Reichl, T. Schliepdiek, and O. Reimuth. 2014. Optimisation and integration of two post combustion capture plants to reduce CAPEX and OPEX. Power-Gen Europe, Cologne, Germany, June 3–5, 2014. Seimens AG, Germany: 1–15.
50. Knudsen, J. N., J. N. Jensen, P. J. Vilhelmsen, and O. Biede. 2009. Experience with CO_2 capture from coal flue gas in pilot-scale: testing of different amine solvents. Energ. Proc. 1:783–790.
51. Knudsen, J. N., J. Andersen, J. N. Jensen, and O. Biede. 2011. Evaluation of process upgrades and novel solvents for the post combustion CO_2 capture process in pilot-scale. Energ. Proc. 4:1558–1565.
52. Rochelle, G. T. 2009. Amine scrubbing for CO_2 capture. Science 325:1652–1654.
53. Freeman, S. A., R. Dugas, D. H. Van Wagener, T. Nguyen, and G. T. Rochelle. 2010. Carbon dioxide capture with concentrated, aqueous piperazine. Int. J. Greenh. Gas Control 4:119–124.
54. Darde, V., K. Thomsen, W. J. Van Well, and E. H. Stenby. 2010. Chilled ammonia process for CO_2 capture. Int. J. Greenh. Gas Control 4:131–136.
55. Reichl, A., G. Schneider, T. Schliepdiek, and O. Reimuth. 2014. CCS for enhanced oil recovery: Integration and optimization of a post combustion CO_2 capture facility at a power plant in Abu Dhabi. Society of Petroleum Engineers. Proceedings of Abu Dhabi International Petroleum Exhibition and Conference, Abu Dhabi, UAE, November 10–13, 2014. SPE 171692. SPE Int.: 1–11.
56. Pires, J. C. M., F. G. Martins, M. C. M. Alvim-Ferraz, and M. Simões. 2011. Recent developments on carbon capture and storage: an overview. Chem. Eng. Res. Des. 89:1446–1460.
57. Zhang, X., X. Zhang, H. Dong, Z. Zhao, S. Zhang, and Y. Huang. 2012. Carbon capture with ionic liquids: overview and progress. Energ. Environ. Sci. 5:6668–6681.
58. Zhang, Y., S. Zhang, X. Lu, Q. Zhou, W. Fan, and X. Zhang. 2009. Dual amino-functionalised phosphonium ionic liquids for CO2 capture. Chem. Eur. J. 15:3003–3011.
59. Samanta, A., A. Zhao, G. K. Shimizu, P. Sarkar, and R. Gupta. 2011. Post-combustion CO_2 capture using solid sorbents: a review. Ind. Eng. Chem. Res. 51:1438–1463.
60. Chester, A. W., and E. G. Derouane. 2009. Zeolite characterization and catalysis: a tutorial. Springer, Dordrecht, The Netherlands.
61. Siriwardane, R. V., M. S. Shen, and E. P. Fisher. 2003. Adsorption of CO_2, N_2, and O_2 on natural zeolites. Energy Fuels 17:571–576.
62. Coriani, S., A. Halkier, A. Rizzo, and K. Ruud. 2000. On the molecular electric quadrupole moment and the electric-field-gradient-induced birefringence of CO_2 and CS_2. Chem. Phys. Lett. 326:269–276.
63. Spigarelli, B. P., and S. K. Kawatra. 2013. Opportunities and challenges in carbon dioxide capture. J. CO2 Utilization 1:69–87.
64. Iwan, A., H. Stephenson, W. C. Ketchie, and A. A. Lapkin. 2009. High temperature sequestration of CO_2 using lithium zirconates. Chem. Eng. J. 146:249–258.
65. Venegas, M. J., E. Fregoso-Israel, R. Escamilla, and H. Pfeiffer. 2007. Kinetic and reaction mechanism of CO_2 sorption on Li_4SiO_4: study of the particle size effect. Ind. Eng. Chem. Res. 46:2407–2412.
66. Dawson, R., A. I. Cooper, and D. J. Adams. 2013. Chemical functionalization strategies for carbon dioxide capture in microporous organic polymers. Polym. Int. 62:345–352.
67. An, H., B. Feng, and S. Su. 2011. CO_2 capture by electrothermal swing adsorption with activated carbon fibre materials. Int. J. Greenh. Gas Control 5:16–25.
68. Mendes, D., A. Mendes, L. M. Madeira, A. Iulianelli, J. M. Sousa, and A. Basile. 2010. The water-gas shift

reaction: from conventional catalytic systems to Pd-based membrane reactors—a review. Asia-Pac. J. Chem. Eng. 5:111–137.
69. Ramasubramanian, K., Y. Zhao, and W. S. W. Ho. 2013. CO2 capture and H2 purification: prospects for CO2-selective membrane processes. AIChE J. 59:1033–1045.
70. Kim, T. J., H. Vrålstad, M. Sandru, and M. B. Hägg. 2013. Separation performance of PVAm composite membrane for CO_2 capture at various pH levels. J. Membr. Sci. 428:218–224.
71. Merkel, T. C., H. Lin, X. Wei, and R. Baker. 2010. Power plant post-combustion carbon dioxide capture: an opportunity for membranes. J. Membr. Sci. 359:126–139.
72. Murali, R. S., S. Sridhar, T. Sankarshana, and Y. V. L. Ravikumar. 2010. Gas permeation behavior of Pebax-1657 nanocomposite membrane incorporated with multiwalled carbon nanotubes. Ind. Eng. Chem. Res. 49:6530–6538.
73. Favre, E. 2011. Membrane processes and postcombustion carbon dioxide capture: challenges and prospects. Chem. Eng. J. 171:782–793.
74. Ho, M. T., G. W. Allinson, and D. E. Wiley. 2008. Reducing the cost of CO_2 capture from flue gases using membrane technology. Ind. Eng. Chem. Res. 47:1562–1568.
75. Olajire, A. A. 2010. CO_2 capture and separation technologies for end-of-pipe applications–a review. Energy 35:2610–2628.
76. Clodic, D., and M. Younes. 2002. A new method for CO_2 capture: frosting CO_2 at atmospheric pressure. Sixth International Conference on Greenhouse Gas Control Technologies, GHGT6. Kyoto, October 2002, 155–160.
77. Tuinier, M. J., M. van Sint Annaland, G. J. Kramer, and J. A. M. Kuipers. 2010. Cryogenic CO_2 capture using dynamically operated packed beds. Chem. Eng. Sci. 65:114–119.
78. Castillo, R. 2011. Thermodynamic analysis of a hard coal oxyfuel power plant with high temperature three-end membrane for air separation. Appl. Energ. 88:1480–1493.
79. Northrop, P. S., and J. A. Valencia. 2009. The CFZ™ process: a cryogenic method for handling high-CO_2 and H_2S gas reserves and facilitating geosequestration of CO_2 and acid gases. Energ. Proc. 1:171–177.
80. Hart, A., and N. Gnanendran. 2009. Cryogenic CO_2 capture in natural gas. Energ. Proc. 1:697–706.
81. Watanabe, Y., and D. O. Hall. 1996. Photosynthetic CO_2 conversion technologies using a photobioreactor incorporating microalgae—energy and material balances. Energ. Convers. Manage. 37:1321–1326.
82. Scott, S. A., M. P. Davey, J. S. Dennis, I. Horst, C. J. Howe, D. J. Lea-Smith, et al. 2010. Biodiesel from algae: challenges and prospects. Curr. Opin. Biotechnol. 21:277–286.
83. Aresta, M., A. Dibenedetto, and G. Barberio. 2005. Utilization of macro-algae for enhanced CO_2 fixation and biofuels production: development of a computing software for an LCA study. Fuel Process. Technol. 86:1679–1693.
84. Toroghi, M. K., G. Goffaux, and M. Perrier. 2013. Observer-based backstepping controller for microalgae cultivation. Ind. Eng. Chem. Res. 52:7482–7491.
85. Abo-Shady, A. M., Y. A. Mohamed, and T. Lasheen. 1993. Chemical composition of the cell wall in some green algae species. Biol. Plantarum 35:629–632.
86. Metzger, P., and C. Largeau. 2005. *Botryococcus braunii*: a rich source for hydrocarbons and related ether lipids. Appl. Microbiol. Biotechnol. 66:486–496.
87. Atabani, A. E., A. S. Silitonga, I. A. Badruddin, T. M. I. Mahlia, H. H. Masjuki, and S. Mekhilef. 2012. A comprehensive review on biodiesel as an alternative energy resource and its characteristics. Renew. Sust. Energ. Rev. 16:2070–2093.
88. Yeang, K.. 2008. Biofuel from algae. Architect. Des. 78:118–119.
89. Hu, Q., M. Sommerfeld, E. Jarvis, M. Ghirardi, M. Posewitz, M. Seibert, et al. 2008. Microalgal triacylglycerols as feedstocks for biofuel production: perspectives and advances. Plant J. 54:621–639.
90. Potts, T., J. Du, M. Paul, P. May, R. Beitle, and J. Hestekin. 2012. The production of butanol from Jamaica bay macro algae. Environ. Prog. Sust. Energ. 31:29–36.
91. Ellis, J. T., N. N. Hengge, R. C. Sims, and C. D. Miller. 2012. Acetone, butanol, and ethanol production from wastewater algae. Bioresour. Technol. 111:491–495.
92. Kim, N.-J., H. Li, K. Jung, N. M. Chang, and P. C. Lee. 2011. Ethanol production from marine algal hydrolysates using *Escherichia coli* KO11. Bioresour. Technol. 102:7466–7469.
93. Yen, H. W., and D. E. Brune. 2007. Anaerobic co-digestion of algal sludge and waste paper to produce methane. Bioresour. Technol. 98:130–134.
94. Sialve, B., N. Bernet, and O. Bernard. 2009. Anaerobic digestion of microalgae as a necessary step to make microalgal biodiesel sustainable. Biotechnol. Adv. 27:409–416.
95. Huang, H., X. Yuan, G. Zeng, J. Wang, H. Li, C. Zhou, et al. 2011. Thermochemical liquefaction characteristics of microalgae in sub- and supercritical ethanol. Fuel Process. Technol. 92:147–153.
96. Singh, S., B. N. Kate, and U. C. Banerjee. 2005. Bioactive compounds from cyanobacteria and microalgae: an overview. Crit. Rev. Biotechnol. 25:73–95.
97. Farrelly, D. J., C. D. Everard, C. C. Fagan, and K. P. McDonnell. 2013. Carbon sequestration and the role of

biological carbon mitigation: a review. Renew. Sust. Energ. Rev. 21:712–727.

98. Wu, H. Y., D. H. Zou, and K. S. Gao. 2008. Impacts of increased atmospheric CO_2 concentration on photosynthesis and growth of micro- and macro-algae. Sci. China C: Life Sci. 51:1144–1150.

99. Israel, A., J. Gavrieli, A. Glazer, and M. Friedlander. 2005. Utilization of flue gas from a power plant for tank cultivation of the red seaweed *Gracilaria cornea*. Aquacult. 249:311–316.

100. Lopez, C. V. G., F. G. A. Fernandez, J. M. F. Sevilla, J. F. S. Fernandez, M. C. C. Garcia, and E. M. Grima. 2009. Utilization of the cyanobacteria *Anabaena* sp. ATCC 33047 in CO_2 removal processes. Bioresour. Technol. 100:5904–5910.

101. Jacob-Lopes, E., L. M. C. F. Lacerda, and T. T. Franco. 2008. Biomass production and carbon dioxide fixation by *Aphanothece microscopica Nägeli* in a bubble column photobioreactor. Biochem. Eng. J. 40:27–34.

102. Ono, E., and J. L. Cuello. 2007. Carbon dioxide mitigation using thermophilic cyanobacteria. Biosyst. Eng. 96:129–134.

103. Weissman, J. C., J. C. Radway, E. W. Wilde, and J. R. Benemann. 1998. Growth and production of thermophilic cyanobacteria in a simulated thermal mitigation process. Bioresour. Technol. 65:87–95.

104. Miyairi, S. 1995. CO_2 assimilation in a thermophilic cyanobacterium. Energ. Convers. Manage. 36:763–766.

105. Dutta, D., D. De, S. Chaudhuri, and S. K. Bhattacharya. 2005. Hydrogen production by Cyanobacteria. Microb. Cell Fact. 4:36.

106. Smeets, E. M. W., I. M. Lewandowski, and A. P. C. Faaij. 2009. The economical and environmental performance of miscanthus and switchgrass production and supply chains in a European setting. Renew. Sust. Energ. Rev. 13:1230–1245.

107. Nanda, S., P. Mohanty, J. A. Kozinski, and A. K. Dalai. 2014. Physico-chemical properties of bio-oils from pyrolysis of lignocellulosic biomass with high and slow heating rate. Energ. Environ. Res. 4:21–32.

108. Mohanty, P., S. Nanda, K. K. Pant, S. Naik, J. A. Kozinski, and A. K. Dalai. 2013. Evaluation of the physiochemical development of biochars obtained from pyrolysis of wheat straw, timothy grass and pinewood: effects of heating rate. J. Anal. Appl. Pyrol. 104:485–493.

109. Nanda, S., P. Mohanty, K. K. Pant, S. Naik, J. A. Kozinski, and A. K. Dalai. 2013. Characterization of North American lignocellulosic biomass and biochars in terms of their candidacy for alternate renewable fuels. Bioenerg. Res. 6:663–677.

110. Nanda, S., A. K. Dalai, F. Berruti, and J. A. Kozinski. 2016. Biochar as an exceptional bioresource for energy, agronomy, carbon sequestration, activated carbon and specialty materials. Waste Biomass Valor. 7:201–235.

111. Ragauskas, A. J., C. K. Williams, B. H. Davison, G. Britovsek, J. Cairney, C. A. Eckert, et al. 2006. The path forward for biofuels and biomaterials. Science 311:484–489.

112. Beerthuis, R., G. Rothenberg, and N. R. Shiju. 2015. Catalytic routes towards acrylic acid, adipic acid and ε-caprolactam starting from biorenewables. Green Chem. 17:1341–1361.

113. Moshkelani, M., M. Marinova, M. Perrier, and J. Paris. 2013. The forest biorefinery and its implementation in the pulp and paper industry: energy overview. Appl. Ther. Eng. 50:1427–1436.

114. Dhabhai, R., S. P. Chaurasia, K. Singh, and A. K. Dalai. 2013. Kinetics of bioethanol production employing mono- and co-cultures of *Saccharomyces cerevisiae* and *Pichia stipitis*. Chem. Eng. Technol. 36:1651–1657.

115. Chung, H., D. R. Zak, and E. A. Lilleskov. 2006. Fungal community composition and metabolism under elevated CO_2 and O_3. Oecologia 147:143–154.

116. Rogers, H. H., S. A. Prior, and S. V. Krupa. 1994. Plant responses to atmospheric CO_2 enrichment with emphasis to roots and rhizosphere. Environ. Pollut. 83:155–189.

117. Cheng, W., and D. W. Johnson. 1998. Elevated CO_2, rhizosphere processes, and soil organic matter decomposition. Plant Soil 202:167–174.

118. Lin, G., J. R. Ehleringer, P. T. Rygiewicz, M. G. Johnson, and D. T. Tingey. 1999. Elevated CO_2 and temperature impacts on different components of soil CO_2 efflux in Douglas-fir terracosms. Glob. Change Biol. 5:157–168.

119. Denef, K., H. Bubenheim, K. Lenhart, J. Vermeulen, O. V. Cleemput, P. Boeckx, et al. 2007. Community shifts and carbon translocation within metabolically-active rhizosphere microorganisms in grasslands under elevated CO_2. Biogeosciences 4:769–779.

120. Carney, K. M., B. A. Hungate, B. G. Drake, and J. P. Megonigal. 2007. Altered soil microbial community at elevated CO_2 leads to loss of soil carbon. Proc. Natl Acad. Sci. USA 104:4990–4995.

121. Montealegre, C. M., C. van Kessel, M. P. Russelle, and M. J. Sadowsky. 2002. Changes in microbial activity and composition in a pasture ecosystem exposed to elevated atmospheric carbon dioxide. Plant Soil 243:197–207.

122. Sartori, F., R. Lal, M. H. Ebinger, and D. J. Parrish. 2006. Potential soil carbon sequestration and CO_2 offset by dedicated energy crops in the USA. Crit. Rev. Plant Sci. 25:441–472.

123. Molina, J. A. E., C. E. Clapp, M. J. Shaffer, F. W. Chichester, and W. E. Larson. 1983. NCSOIL, a model

of nitrogen and carbon transformations in soil: description, calibration, and behavior. Soil Sci. Soc. Am. J. 47:85–91.
124. Wei, X. R., G. X. Wang, P. Massarotto, S. D. Golding, and V. Rudolph. 2007. A Review on recent advances in the numerical simulation for coalbed-methane-recovery process. SPE Reserv. Eval. Eng. 10:657–666.
125. Ulrich, G., and S. Bower. 2008. Active methanogenesis and acetate utilization in Powder River Basin coals, United States. Int. J. Coal Geol. 76:25–33.
126. Strąpoć, D., F. W. Picardal, C. Turich, I. Schaperdoth, J. L. Macalady, J. S. Lipp, et al. 2008. Methane-producing microbial community in a coal bed of the Illinois basin. Appl. Environ. Microbiol. 74:2424–2432.
127. Flores, R. M., C. A. Rice, G. D. Stricker, A. Warden, and M. S. Ellis. 2008. Methanogenic pathways of coal-bed gas in the Powder River Basin, United States: the geologic factor. Int. J. Coal Geol. 76:52–75.
128. Penner, T. J., J. M. Foght, and K. Budwill. 2010. Microbial diversity of western Canadian subsurface coal beds and methanogenic coal enrichment cultures. Int. J. Coal Geol. 82:81–93.
129. Guo, H., Z. Yu, R. Liu, H. Zhang, Q. Zhong, and Z. Xiong. 2012. Methylotrophic methanogenesis governs the biogenic coal bed methane formation in Eastern Ordos Basin, China. Appl. Microbiol. Biotechnol. 96:1587–1597.
130. Singh, D. N., A. Kumar, M. P. Sarbhai, and A. K. Tripathi. 2012. Cultivation-independent analysis of archaeal and bacterial communities of the formation water in an Indian coal bed to enhance biotransformation of coal into methane. Appl. Microbiol. Biotechnol. 93:1337–1350.
131. Pfeiffer, R. S., G. Ulrich, G. Vanzin, V. Dannar, R. P. DeBruyn, and J. B. Dodson. 2008. Biogenic fuel gas generation in geological hydrocarbon deposits. United States Patent No. 7,426,960 B2.
132. Bennett, B., and S. R. Larter. 2000. Quantitative separation of aliphatic and aromatic hydrocarbons using silver ion-silica solid-phase extraction. Anal. Chem. 72:1039–1044.
133. Strapoć, D., M. Mastalerz, K. Dawson, J. Macalady, A. K. Callaghan, B. Wawrik, et al. 2011. Biogeochemistry of microbial coal-bed methane. Annu. Rev. Earth Pl. Sc. 39:617–656.
134. Shimizu, S., M. Akiyama, T. Naganuma, M. Fujioka, M. Nako, and Y. Ishijima. 2007. Molecular characterization of microbial communities in deep coal seam groundwater of northern Japan. Geobiology 5:423–433.
135. Stephen, A., A. Adebusuyi, A. Baldygin, J. Shuster, C. G. Southam, C. K. Budwill, et al. 2014. Bioconversion of coal: new insights from a core flooding study. RSC Adv. 4:22779–22791.
136. Baldygin, A., D. S. Nobes, and S. K. Mitra. 2014. Water-alternate-emulsion (WAE): a new technique for enhanced oil recovery. J. Pet. Sci. Eng. 121:167–173.
137. IPCC (Intergovernmental Panel on Climatic Change). 2000. Land use, land-use change, and forestry. Cambridge Univ. Press, Cambridge, UK.
138. Nanda, S., R. Azargohar, J. A. Kozinski, and A. K. Dalai. 2014. Characteristic studies on the pyrolysis products from hydrolyzed Canadian lignocellulosic feedstocks. Bioenerg. Res. 7:174–191.
139. Lehmann, J. 2007. Bio-energy in the black. Front. Ecol. Environ. 5:381–387.
140. Renner, R. 2007. Rethinking biochar. Environ. Sci. Technol. 41:5932–5933.
141. Lehmann, J.. 2007. A handful of carbon. Nature 447:143–144.
142. Lehmann, J., J. Gaunt, and M. Rondon. 2006. Biochar sequestration in terrestrial ecosystems - a review. Mitig. Adapt. Strategies Glob. Chang. 11:403–427.
143. Zandersons, J., J. Gravitis, A. Kokorevics, A. Zhurinsh, O. Bikovens, A. Tardenaka, et al. 1999. Studies of the Brazilian sugarcane bagasse carbonisation process and products properties. Biomass Bioenerg. 17:209–219.
144. Mathews, J. A. 2008. Carbon-negative biofuels. Energ. Policy 36:940–945.
145. Major, J., J. Lehmann, M. Rondon, and C. Goodale. 2010. Fate of soil-applied black carbon: downward migration, leaching and soil respiration. Glob. Change Biol. 16:1366–1379.
146. Stewart, C., and M. A. Hessami. 2005. A study of methods of carbon dioxide capture and sequestration—the sustainability of a photosynthetic bioreactor approach. Energ. Convers. Manage. 46:403–420.
147. Herzog, H. 2001. What future for carbon capture and sequestration? Environ. Sci. Technol. 35:148–153.
148. Cox, P. M., R. A. Betts, C. D. Jones, S. A. Spall, and I. J. Totterdell. 2000. Acceleration of global warming due to carbon-cycle feedbacks in a coupled climate model. Nature 408:184–187.
149. Fung, I. Y., S. C. Doney, K. Lindsay, and J. John. 2005. Evolution of carbon sinks in a changing climate. Proc. Natl Acad. Sci. USA 102:11201–11206.
150. Khoo, H. H., and R. B. H. Tan. 2006. Life cycle investigation of CO_2 recovery and sequestration. Environ. Sci. Technol. 40:4016–4024.
151. Xie, X., and M. J. Economides. 2009. The impact of carbon geological sequestration. J. Nat. Gas Sci. Eng. 1:103–111.
152. Michael, K., A. Golab, V. Shulakova, J. Ennis-King, G. Allinson, S. Sharma, et al. 2010. Geological storage of CO_2 in saline aquifers—a review of the experience from existing storage operations. Int. J. Greenh. Gas Control 4:659–667.

153. Yang, F., B. J. Bai, D. Z. Tang, S. Dunn-Norman, and D. Wronkiewicz. 2010. Characteristics of CO_2 sequestration in saline aquifers. Petrol. Sci. 7:83–92.
154. Shoji, K., and I. S. F. Jones. 2001. The costing of carbon credits from ocean nourishment plants. Sci. Total Environ. 277:27–31.
155. Anderson, L. G., T. Tanhua, G. Bjork, S. Hjalmarsson, E. P. Jones, S. Jutterstrom, et al. 2010. Arctic ocean shelf-basin interaction: an active continental shelf CO_2 pump and its impact on the degree of calcium carbonate solubility. Deep Sea Res. 57(Pt 1):869–879.
156. Raven, J. A., and P. G. Falkowski. 1999. Oceanic sinks for atmospheric CO_2. Plant, Cell Environ. 22:741–755.
157. Jiao, N., G. J. Herndl, D. A. Hansell, R. Benner, G. Kattner, S. W. Wilhelm, et al. 2010. Microbial production of recalcitrant dissolved organic matter: long-term carbon storage in the global ocean. Nature 8:593–599.
158. van der Zwaan, B., and K. Smekens. 2009. CO_2 capture and storage with leakage in an energy-climate model. Environ. Model. Assess. 14:135–148.
159. Doney, S. C., V. J. Fabry, R. A. Feely, and J. A. Kleypas. 2009. Ocean acidification: the other CO_2 problem. Ann. Rev. Mar. Sci. 1:169–192.

4

Low-grade heat recycling for system synergies between waste heat and food production, a case study at the European Spallation Source

Thomas Parker[1] & Anders Kiessling[2]

[1]WA3RM AB, Lund, Sweden
[2]Swedish University of Agricultural Sciences, Uppsala, Sweden

Keywords
Agriculture, aquaculture, cooling, horticulture, temperature, waste heat

Correspondence
Thomas Parker, WA3RM AB, Sandgatan 14F, 223 50 Lund, Sweden.
E-mail: thomas@wa3rm.se

Funding Information
Part of this work was substantially assisted by EU Grant Agreement No.: 312453 "EuCARD-2" for Enhanced European Coordination for Accelerator Research & Development

Abstract

At present food production depends almost exclusively on direct use of stored energy sources, may perhaps they be nuclear-, petroleum-, or biobased. Arable land, artificial fertilizers, and fresh water resources are the base for our present food systems, but are limited. At the same time, energy resources in the form of waste heat are available in ample quantities. The European Spallation Source (ESS) will require approximately 270 GWh of power per year to operate, power that ultimately is converted to heat. This multidisciplinary case study details an alternative food production cooling chain, using low-grade surplus heat, and involving fermentation, aquaculture, nutrient recapture, and greenhouse horticulture including both use of low-grade surplus heat and recycling of society's organic waste that is converted to animal feed and fertilizer. The study indicates that by combining the use of surplus energy with harvest of society's organic side flows, for example, food waste and aquatic-based cash crops, sustainable food systems are possible at a level of significance also for global food security. The effects of the proposed heat reuse model are discussed in a system perspective and in the context of the UNSCD indicator framework. The potential sustainability benefits of such an effort are shown to be substantial and multifaceted.

Introduction

The opportunity to recycle low-grade heat

In recent years, there has been a substantially increased research effort into the use of low-grade waste heat. The driver behind this interest seems to be the combination of concern of the climate impact from energy use and the substantial supply of low-grade heat. Low-grade heat is plentiful, because it is a by-product of thermal power production as well as various industrial processes in sectors such as metals and pulp and paper. In the United Kingdom, 11.4 TWh of recoverable heat was found to be wasted each year [1]. However, this figure only represents the wastage where there is a technically viable use available. The total amount of wasted heat is 48 TWh per year [2]. Similarly, "In the USA, over two-thirds of the primary energy supply is ultimately rejected as low-grade waste heat" [3]. Sweden, considered a world leader in heat recycling with its well-developed district heating networks, reuses 4 TWh of 9.5 available industrial waste heat [4]. This figure does not include the considerably larger waste heat streams from nuclear power.

An emerging source of low-grade waste heat may be data centers. In these, "temperatures as high as 60°C are sufficient to cool microprocessors" and "switching to liquid cooling [is] inevitable" [5]. This would lead to two improvements, greater efficiency in the data center and the possibility to utilize the waste heat.

Certainly, the need to address the climate impact of world energy supply is well established. The concept of "energy poverty" also pinpoints a need to address energy efficiency from a viewpoint of economy and equity. The recovery of low-grade heat has been shown to be able to play a role in this and the opportunities for doing so seem to be increasing [6]. A low-temperature district heating network greatly facilitates heat recovery from industrial waste heat and leads to more efficient industry, cheaper heat, and lower emissions [7].

Purpose of the research

There is thus significant indication of an opportunity to create sustainability benefits by recycling low-grade heat. The purpose of research is to contribute to knowledge about how low-grade waste heat can be used in ways that are both practicable and sustainable. Target audiences for this article are sustainability managers, energy managers, government bodies involved in planning and energy systems, and researchers in the field.

To achieve the research goal, based on the identified case, the following research questions were posed:

1. What uses for industrial waste heat have been identified with development potential?
2. What are the identified sustainability benefits of the identified heat-recycling initiatives?
3. What potential sustainability benefits and costs might be associated with the heat recycling and how may they be evaluated?

Methods

This research is based on an in-depth case study. The organization in the case, the European Spallation Source (ESS) is large-scale, multinational research facility, a type of institution often called "research infrastructure." The study of research infrastructure offers some advantages. As organizations dedicated to facilitating scientific endeavor, research infrastructure tends to be default support research and to be open to the study. They tend to have an academic culture and publish design reports and other documents, and work with peer review as a management process [8]. Energy issues in research infrastructure have in recent years also attracted considerable interest [9].

The case is of special interest because ESS has committed to recycling its waste heat, and doing so in an as efficient way as possible. The commitment is strong, having been made between partner governments.

To achieve its goals, the ESS has formed a close collaboration with the Swedish University of Agricultural Sciences (SLU) and other interested parties, and has developed a proposal, based on the biological systems to reuse low-grade heat, which is detailed in the case study.

The case is analyzed in the perspectives of the interaction of the involved systems and sustainability impacts of the case proposal in these systems. In the systems perspective, the point of departure is the effect in the case on the energy system, and the analysis continues with connections to food and nutrient systems and water.

Many tools and indices to assess sustainability have been proposed, and these proposals have in turn been analyzed. In one such sustainability indices, the authors conclude that "We show that these indices fail to fulfill fundamental scientific requirements making them rather useless if not misleading with respect to policy advice" [10].

Indices do not express objective truths, but instead are powerful expressions of a chosen set of values [11]. "Indicators arise from values (we measure what we care about), and they create values (we care about what we measure)" [12] and further "from a scientific point of view, there cannot be such a thing as one comprehensive measure or index of sustainability" [13, 14].

Therefore, the sustainability assessment will focus on identifying the relevant categories of sustainability impacts and discussing the case from a systems perspective in these categories. Rather than attempting to assess the proposal into an index figure, the case is discussed in the context of each relevant theme in the framework put forward by the United Nations Commission for Sustainable Development [15].

Structure of the article

This first section of this article serves to introduce the issues and thereafter to present the purpose of the research and the methods employed to attain them. The rest of the article is structured as follows: "Use of Low-Grade Heat in the Literature" section contextualizes the study by giving a brief overview of uses of low-grade heat from the literature. Section "Case Study: The European Spallation Source" thereafter presents the case study, starting with the organization studied, its goals, and operating conditions as well as design solutions implemented at the point of study. Relevant local market and climate conditions are also briefly presented. Section "Proposal: waste heat for food production with a nutrient loop" contains a presentation of a proposal for improved heat recycling at lower temperatures that included integration to horticulture and aquaculture. Section "Analysis" is the analysis of the findings in the case, starting with the general applicability of the case study to industry, thereafter applying a systems perspective and lastly a sustainability perspective. And the final section presents the main conclusions of the study.

Use of Low-Grade Heat in the Literature

Potential uses of low-grade heat

The simplest use for low-grade waste heat is for space heating. Today, temperatures as low as 40°C can easily be used for heating purposes, either via ventilation or floor heating. The environmental benefits of this can be substantial, if the waste heat replaces burning of fossil fuels. Looking at waste heat in the United Kingdom, "one-third of all fossil fuels consumed in the UK to produce low-grade heat for buildings" and further that "district heating schemes can provide cost-effective and low-carbon energy to local populations" [16]. The word "local" is significant. Heating, as opposed to electrical power, can be relatively easily stored, but is difficult to transport. Therefore, "direct heat use will depend on whether [a] potential user can be found" [6].

Aside from residential space heating, low-grade heat may also be supplied to greenhouses. However, as a cooling source for industry, greenhouses are viable only in winter, or in very northern climate areas [17].

Heating demand for greenhouses in Sweden is around 0.5 TWh [18, 19]. Nonetheless, reducing the sustainability impact of energy use would be a significant improvement of the sustainability performance of greenhouses. Also, "energy is typically the largest over-head cost in the production of greenhouse crops." Counting indirect energy use, fertilizer is one of the most energy-consuming parts of greenhouse operation, accounting for 21% of energy use [18].

Greenhouses contribute to sustainable development by vastly increasing the yield for a given area. The increase can be by a factor of 10–20 times compared to outdoor horticulture [18].

An intriguing possibility is to generate electrical power with waste heat. The most common proposal for this to make use of the organic Rankine cycle (ORC), but many other proposals exist, such as Stirling engines or condensing boilers [1, 20]. Suppliers of ORC systems claim to be able to produce power at temperatures such as 80°C and even lower, but typically require a heat sink with a temperature difference to the supply of at least 40°C. Moreover, at these extreme levels the production is not very economical. Other systems tend to demand higher temperatures and/or higher temperature differences. According to Fang et al. [7], industrial uses of waste heat include desalination and power generation, but are difficult at temperatures under around 200°C.

It would seem more advantageous to find uses of heat not requiring conversion to mechanical or electrical power. There is a significant difference between *recovery* and use as heat or *upgrade* to work, electricity or cooling [3]. Aside from space heat, refrigeration and desalination allow direct use without conversion to mechanical work [5]. Desalination can be achieved at as low temperatures as 45–50°C using "near vacuum level pressures" [6]. Even freezing is possible [21]. The much-awaited hydrogen economy would open up new possibilities, such as biohydrogen via dark fermentation at 70°C [22], or biohydrogen and biomethane at 37°C [23]. Additional uses include bacteria growth, typically at 37–38°C, biogasification, drying biomass, and production of a variety of substances, including ammonia, hydrogen, and pure water [6].

The drying process can have a crucial impact on the total energy efficiency of the use of biomass as an energy carrier, because the drying requires significant amounts of energy. Similarly, in digestion processes, digestibility has been shown to increase by 5–10% when the temperature is increased from 35°C to 50°C, but the energy use for heating was higher than the gain [24]. Production of protein meals for constructed animal feed from biomass sources as fish, microbes, macroalgae, plants, and insects require energy for drying. For example, meal production from as varying sources as yeast or fish requires removal of more than 70–80% of the weight in water [25]. Traditionally, high-temperature systems with temperatures over 100°C have been used, but more recently it has become clear that lower drying temperatures improve the quality of the product [26]. Today more than 27 million tons of fish is processed into fishmeal annually [27, 28] requiring removal of nearly 20 million tons of water. Considering the prognosis of more than 90 million ton of farmed fish to 2030 [28] and a replacement of soy as the major protein source in their diet with single cell protein (as yeast, microalgae etc.), macroalgae, feed mussel meal, and/or insects, more than 300 million tons of water needs to be evaporated by low-temperature drying techniques in order to form a transportable commodity to be used in the aquatic feed production. Even if fish has an uniquely high protein need, the amounts will be similar or even larger in the terrestrial farmed animals considering their larger volumes and that also in these animals the use of human grade protein sources (mainly soy and corn) must be replaced by alternative sources like the ones mentioned earlier and thereby require more energy for evaporating water than presently is the case. Considering that 1 kg of water requires 2.3 MJ for evaporation, it becomes clear that alternative techniques to the present ones based on fossil energy are urgently needed.

Production of fish and microalgae in more closed systems will require heating, especially in temperate climates. The need of energy will vary with farming temperature, exchange of water, and ventilation. Salmon requires 14°C as the optimal temperature, while tilapia prefer close to 30°C. Most fish will have a Q10 of more than a doubling and, for example, will salmon smolt double their growth already when temperature is increased from 9°C to 14°C [29, 30]. On the other hand, fish, in parity with all biological organisms, are more sensitive to a temperature above their optimum than to a lower one [31]. In a complete flow through system roughly 70 m^3 of water is needed per kg of salmon production, while about 2/3 is needed in less oxygen-demanding and CO_2-resistant species [32]. Given input of oxygen and CO_2 stripper, this requirement could be reduced to 50% and with high-technology filtration, protein skimming, and biological filters a reduction to 2.5% is possible [33]. However, in most food production system an exchange below 15% of farming volume per day is unusual due to quality reason. A quick calculation then gives with hand that a standing biomass of, for example, 1000 ton, with a density of 50 kg/m^3, will require a total farming volume of 20,000 m^3. An ambient water temperature of 10°C and a farming temperature of, for example, 25°C would require at an exchange of 15% heating of 3000 m^3 at 15°C per day.

The importance of temperature

Temperature is fundamental to heat recycling: "The temperature of the low-grade heat stream is the most important parameter, as the effective use of the residual heat or the efficiency of energy recovery from the low-grade heat sources will mainly depend on the temperature difference between the source a suitable sink, for example another process or space heating/cooling" [6].

Within the literature discussing recycling and use of low-grade heat, there is variance in the definition "low grade," including the "widely accepted threshold temperature": 250°C [6, 20], 260°C [1], 60°C, and 120°C [3], and a typical heat from a solar collector (70°C) [21].

Case Study: The European Spallation Source

The European Spallation Source

The European Spallation Source (ESS) is a large-scale research facility in construction in Lund, Sweden. The facility will supply neutrons and a suite of neutron instruments for use in research in materials, life science, energy, and other disciplines. The facility will generate a far stronger neutron flux than existing facilities. To generate the flow of neutrons, a linear proton accelerator, the most powerful in the world, will propel bunches of 10^{12} protons into a target in the form of a large wheel of tungsten. In this target, the spallation process takes place, generating 30 neutrons for every proton.

Neutrons, being neutral particles, can penetrate into materials and can be used to create images of the insides of materials and substances on a nanometer scale and with nanosecond resolution. Neutrons are particularly useful for investigating light atoms, such as hydrogen, carbon, and oxygen found in organic molecules. This makes neutrons an important tool within life sciences, sustainability, and energy research. Within energy specifically, neutrons can help study the movement and structure of ions in batteries and fuel cells. The storage of hydrogen in metal substrates is another active research area, as is carbon capture and storage. With somewhat different research methods, neutrons can also be used to investigate photovoltaics and photosynthesis. The ESS and similar facilities can be used for in situ studies, typically looking at ongoing combustion processes to explore mechanical and chemical process improvements. A highly specialized area of research is superconductivity. This research makes use of another property of neutrons, their magnetic spin, to explore how magnetic properties of materials change with temperature.

Development of a sustainability strategy

The decision to build ESS in Lund was preceded by a competition between countries to host the facility. It was in this competition that the host governments of Sweden and Denmark committed to building a sustainable research facility, by implementing and energy strategy called Responsible, Renewable, and Recyclable. This meant that the facility was to be energy efficient, use energy from renewable sources, and recycle the waste heat resulting from activities. Importantly, this trio of goals was given as a hierarchy, so that energy efficiency was of higher priority than heat recycling [34].

The three parts of the energy strategy were given specific requirements in an Energy Policy [35]. The target for energy efficiency was set for a maximum of 270 GWh total annual energy use at full operation. The target for Renewable was that all energy used would be from renewable sources, and for Recyclable, it was that all "recuperated" waste heat would be reused. "Recuperated" meant heat captured in cooling systems.

ESS Scandinavia, as the Swedish and Danish bid to host was called, also proposed a shift to an almost completely superconducting linear accelerator, which was a significant gain in overall efficiency at ESS. A superconducting accelerator does not suffer from losses due to resistance in the accelerator, but requires cryogenics to chill the accelerator to approximately 2 K, in itself an energy-intensive process. Despite this, superconducting was still a net gain. This technological leap also had an important effect on heat recycling, as the cryogenics is required to run constantly, in order to preserve the cryogenic helium and to avoid thermal expansion in the accelerator, which would require time-consuming retuning. Therefore, a superconducting accelerator requires a constant minimum cooling, and thus supplies a constant minimum flow of heat.

The original energy concept envisioned heat recycling to the local district heating system. This system supplies Lund, as well as neighboring townships with a total of around a TWh of heat per year. Additionally, the operator has embarked on a project to connect this system with more distant heating systems in neighboring cities. However, district heating systems typically operate at temperatures of 80–120°C, whereas cooling systems for accelerator-based research facilities typically have cooling loops at two levels, one cooling-tower level of 30–40°C and one chilled water level of 5–20°C. Technically, this gap could easily be bridged with the use of heat pumps, but to do so for the full heat load at ESS would have directly conflicted with the "Responsible" goal of energy efficiency.

Since before the ESS Scandinavian proposal, ESS Scandinavia, and subsequently ESS has been collaborating in various agreements with the district heating operator Lunds Energi (now Kraftringen) to pursue a sustainable research facility. Throughout this long-term collaboration Kraftringen has been pursuing an independent effort to reduce temperatures in it local district heating system to reduce losses. As of December 2013, the two parties have reached a formal agreement to connect ESS to the district heating system. This agreement requires ESS to provide a temperature of 80°C, which is deemed sufficient for heating needs. The ESS receives back a temperature of under 50°C.

The evolution of the ESS energy strategy has been analyzed in sustainability strategy research [36].

Energy inventory

ESS has implemented a program to raise cooling temperature levels, in order to make use of the district heating system directly as cooling, an effort that has led to attention in the field [9]. High-temperature cooling is being implemented primarily for the klystrons providing the accelerating power and for the helium compressors providing the cryogenic cooling. Nevertheless, a significant amount of cooling is still necessary at lower temperature levels. The energy inventory, a biannual exercise conducted at ESS shows the projected energy use as well as the cooling needs at the three temperature levels.

The latest available energy inventory [37] shows a total annual power use of 265 GWh, of which 60 GWh is for the heat pumps that provide the lower temperature cooling and eject heat to the district heating system. The estimated total recovered heat amounts to 253 GWh per year, divided into temperature levels according to Table 1.

Table 1. The cooling temperature levels at ESS.

Temperature levels	Supply temperature (°C)	Return temperature (°C)	Part of total (%)
Low	5–15	30–35	30
Medium	30–35	40–50	35
High	40–50	75–80	35

The possible savings from recycling heat at lower temperatures than the 80°C required for the district heating system are thus significant.

Sustainability issues aside from energy at ESS

The indirect effect of energy was considered the most important and variable sustainability issue at ESS, but of course there were others as well. The facility would generate substantial volumes of radioactive waste, albeit mostly at rather low levels. Radiation protection was an important issue. Radioactive waste handling and radiation safety issues were dealt with a regulatory framework.

An issue that came up in relation to local inhabitants and in conjunction with the regulatory process was the use of 60 ha of prime agricultural land for the construction of the facility. This had led to some local opposition in the licensing process from the local farmer's organization and a local nature conservation organization. Placing this large facility on such excellent soil was perceived as unsustainable and a threat to future food security.

Indicative prices in the case

A full business case for heat recycling alternatives was not yet available, but some indicative prices could be uncovered. The long-term gross power price on the Nordic market is estimated at 5 ¢/kWh (eurocents per kilowatt hour). ESS is exempt from energy tax, but this would otherwise amount to 3 ¢/kWh. The average price paid for heat recovered to a district heating system in Sweden was 2 ¢/kWh. The average price of district heating to large companies in Sweden was 7 ¢/kWh. At an estimated COP for a heat pump to cool at 40°C and eject heat to district heating at more than 80°C would be around 4, indicating a cost of electrical power for the process of ¼ times 5 ¢ or 1.25 ¢/kWh.

Climate conditions in the case

The conventional cooling solution for research infrastructure facilities like the ESS is to either make use of a local body of water, if available and allowed, or to use cooling towers. Heating a local ecosystem has an associated impact on that system. Cooling towers consume electrical power and substantial amounts of water and chemicals.

A closed loop system based on heat pumps requires more energy than cooling towers, but does not consume the cooling water or the chemicals, and much of the chemical use is avoided completely as the system is not open to contamination.

Specific conditions in Sweden, compared to global averages, are that the climate is cool, so that heating is required for most of the year, whereas cooling is not in great demand, and that the supply of fresh water is ample.

Collaboration and comparison with other facilities

A study conducted within the EU-sponsored EuCARD2 (Grant Agreement 312453) project for accelerator development examines 12 large-scale research facilities (Research Infrastructure), of which 10 were in operation and two in construction. The average annual energy consumption for the facilities (including estimates for the two in construction) was 180 GWh per year, with considerable variation in the group. Discounting the outlier in each end, the average fell to somewhat under 100 GWh. The cooling requirement varied from 40% to 60% of the electricity use. Cooling at operating facilities was at low temperature (up to approximately 40°C). The facilities in construction had included a high-temperature (up to approximately 80°C) cooling loop for part of the cooling demand [38].

The study examines a number of technologies for reuse of surplus heat, divided into high-temperature technologies, meaning those requiring 80°C or more, and low-temperature technologies. The high-temperature technologies examined were district heating, heat-driven cooling, and power production using the organic Rankine cycle (ORC). Both the heat-driven cooling technologies examined and the ORC required cooling to function and thereby produced a flow of lower temperature heat. These options were therefore only of interest for facilities with a low-temperature heat sink available. District heating, on the other hand, was only of interest for facilities located close to a significant heat demand, preferably with an existing infrastructure for distribution.

The low-temperature option studies included low-temperature district heating, heat storage, food/fodder production, biological/chemical purification/separation techniques, wastewater treatment, and ground heating (e.g., for ice and snow removal).

The EuCARD2 project involves 40 partners from 15 European countries. The energy efficiency effort within this collaboration stems from "the need to increase the efficiency of energy use during operation for cost and sustainability reasons is common in all accelerator facilities in research and industry" [39]. Another European collaboration within a similar area is the workshop series "Energy for Sustainable Science at Research Infrastructure" hosted by CERN, ESS, and ERF (European Association of National Research Facilities) (https://indico.cern.ch/event/245432/, http://europeanspallationsource.se/energy-workshop). After the first workshop, the hosts published an executive summary highlighting the value for society of the efforts at research infrastructure in energy and sustainability management. The value was created first by the direct effects in energy efficiency at the facilities, but potentially more important effects could be created by using research infrastructure as innovation hubs, testing grounds, and training grounds, roles that research infrastructure is created to fulfill [40].

Proposal: waste heat for food production with a nutrient loop

Based on the estimated heat flows at various temperatures, the value of heat sold for district heating purposes, and the estimated cost of electricity to drive the heat pumps, the estimated value of heat at the different temperature levels can be derived. This varies over the year with the value of the heat and electricity prices and may be a positive or negative value. These calculations form an economic basis for the development of alternative uses of waste heat.

The first such use to be developed in the case was for onsite space-heating needs. Space heating can easily be achieved in buildings so designed with water temperatures of 40°C. ESS has developed an internal heat distribution network with a supply temperature of 55°C and return 25°C. The annual heat use was estimated at 5 GWh. The remaining heat would therefore amount to 265 GWh.

Since existing demand required heat pumps to augment temperature, and internal demand was limited, ESS sought opportunities to create new demand, by offering the heat to users that would establish nearby specifically to utilize the offered heat. This was done in an open call for proposals published on 10 of September 2013.

Biological systems offer an opportunity to make use of heat. Fish, plants, algae all have in common that, within limits, growth is stimulated by an increase in temperature. As one example, an increase in temperature from 8.6°C to 13.7°C has been shown in a specialized research facility to double the growth rate in salmon smolt [29].

An existing use of surplus heat for biological systems exists only 125 km northeast of ESS, at the Elleholm greenhouse facility, which at 8000 m^2 greenhouse area is Sweden's second largest producer of tomatoes. Elleholm uses waste heat from nearby Södra Cell, a pulp and paper plant.

The details of the Elleholm case differ from ESS, in that the supplying facility at all times has a significant

Table 2. Identified uses for excess heat under 60°C in a cooling chain from heat to food.

Temperature (°C)	Food and fodder production process
40–60	Low-temperature drying
32–40	Fermentation of microbes (yeast, bacteria, microalgae)
22–32	Warm water fish farming (tilapia, shrimp, perch, turbot)
18–22	Green house hydroponics (tomatoes, cucumbers)
10–18	Cold water fish farming (salmonids, white fish, sturgeon, crayfish)

excess of heat of prime district heating temperature. There is therefore no incentive to explore low-temperature heating systems. Despite this, the Elleholm management has estimated minimum temperature needed at 38°C.

In collaboration with the Swedish University of Agricultural Sciences, SLU, and based on the results of the abovementioned open call, ESS has identified a number of potential biological uses of its excess heat, shown in Table 2.

A vital aspect of the SLU-ESS proposal is that the components of the cooling chain also could be linked in a nutrient chain. The heat recycling at the Elleholm facility is a significant gain in sustainability, but the facility still uses commercial fossil fertilizer, which represents a significant indirect energy use. If a greenhouse facility were colocated with fish farming, the fish excrements could be directly used as fertilizer in a hydroponic system. Fish fodder could be made from yeast, based on a substrate of food waste and agricultural waste (including plant waste from the greenhouses). This would require a drying process to create fodder, which could also be achieved with low-grade waste heat.

A graphic of the SLU-ESS initial proposal taken from the ESS Energy Design Report [41] is shown in Figure 1.

Analysis

Generality of the case

The case of the European Spallation Source shows an example of the phenomenon that surplus heat has become an economic and environmental cost. The cost is first, a direct cost for operating cooling systems, and second, the cost of a lost opportunity, and third a presently undefined cost for the ecosystem in coping with an increased ambient temperature due to the release of large volumes of cooling water. The case thus shows some characteristics that may be common to energy-intensive activities, these being that (i) large quantities of excess heat are produced, (ii) well-managed, the heat can be conserved at a temperature that can be useful for heat and biological processes, and (iii) conventional management of this resource represents a cost burden to the organization. In EU it is estimated that this low-temperature heat loss equals 500 billion Euro in petrol equivalents [25].

Identification and interaction of relevant physical systems

In the case, a hierarchy of energy forms was established by the relative prices of heat and electricity, and the electricity demand for heat pumps to supply appropriate temperatures. The electrical power system represented the

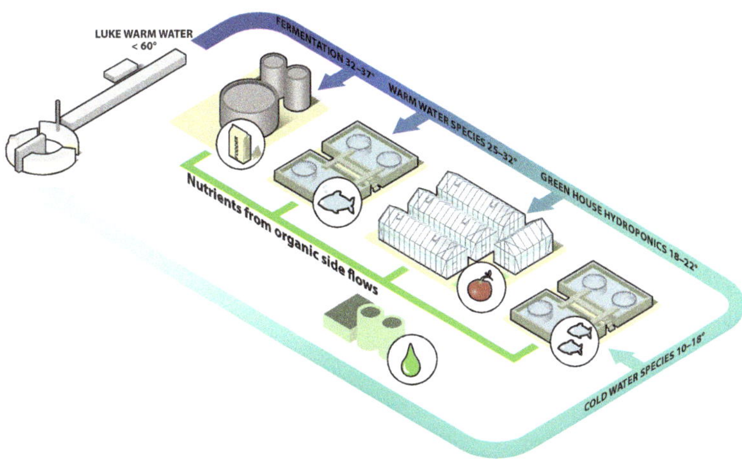

Figure 1. UNCSD theme indicator framework [15].

highest value systems, and cooling systems of various temperatures represented falling value with temperature level. This is a hierarchy of monetary value, but the same outcome would result from an analysis based on the second law of thermodynamics. The theoretical limits in Carnot cycle conversions function can be used to assign relative values to heat of various temperatures compared to electrical power [42].

A low-temperature heat-recycling scheme is therefore also a low-value scheme. This can enable uses that require only low-grade heat, but are not competitive if they must purchase heat at the high-value price.

Food is energy

Agriculture plants convert solar energy into energy for human consumption in the form of food. However, modern agriculture methods depend heavily on fossil fuels. Fertilizer represents a significant, indirect part of energy consumption for food production. For example, in Europe, farmers use about 10.5 million tons of nitrogen fertilizer 2010 and the trend is an increase. Phosphate fertilizer 2011/2012 was 23 million tons. Of this, EU imports roughly 25% (IFA, Fertilizer Europe.Com, 2014). The energy consumption to produce fertilizer varies considerably with the fertilizer type; nitrogen fertilizer requires much more than phosphorus or potash, but in order of magnitude this may represent an energy consumption of 100–150 TWh, or an eighth of a PWh. On the other hand, mining of organically available phosphorus, besides fossil fuels, do include a fraction of heavy metals, especially cadmium and aluminum, adding to its environmental load [43]. The total electrical consumption of Sweden in 2012 was 142 TWh; replacing mineral fertilizer is therefore a significant energy efficiency gain for agriculture.

Peak oil use, peak land, and peak fisheries

The term "peak oil" is used to describe a moment when the supply of oil would begin to decline, and when that might happen has been periodically much debated. More recently, "peak oil consumption" is being discussed, meaning the event that oil consumption would begin to decline.

It is clear that the event "peak arable land" has repercussions for future food supply. This moment might already have past. Loss of arable land is caused by degradation, and by constraints imposed by climatic, environmental, and human activity. The United Nations' Food and Agricultural Organization (FAO) track land degradation and publish a "Global Land Assessment of Degradation." If indeed "peak arable land" has been passed, it follows that all increased food production must come from greater output per land unit.

Furthermore, it is clear that "peak wild fish harvest" has passed. Nearly 85% of our wild fish stocks are near, at, or over its maximum harvest [27]. Increased fish consumption must therefore come from aquaculture.

Food and fodder

Within the world's food production system, basic food for human consumption competes with animal feed, and to a lesser extent also for energy production. In animal feed, 47% of soy and 60% of corn produced in the United States is used and at global scale more than 40% of these crops are used for animal feed [28, 43]. In 2007, EU produced less than 1 million tons of soy per year, and imported around 25 million ton of soybean meal (EU-27) (http://www.indexmundi.com/agriculture/?country=eu&commodity=soybean-meal&graph=imports). Today, only around 6% of soybeans are eaten directly as whole beans or in products like tofu and soy sauce [44]. Approximately 75% of all produced soybean is used for animal feed [44, 45]. To produce this amount of soy almost 15 million ha of agricultural land area is needed, nearly equal the total agricultural land of Germany [45]. In fact, soy fields now cover more than 1 million square kilometers of the world – the total combined area of France, Germany, Belgium, and the Netherlands [44]. Also, approximately 80% of that soy is genetically modified [44]. Based on the present increase in animal-derived food products, FAO [28] estimates that world soy production will double to 2050. Ever since soy production began increasing in South America in the 1960s, soy has been associated with clearance of some of the world's most crucial ecosystems, such as the Amazon and Cerrado, leading to loss of biodiversity. This loss of valuable forests and other native vegetation means that the carbon storage services they provide are lost forever, contributing to global climate change. Soy production is also linked with unsafe and excessive use of pesticides, violation of land rights, and unfair labor conditions [44]. The import of soy to EU has remained more or less stable over the last 15 years (EU-27) (http://www.indexmundi.com/agriculture/?country=eu&commodity=soybean-meal&graph=imports). In addition to plant-based nutrients, about a quarter of world catch of fish is used as animal feed [27] underlining the enormous amount of high-quality nutrients of human food quality presently used in animal feeds.

Horticulture is water intensive; the plants consume water and much is also lost in evaporation, one of the issues driving the idea of closed greenhouses. In the case, water usage was seen as an important sustainability advantage for location of greenhouses in Sweden, as the supply of fresh water is ample and the cost of often negligible. The cool climate and the resulting need for heat in greenhouses was a competitive disadvantage, to the extent that Sweden's horticulture industry was in a steady state of decline. Only 10% of tomatoes in Sweden are produced domestically [46]. An inexpensive, sustainable heat source would therefore potentially shift greenhouse production to a place with an abundance of water and thereby lessen the pressure on scarce resources elsewhere.

The indicative prices uncovered show that value of cutting out the middle man in the district heating system and establishing a direct relationship between the waste heat producer and the consumer was as much as 5 ¢/kWh. Additionally, a supply and demand at 40°C rather than 80°C would mean a savings of 1.25 ¢/kWh. With a total heat supply of 250 GWh per year, this would mean a total added value of 16 million Euros a year, quite likely enough to negate the competitive disadvantage of the heating need in Sweden.

Microbes and the protein chain

In cells, RNA relays the information of DNA to the protein synthesis. Microbes have high levels of RNA (10–15%) because of the high level of protein synthesis. Living cells metabolize the N (nucleotides) in RNA to uric acid. In mammals uric acid leads to kidney stones and gout [47]. Fish, however, have retained their ability to eat microbes and have no problem with uric acid [47–50]. Using microbes to produce fish fodder offers an opportunity to profitably use the fantastic growth rates of microbes. Microalgae tend to multiply at a rate of once per day, that is, a daily rate of 2^1. Yeast doubles every 2 h, that is, 2^{12}, whereas, for example, *Escherichia coli* bacteria the pace is as high as every 20 min during the exponential growth phase, leading to a daily rate of 2^{72}.

Taking the example of yeast, a study at Swedish yeast factory demonstrated a in favorable conditions a start culture of 10 mg of yeast developing into 150 tons in a week given free access to short carbon chains and ample supply of minerals.

Protein is important because around 40% of fish fodder is protein, in the average (tilapia/carp 30%, marine species 50%, salmon 40%). Table 3, below, details some of the main similarities and differences between protein chains for food production based on soy compared with yeast. For fodder production using yeast, the dominant energy use is for the drying process.

Table 3. Protein chain comparison with energy use.

Protein chain	Soy	Yeast
Product	Soy meal	Protein meal
Protein content	50% dry matter	50% dry matter
Suitability	Terrestrial farmed animal[1] and humans	Fish and shrimp[2]
Main energy use	Farm/harvest, processing, transport	Drying
Energy use, kWh/kg	2.8	1.46
Energy supply	Fossil based	Surplus low-grade heat

[1]Not suitable for most fish, but if used to fish needs further processing into soy concentrate which require further energy in alcohol extraction and heat treatment to reduce antinutrient and endothelial inflammatory factors. In this process protein is concentrated from under 50% dry matter to over 65% demanding a parallel increase in amount of soy bean raw material per kg of concentrate, that is, in production energy.
[2]Suitable for monogastric terrestrial farmed animals at low inclusion level.

Creation of new dependency

From a systemic view, recycling is usually seen as an inferior option to prevention. This is also the case in the ESS program, where the "Responsible" goal of energy efficiency was the superior goal. A related issue is the question of whether heat recycling might cause a dependency on a fossil-based and/or wasteful process. For example, in this case, further efficiency gains were envisioned in the accelerator.

In comparison, heat recovery from incineration of waste streams or from electrical power production from fossil fuel sources may be seen as an unsustainable subsidy to these practices and as a risk if the heat source should cease.

In this case, ESS, backed by 17 democratic governments and a major potential source of new knowledge enabling new sustainability solutions, may be seen as a sustainable activity in its own right. The explicitly planned life span of 40 years also adds a measure of certainty of supply.

More generally, source dependency is lessened by lower temperature level. This is because the lower the temperatures, the more easily the heat source can be replaced by solar or geothermal heat. Therefore, heat recycling at low temperature can be seen as an enabler for restructuring of the energy system.

Sustainability assessment

The UNSCD sustainability framework [15] is shown in Figure 2. The framework is established for indicators on for a nation or region, not for a specific technology. It is chosen here as a reflection of a consensus view of the main international sustainability challenges.

In the CSD framework, the Health theme includes the subtheme Drinking Water. Climate Change is a subtheme of Atmosphere. Agriculture is a subtheme of Land and includes the indicators Arable and Permanent Crop Land Area and Use of Fertilizers. Oceans, Seas, and Coasts

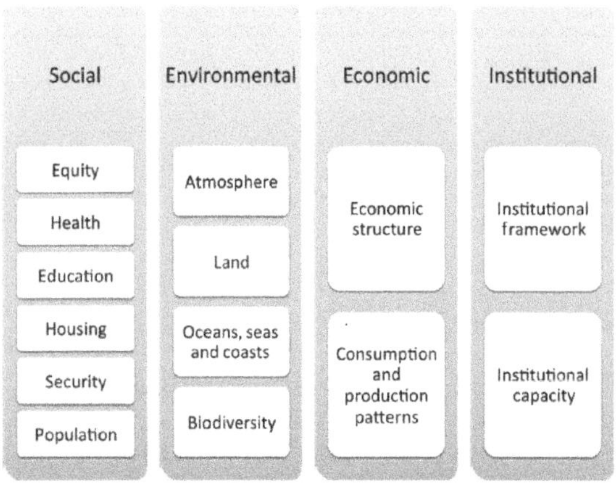

Figure 2. The proposed cooling chain for ESS with food and fuel production (*source*: [41]).

include issues of the impacts of nutrient flows into bodies of water as well as fishing yields. Among the Economic themes, Energy Use is a subtheme of Consumption and Production Patterns as is Waste Generation and Management, which includes the indicator Waste Recycling and Reuse.

The following themes, subthemes, and indicators are therefore relevant as categories of analysis to the case: groundwater and coastal waters; fishing, arable, and permanent crop land area; use of fertilizers; energy use; climate change; and waste recycling and reuse. The Institutional themes are specific to the national or regional level, and not applicable to a specific technology.

To ensure relevancy, a comparison was made with an assessment method developed specifically for energy [51]. As a result, four additional categories were added: resource depletion, cost/benefit, security and diversity of supply, and public acceptability. A summary of the assessments according to these categories is shown in Table 4.

The positive climate effect of replacing red meat with fish has been quantified in studies. At global level, meat production accounts for 18% of released climate gases [27, 28]. Gonzales et al. [52] in parity with Pelletier and Tyedemers [53] show a much lower energy use and release of CO_2 equivalents in producing 1 kg fish compared to red meat. Naturally this varies with species and production system. For example, Troell et al. [54], Tyedemers et al. [55], and Pelletier and Tyedemers [53] showed that cultured carp yielded an "industrial energy" return in edible food of 94%, while fishery harvested fish reached 8% and farmed salmon using fabricated diet based on fish and plant protein meals varied between 8% and 17% pending conventional or organic sources, while feed lot beef reached 2.5%. The differences mainly lie in the feed compartment. In these studies, the origin of the energy was not considered.

The increased predictability of supply from land-based fish farms is because unpredictable factors such as weather, diseases, etc., will be possible to control to a much higher extent in a closed system.

Main Conclusions

In this case, the value of energy quality is demonstrated. The preservation of temperature in cooling system was

Table 4. Summary of sustainability assessment by category.

Theme	+/–	Motivation
Groundwater and coastal waters	+	On-land closed-system fish farms can materially reduce the impact on inland and coastal waters from open-water fish farms.
Fishing	+	On-land aquaculture has the potential to alleviate pressure on wild stocks, allowing these to recover.
Arable and permanent crop land area	+	The production from greenhouses is 40–80 times higher (in monetary value) than from farmland per unit area.
Use of fertilizers		By combining the appropriate amount of fish farming with horticulture, nutrient flows from fish excrements can replace commercial fertilizer in greenhouses, provided the feed is based on nonplant materials.
Energy use	+	Use of waste heat for greenhouses and land-based fish farms can substantially reduce their energy use.
Climate change	+	A positive climate change effect comes from reduced energy use, reduced fertilizer use, and a replacement of red meat by fish due to avoided methane release and great energy efficiency.
Waste recycling and reuse	+	Waste nutrient streams are proposed as a basis for producing fish fodder. If implemented, this would be a significant valorization of a waste stream. However, technical hurdles remain.
Resource depletion	+	Fossil fuel, fresh/ground water, and fossil nutrient use is reduced.
Security and diversity of supply	+	The diversity of energy supply would increase. Heat storage capacity would be necessary and could contribute to stability. Diversity and predictability of food supply would increase.
Public acceptability	–	Industrial-scale fish and greenhouse farming may be considered unsightly. Light pollution may be an issue. Traditional farming and fishing is culturally ingrained. Animal feed based on recycled nutrients need strict food security control.

shown to add significant value. Conversely, using as low quality as possible can significantly lower costs. The indicative figures provided in the case, for example, 40°C for heating, 60°C for cooling data centers, and 70–80°C for cooling power electronics corroborate some earlier research. With additional verification, these benchmarks could inform future energy performance efforts.

Heat recycling was shown to be a considerable enabler for food production. The analysis indicates that waste heat resources available that are comparable to those in the case are abundant. If these can be combined with waste nutrient streams and converted to food and energy, there is potential for noticeable impact on global food supply. Additionally, if the efforts to integrate nutrient and cooling chains are successful, a large-scale rollout of this technology could supplant substantial amounts of fossil fertilizer and thus significantly lower the environmental impact of food supply.

Making use of low-grade heat in biological systems may also enable future development by lowering the threshold for renewable heat sources such as solar and geothermal.

The study is limited to a single case, and although the studied case organization is well established, the energy processes described are still being designed and will not be observable for some years. Further studies are therefore urgently needed, particularly studies of real energy flows and demonstration facilities for food production.

Conflict of Interest

None declared.

References

1. Law, R., A. Harvey, and D. Reay. 2013. Opportunities for low-grade heat recovery in the UK food processing industry. Appl. Therm. Eng. 53:188–196.
2. Element Energy, Ecofys, Imperial College, P. Stevenson, and R. Hyde. 2013. The potential for recovering and using surplus heat from industry. Department of Energy & Climate Change, London.
3. Little, A. B., and S. Garimella. 2011. Comparative assessment of alternative cycles for waste heat recovery and upgrade. Energy 36:4492–4504.
4. Swedish Government Official Reports. 2005. SOU 2005:033 Fjärrvärme och kraftvärme i framtiden. Fritzes offentliga publikationer.
5. Zimmermann, S., M. K. Tiwari, D. Poulikakos, I. Meijer, S. Paredes, and B. Michel. 2012. Hot water cooled electronics: exergy analysis and waste heat reuse feasibility. Int. J. Heat Mass Transf. 55 no. 23–24:6391–6399.
6. Ammar, Y., S. Joyce, R. Norman, Y. Wang, and A. P. Roskilly. Jan. 2012. Low grade thermal energy sources and uses from the process industry in the UK. Appl. Energy 89:3–20.
7. Fang, H., J. Xia, K. Zhu, Y. Su, and Y. Jiang. 2013. Industrial waste heat utilization for low temperature district heating. Energy Pol. 62:236–246.
8. Parker, T. 2013. The view from below – a management system case study from a meaning-based view of organization. J. Clean. Prod. 53:81–90.
9. Parker, T. 2011. Sustainable energy: cutting science's electricity bill. Nature 480:315.
10. Böhringer, C., and P. E. P. Jochem. 2007. SURVEY: measuring the immeasurable – a survey of sustainability indices. Ecol. Econ. 63:1–8.
11. Parker, T. 1998. Total cost indicators – operational performance indicators for managing environmental efficiency. International Institute of Industrial Environmental Economics, Lund.
12. Meadows, D. 1998. Indicators and information systems for sustainable development – a report to the Balation Group. The Sustainability Institute, Hartland, WI.
13. 06/02001 Will the information society be sustainable? Towards criteria and indicators for a sustainable knowledge society. Spangenberg, J. H. International Journal of Innovation and Sustainable Development, 2005, 1, (1–2), 85–102. *Fuel Energy Abstr.* 47: 300, 2006.
14. Singh, R. K., H. R. Murty, S. K. Gupta, and A. K. Dikshit. Jan. 2012. An overview of sustainability assessment methodologies. Ecol. Indic. 15:281–299.
15. UNCSD. 2001. Indicators of sustainable development: framework and methodologies. Division for Sustainable Development, UN Department of Economic and Social Affairs, Background Paper No. 3.
16. Swithenbank, J., K. N. Finney, Q. Chen, Y. B. Yang, A. Nolan, and V. N. Sharifi. 2013. Waste heat usage. Appl. Therm. Eng. 60:430–440.
17. Leffler, R. A., C. R. Bradshaw, E. A. Groll, and S. V. Garimella. 2012. Alternative heat rejection methods for power plants. Appl. Energy 92:17–25.
18. Vadiee, A., and V. Martin. 2014. Energy management strategies for commercial greenhouses. Appl. Energy 114:880–888.
19. Statens Jordbruksverk. 2012. The 2011 horticultural census. Statens Jordbruksverk, Stockholm.
20. Walsh, C., and P. Thornley. 2013. A comparison of two low grade heat recovery options. Appl. Therm. Eng. 53:210–216.
21. Le Pierrès, N., D. Stitou, and N. Mazet. 2007. New deep-freezing process using renewable low-grade heat: from the conceptual design to experimental results. Energy 32:600–608.
22. Markowski, M., K. Urbaniec, A. Budek, M. Trafczyński, W. Wukovits, A. Friedl, et al. 2010. Estimation of

energy demand of fermentation-based hydrogen production. J. Clean. Prod. 18(Suppl. 1):S81–S87.
23. Liu, D. 2008. Bio-hydrogen production by dark fermentation from organic wastes and residues. Ph.D. Thesis, Department of Environmental Engineering Technical University of Denmark.
24. Lakaniemi, A.-M., O. H. Tuovinen, and J. A. Puhakka. 2013. Anaerobic conversion of microalgal biomass to sustainable energy carriers – A review. Biorefineries 135:222–231.
25. Langeland, M., A. Kiessling, and O.-I. Lekang. 2014. Baltic Aquaculture Innovation Centre (BIC), Aquaculture Center East, 1.
26. Halver, J. E., and R. W. Hardy. 2002. Fish nutrition, 3rd ed. Elsevier Science and Academic Press, San Diego, CA.
27. FAO. 2014. The state of World Fisheries and Aquaculture. Opportunities and challenges. Food and Agriculture Organisation of the United Nations, Rome.
28. FAO. 2014. Sustainable fisheries and aquaculture for food security and nutrition. Food and Agriculture Organisation of the United Nations, Rome.
29. Terjesen, B. F. 2012. Forskning på miljökrav og produktionsmetoder I RAS for Atlantisk laks. *in Proceedings*, Sunndalsöra, Norway.
30. Terjesen, B. F., S. T. Summerfelt, S. Nerland, Y. Ulgenes, S. O. Fjæra, B. K. Megård Reiten, et al. 2013. Design, dimensioning, and performance of a research facility for studies on the requirements of fish in RAS environments. Aquaculture 54:49–63.
31. Randall, D., W. Burggren, and K. French. 2001. Eckert: animal physiology, 5th ed. W.H. Freeman, New York, NY.
32. Lekang, O. I. 2013. Aquaculture hatchery water supply and treatment systems. Pp. 3–22 *in* X. Allan and X. Burnell, eds. Advances in aquaculture hatchery technology. Woodhead publishing, Cambridge, U.K.
33. Bergheim, A., H. Thorarensen, A. Jøsang, O. Alvestad, and F. Mathisen. 2013. Water consumption, effluent treatment and waste load in flow-through and recirculating systems for salmonid production in Canada – Iceland – Norway *in Proceedings*.
34. Malm, M., K. McFaul, P. W. Carlsson, C. Vettier, and C. Carlile. 2008. Focus Lund. The ESS Scandinavia submission to the ESFRI Working Group on ESS siting. European Spallation Source Scandinavia.
35. Parker, T. 2013. Energy policy. European Spallation Source ESS AB. 13-June-2013.
36. Peck, P., and T. Parker. 2015. The "sustainable Energy Concept" – making sense of norms and co-evolution within a research facility's energy strategy. J. Clean. Prod. doi: 10.1016/j.jclepro.2015.09.121.
37. Lindström, E. 2014. Energy inventory. European Spallation Source ESS AB, 24-Feb-2014.
38. Torbentsson, J. 2014. Cooling related inventory. EuCARD2 Consortium, Lund, Sweden, Deliverable Report EuCARD2-Del-D3-1, 2014.
39. Stadlmann, J., R. Gehring, E. Jensen, T. Parker, and P. Seidel. 2014. Energy efficiency of particle accelerators – a networking effort within the EuCard2 program, presented at the 5th International Particle Accelerator Conference, Dresden, Germany, Pp. 4016–4018.
40. Bordry, F., T. Parker, and C. Rizzuto. 2011. Main findings of the first joint workshop on Energy management for large-scale research infrastructures, presented at the Energy for Sustainable Science workshop, Lund, Sweden.
41. Indebetou, F. 2013. Business plan energy recycling. Pp. 104–117 *in* T. Parker, ed. ESS energy design report. ESS, Lund, Sweden.
42. Lebrun, P. 2014. Energy consumption and savings potential of CLIC, presented at the 55th ICFA Advanced Beam Dynamics Workshop on High Luminosity Circular e+e- Colliders – Higgs Factory, Beijing.
43. Brown, L. R. 2012. Full planet empty plates. The new geopolitics of food scarcity. W.W. Norton and Company, New York, NY.
44. WWF Soy Report Card. 2014. Assessing the use of responsible soy for animal feed in Europe. World Wildlife Found, International, 56 pp.
45. Gelder, J. V., K. Kammeraat, and H. KroesSoy. 2008. Consumption for feed and fuel in the European Union. Friends of the Earth Netherlands, 22 pp.
46. Ekelund, L. L., L. Johnson, S. Lundqvist, B. Persson, H. Sandin, H. Schroeder, A. Sundin, I. Christensen, G. Larsson, and L.-L. Björkman. 2012. Branschbeskrivning Trädgård – område hortikultur, utemiljö och fritidsodling, Sveriges lantbruksuniversitet, Fakulteten för landskapsplanering, trädgårds- och jordbruksvetenskap, Omvärld Alnarp 2012, ISBN:978-91-576-9114-9.
47. Rumsey, G. L., R. A. Winfree, and S. G. Hughes. 1992. Nutritional value of dietary nucleic acids and purine bases to rainbow trout (*Oncorhynchus mykiss*). Aquaculture 108:97–110.
48. Kinsella, J. E., B. German, and J. Shetty. 1985. Uricase from fish liver: isolation and some properties. Comp. Biochem. Physiol. B 82:621–624.
49. Andersen, Ø., T. S. Aas, H. Takle, S. van Nes, B. Grisdale-Helland, and B. F. Terjesen. 2006. Purine-induced expression of urate oxidase and enzyme activity in Atlantic salmon (*Salmo salar*). FEBS J. 273:2839–2850.
50. Langeland, M., A. Vidakovic, J. Vielma, J. Lindberg, A. Kiessling, and T. Lund. 2015. Digestibility of microbial and mussel meal for Arctic charr (*Salvelinus alpinus*) and Eurasian perch (*Perca fluviatilis*). Aquac. Nutr. *In press.*
51. Santoyo-Castelazo, E., and A. Azapagic. 2014. Sustainability assessment of energy systems: integrating environmental, economic and social aspects. J. Clean. Prod. 80:119–138.

52. Gonzales, A., B. Frostell, and A. Carlsson-Kanyama. 2011. Protein efficiency per unit energy and per unit greenhouse gas emissions: potential contribution of diet choices to climate change mitigation. Food Policy 36:562–570.
53. Pelletiers, N., and P. Tyedemers. 2007. Feeding farmed salmon: is organic better? Aquaculture 272:399–416.
54. Troell, M., P. Brunsvik, N. Kautsky, and P. Ronnback. 2004. Aquaculture and energy use. Pp. 97–108 *in* C. Cleveland, ed. The encyclopedia f energy Vol. 1. Elsevier, St. Louis, MO.
55. Tyedmers, P. H., R. Watson, and D. Pauly. 2005. Fueling global fishing fleets. Ambio 34:635–638.

A study of the annual performance of bifacial photovoltaic modules in the case of vertical facade integration

Bruno Soria[1], Eric Gerritsen[1], Paul Lefillastre[1] & Jean-Emmanuel Broquin[2]

[1]Photovoltaic Modules Laboratory, CEA-INES, 50 avenue du Lac Léman, Le Bourget-du-Lac F-73375, France
[2]CNRS, IMEP-LACH, Grenoble Alpes University, Grenoble F-38000, France

Keywords
Bifacial, facade, half-cells, indoor, N-type, outdoor, ray-tracing, textured glass

Correspondence
Bruno Soria, CEA-INES, Photovoltaic Modules Laboratory, 50 avenue du Lac Léman, Le Bourget-du-Lac F-73375, France.

E-mail: brunosoria@ntymail.com

Funding Information
Study funded by CEA-INES.

Abstract

Despite the apparent benefits of bifacial modules, their application still suffers from a lack of visibility on the performance gain that they can actually provide. In this work, we consider the specific application of vertically oriented bifacial modules, notably for facade integration. We have developed a methodology to evaluate the annual electrical performance of bifacial modules based on three tools. First, a double illumination characterization setup is used in a solar simulator for comparing module architectures. Then, a reduced scale outdoor test bench allows us to evaluate bifacial module performance in a variety of configurations. Finally, a ray-tracing model validated with short-term outdoor data leads to the determination of the annual performance gain. This methodology allowed us to find optimal performance according to the most important parameters of application and module. Specifically, a module architecture using half-cut cells, a parallel cell interconnection, and textured glasses have been analyzed with respect to their influence on the resistive losses which increase in dual side illumination as well as to their influence on the effect of non-uniform and diffuse irradiance on the backside of the module. This work enabled us to give directions for innovative full-size module architectures.

Context and Aim of the Study

Bifacial photovoltaic (PV) cells which are able to convert solar radiation from both sides have been developed since the 1960s, but their terrestrial application only started around the 1980s [1, 2]. Even though some modeling work is performed at the megawatt scale [3], only few large bifacial plants have been set up so far, probably due to a lack of confidence from investors in the actual gain of bifacial technology. Indeed, actual commercial energy prediction softwares do not take into account the complex opto-geometrical environment of the modules (generating non-uniform and diffuse backside albedo for instance), and the experimental tests performed so far have often been short term, at small scale or in specific configurations only. New methodologies which consider the complex opto-geometrical environment of bifacial modules are however in development in the research community [4, 5].

Vertical facade integration of PV modules attracts interest in the building integrated photovoltaics (BIPV) community because PV facades have a large collection area without additional footprint and their vertical installation makes them less sensitive to snow, dust, bird droppings, or tree leaves. However, they suffer from non-optimal orientation and module heating due to integration onto a wall. Using bifacial modules in a double skin configuration can minimize these drawbacks. Figure 1 shows how the gap between module and inner wall allows backside albedo and natural or forced ventilation to give additional power to the module. Additionally, there are seasonal advantages like a higher PV production and heating inside the building during winter as well as vertical sun shading during summer.

The aim of this article is to evaluate and to maximize the annual performance of a bifacial module in a double skin configuration compared to a monofacial one. For this purpose, we developed a methodology based on indoor characterization with a double illumination setup in solar simulator, on outdoor characterization with an adjustable test bench of the application, and on ray-tracing simulation with an optical model of the application. The study has been performed on

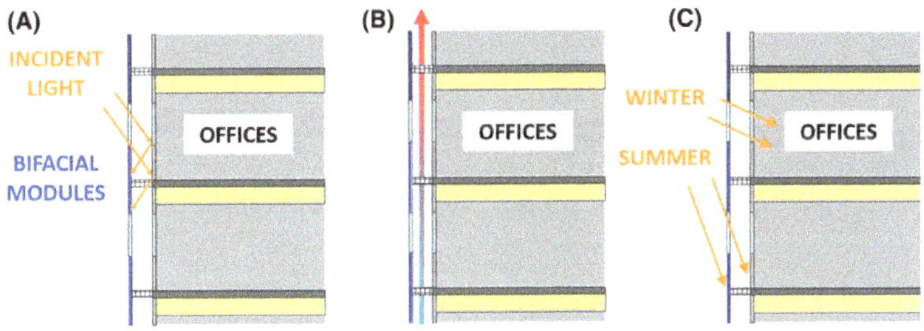

Figure 1. Advantages of using bifacial modules on a vertical facade in a double skin configuration: a backside albedo (A), natural or forced ventilation (B), and seasonal advantages (C).

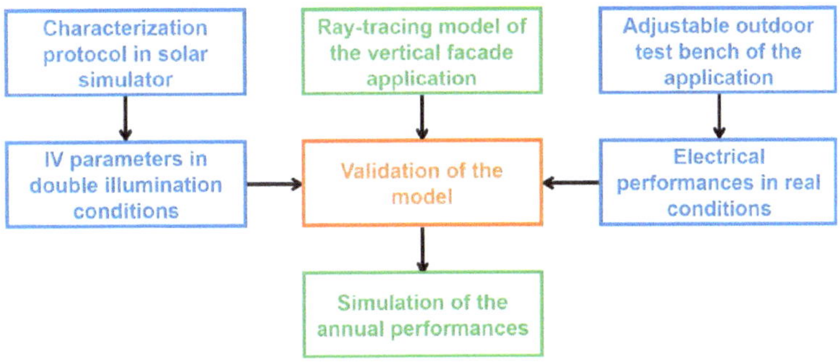

Figure 2. Schematics of the methodology used for evaluating the annual performance of bifacial modules in a vertical facade application (experimental part in blue, simulation part in green, and validation in orange).

a reduced scale application using mini-modules. The bifacial gain is dependent on the opto-geometrical parameters of the application (double skin distance, reflective properties of the inner wall) and on the module architecture (electrical and optical parameters). Therefore, these two categories of parameters have been studied separately. The results enabled us to propose some design rules of the application and perspectives for the optimal bifacial module architecture.

Methodology of the Study

We used mini-modules with four N-type bifacial cells connected in series, encapsulated between two glasses in EVA-HLT (ethylene vinyl acetate - high-light transmission from STR). The type of cells used in this study has a back to front power ratio of 0.9 and has been reported in detail in [6]. Figure 2 shows the different steps implemented for the simulation of the annual performance and the following subsections will describe the three tools used in this methodology.

Double illumination characterization setup

Standard characterization protocols for bifacial modules in solar simulators usually consist in measuring each side of the module independently using a black mask on the opposite side. Here, we adapted a double illumination setup existing at the cell level [7–9] for our mini-modules. Figure 3 shows the setup we designed. The bifacial module is placed between two aluminum mirrors (Alanod MIRO® 4200GP, Ennepetal, Germany) with an angle between mirror and module of $\psi = 44.1°$ optimized with a ray-tracing model of our PASAN solar simulator performed with TracePro® [10]. Metallic grid filters with several pitches are used for attenuating light on the backside of the module in order to mimic different albedo conditions.

With this setup, we used a protocol with nine main steps depending on the PASAN intensity filters and on our metallic filters (see Table 1). We performed a reproducibility study in order to quantify the 3σ relative uncertainty of every current–voltage (IV) parameter for each step. Figure 4 shows that this uncertainty is comprised in the ±3% range as in Ohtsuka's study [7]. The uncertainty is high at step 1 probably because the voltage range of the IV sweeping (kept equal for each step of our protocol) is not fully adapted for low irradiance measurements. It is also high at steps 7, 8, and 9 because the manual change of filters can slightly modify the alignment between the PASAN source and the module. Finally, step 5 shows the lowest uncertainty as this is the calibration

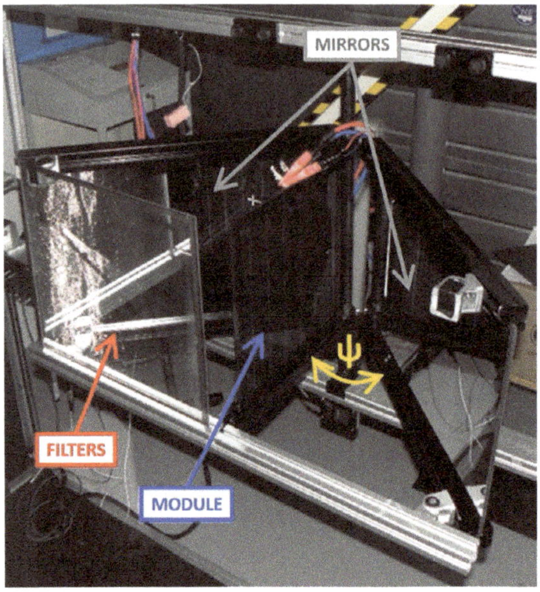

Figure 3. Picture of the double illumination setup positioned in our PASAN solar simulator.

step: here, the front short circuit current measured with our setup ($I_{sc\text{-}front}$) is compared to the I_{sc} measured in standard test conditions (STC) with a black mask on the back ($I_{sc\text{-}front\text{-}STC}$) and the PASAN light source intensity is adjusted in order to compensate mirrors absorption (about 5%).

Figure 5 shows the spectral intensity measured with a spectrometer in the standard STC measurement case (in blue) and pondered with the spectral reflectivity of the aluminum mirror with our calibrated setup (in red). This shows that the spectral intensity of the PASAN source belongs to the A+ class (deviation compared to AM1.5G below ±12.5%) as long as the calibration step which compensates the mirror absorption is performed.

Table 1. The nine main steps of our protocol. Additional steps consist in measuring the backside of the module (black mask on the frontside) in order to measure the optical transmission of the metallic filters.

Protocol step number	Irradiance on the reference cell (below our setup)	Percentage of light on the back of the module
1	100 W/m^2	0% (mask)
2	200 W/m^2	0% (mask)
3	400 W/m^2	0% (mask)
4	700 W/m^2	0% (mask)
5	1000 W/m^2	0% (mask)
6	1000 W/m^2	23% (filter)
7	1000 W/m^2	47% (filter)
8	1000 W/m^2	78% (filter)
9	1000 W/m^2	100% (no filter)

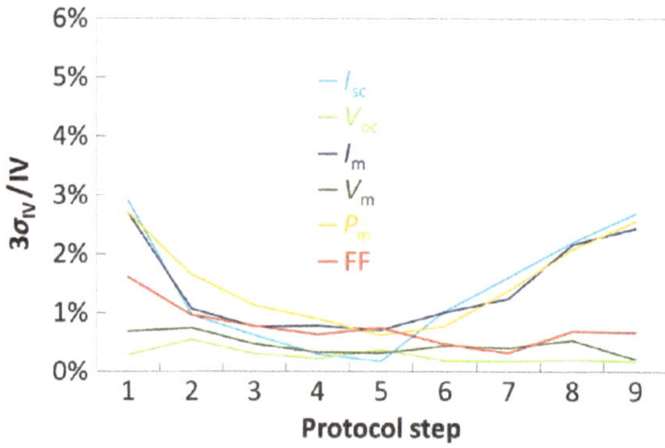

Figure 4. 3σ relative uncertainty measured for each step of the protocol as well as each IV parameter (σ being the standard deviation).

Using our setup, we ensured that all the IV parameters follow the same tendency independently of the illumination mode (frontside only, backside only, double side) and identified the resistive losses which increase when the module receives higher irradiance. Figure 6A shows that the power in double illumination ($P_{m\text{-}bi}$) gets lower than the linear tendency, which is the sum of frontside and backside powers ($P_{m\text{-}front}$ + $P_{m\text{-}back}$) as the total irradiance on the module $I_{sc\text{-}bi}/I_{sc\text{-}front\text{-}STC}$ increases. Figure 6B details this behavior plotting the gain between $P_{m\text{-}bi}$ and ($P_{m\text{-}front}$ + $P_{m\text{-}back}$). This shows that the resistive losses reach −5% at 2 sun. Note that for front characterization at low irradiances, the gain is actually positive because of the logarithmic behavior of the voltage (see the V_m parameter evolution in blue).

Outdoor test bench

Our outdoor study has been performed on an adjustable reduced scale test bench inspired by a real building facade. From November 2012 until March 2013, we studied the influence of the opto-geometrical parameters of the application, namely the diffuse reflector on the inner wall and its distance to the module. Figure 7A shows the bifacial module in the middle, a monofacial reference on the right (bifacial cells from the same batch encapsulated with EVA-HLT between a glass and a white backsheet), and a reflector behind. In this article, the measure for length will be quantified in terms of number of *c*: c = 36 cm being the length of our modules, $3c \times 2c$ the area of the reflector, and $3c \times 3c$ the aperture of the opto-geometrical environment. From April 2013 until December 2013, we studied the influence of different module architectures. Figure 7B shows two bifacial modules (a standard architecture on the left and novel one on

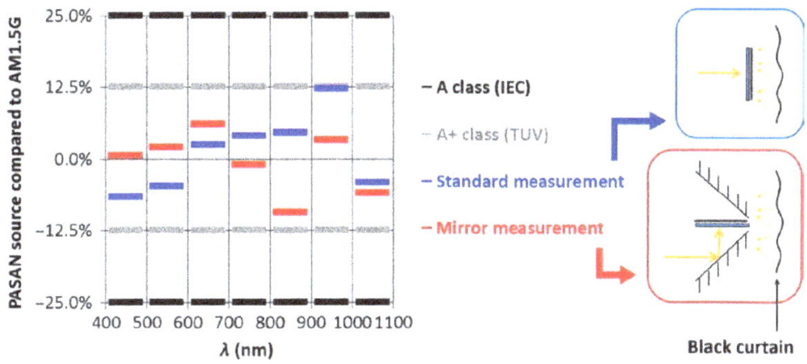

Figure 5. Spectral repartition of the PASAN source intensity compared to AM1.5G reference. The blue schematics shows a standard STC measurement and the red one shows a measurement with our double illumination setup. Black curtain reflections are also shown behind the setup.

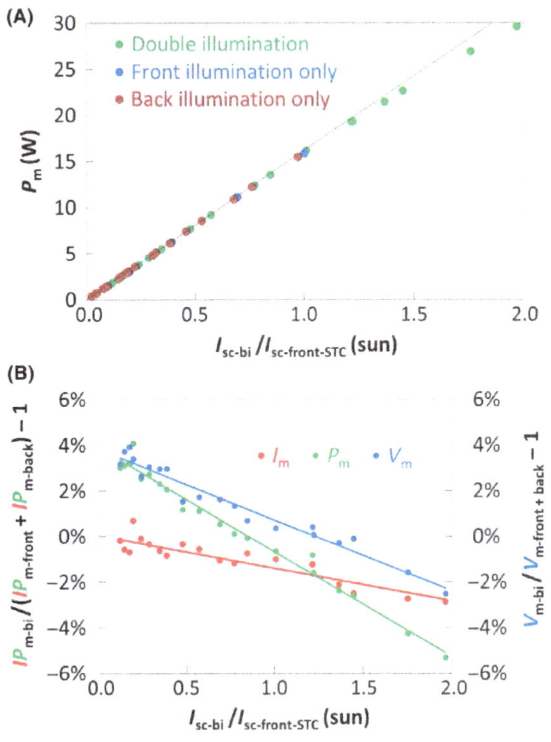

Figure 6. (A) Power P_m measured for different illumination modes and linear tendency showing the sum of frontside and backside powers ($P_{m\text{-front}}$ and $P_{m\text{-back}}$). (B) Gain between the double illumination power $P_{m\text{-bi}}$ and ($P_{m\text{-front}} + P_{m\text{-back}}$) and its separation into the current contribution I_m and the voltage contribution V_m.

the right), the monofacial reference below, and a reflector behind. The whole setup is equipped with frontside and backside irradiance sensors (in yellow and orange) as well as calibrated temperature sensors on the back of the modules. These are T-type sensors glued with epoxy and protected from light using aluminum adhesive. The IV parameters have been followed over the year with a Daystar MT5 multi-tracer. These data have been used to compare the temperatures measured with thermocouples and the ones calculated using outdoor and indoor IV parameters. Equation 1 shows the calculation of the temperature T as a function of the open circuit voltage V_{oc} and the voltage temperature coefficient β (−0.31%/°C, assumed independent of the irradiance G [11]). V_{oc}(T, G) is directly measured outdoor and V_{oc}(25, G) is calculated from I_{sc}(T, G) also measured outdoor. Note that we assume $I_{sc}(25, G) = I_{sc}(T, G)$, that is to say a zero current temperature coefficient, otherwise the equation would become non-linear.

$$T = 25 + \frac{V_{OC}(T, G) - V_{OC}(25, G)}{\beta \cdot V_{OC}(25, 1000)} \tag{1}$$

Figure 8 shows the energy yield of the monofacial module in terms of kilowatt-hour per kilowatt-peak over the period of test. Minimum production appears near summer solstice due to non-optimal orientation of the module with respect to solar positions (high azimuth angles on the morning and evening and high elevation angles around noon), and near winter solstice due to short duration of irradiation over the day. This results in a maximum production achieved around spring and autumn solstices.

Annual performance simulation

Ray-tracing model

We used the ray-tracing software TracePro® to create a simple optical model of our test bench (see Figure 9A). The solar disk sources can be set for any position in the sky with solid angle of the sun not being taken into account. Moreover, an isotropic semi-hemispherical background source is defined with an importance

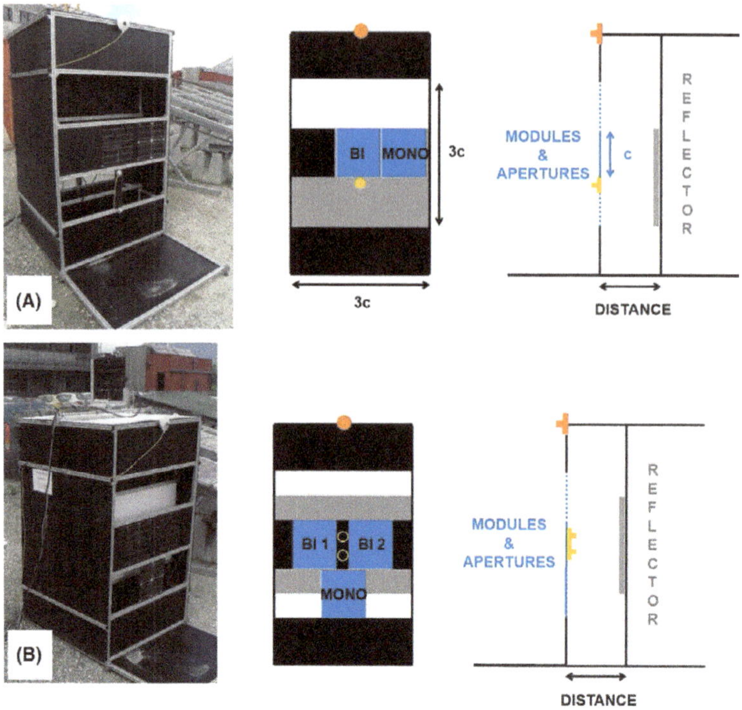

Figure 7. Picture and schematics of our reduced scale test bench for studying the opto-geometrical parameters (A) and the module parameters (B). Filled orange circle denotes the front pyranometer CMP3, while the filled and empty yellow circles denote the SPlite pyranometers for front and rear, respectively. All pyranometers are installed vertically and front pyranometers face south.

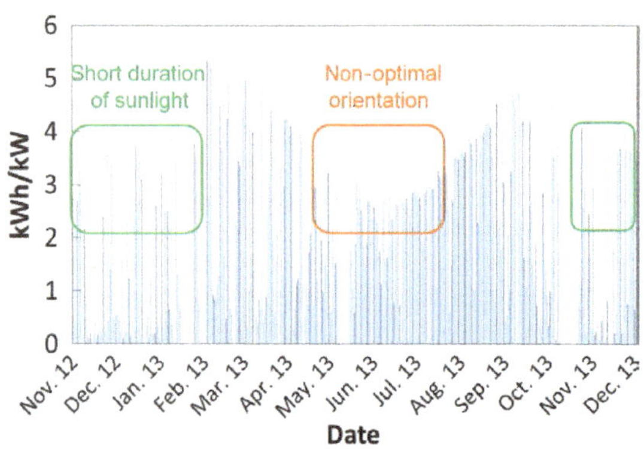

Figure 8. Daily performance of the monofacial reference module over the whole period of experimental tests.

sampling applied on the aperture area $3c \times 3c$. We limit the model to the opto-geometry of the system and do not take into account spectral effects. Additionally, we did not take into account the ground albedo in front of our test bench in order to be able to use the importance sampling tool. Also, the reflective losses, semi-transparency of the module as well as global irradiances below 10 W/m^2 have not been integrated in the model. Note that Reich went down to 8 W/m^2 [12] and Sprenger to 5 W/m^2 [13]. With this model, we can set any diffuse to global ratio and any sun position in order to simulate the spatial and angular distribution of hourly irradiance on the backside of the bifacial module. We defined a typical day for each season of the year as shown on Figure 9B. The input data is taken from PVsyst database [14] and averaged over ±1.5 month around each solstice. Note that the number of rays have been chosen in order to ensure sufficient uniformity of the irradiance as well as limited time of simulation.

Optical to electrical conversion at 25°C

Once the model is designed, we use indoor flash test measurements (front and back independently measured) for converting irradiance data into electrical power data at 25°C. The spatial distribution of the irradiance is taken into account associating indoor STC measurements (Fig. 10A) with the least irradiated cell in the outdoor test bench simulations (Fig. 10B). The conversion is done using a logarithmic relationship between efficiency and irradiance ($\eta = a_1 \ln(G) + b_1$). The angular distribution

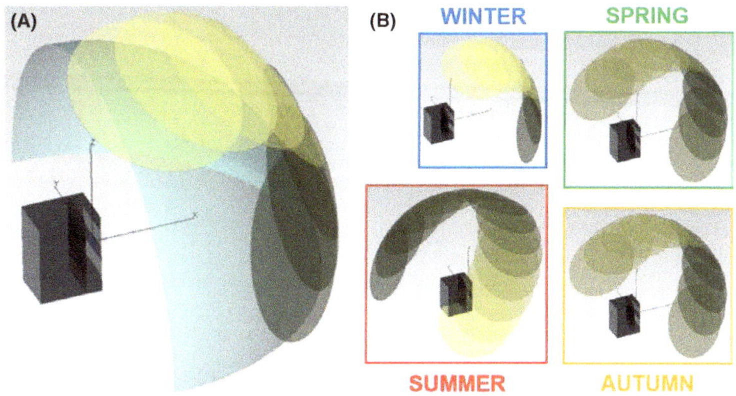

Figure 9. (A) Optical model of the test bench with direct (solar disks) and diffuse (isotropic sky) sources irradiating the test bench. (B) Solar positions for each typical day.

is taken into account associating indoor angle measurements (Fig. 10C) with the incidence angles of light rays in the outdoor test bench simulations (Fig. 10D). The conversion is done using a cosinus relationship between power and angle ($P_m = c_1 \cos(\theta)$). These conversions are critical hypothesis as we associate irradiance data from input meteorological data based on the measurement device spectral and angular responses with irradiance data based on the monitor cell spectral response at normal incidence and AM1.5G indoor spectrum.

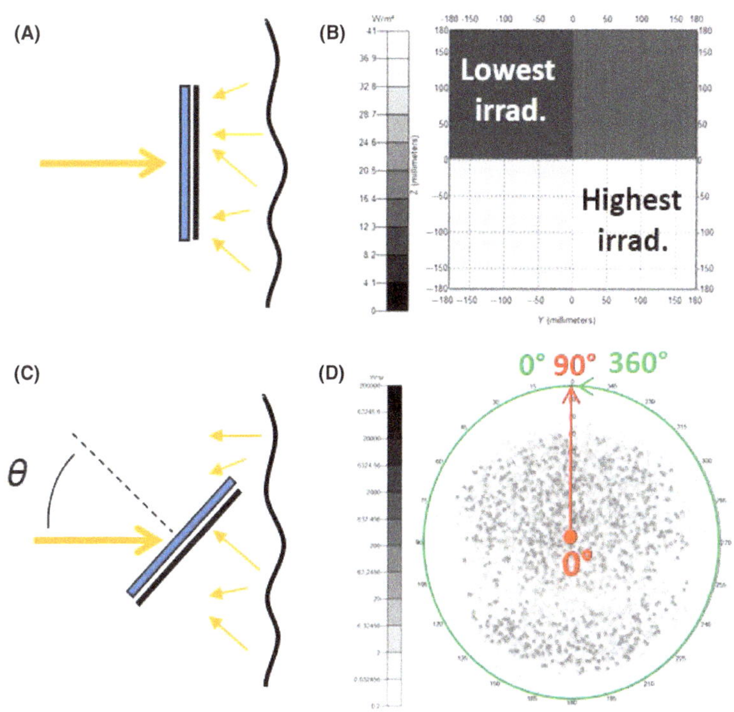

Figure 10. Standard indoor measurement (A) correlated with an example of spatial distribution of outdoor irradiance data (B) each of the four pixels corresponds to a cell, the darker the less irradiated. Indoor angle measurements (C) correlated with an example of angular distribution of outdoor irradiance data (D): red straight axis from 0° to 90° shows the incidence angle of a ray and green circular axis from 0° to 360° maps the direction of incidence of a ray, the darker the more irradiated in this case.

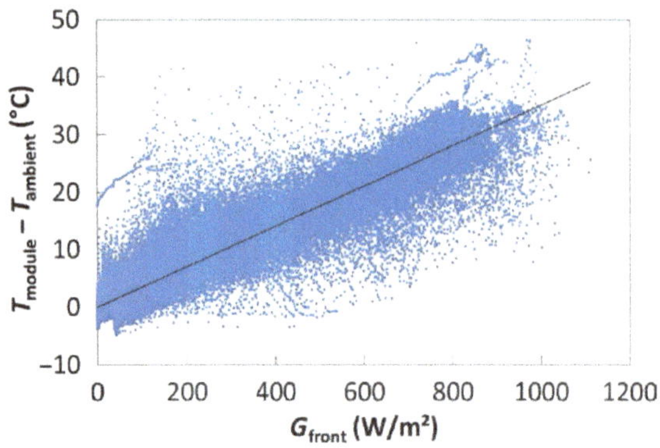

Figure 11. Temperature difference between monofacial module and ambient as a function of the front irradiance in minute data over the year of test, and its linear approximation.

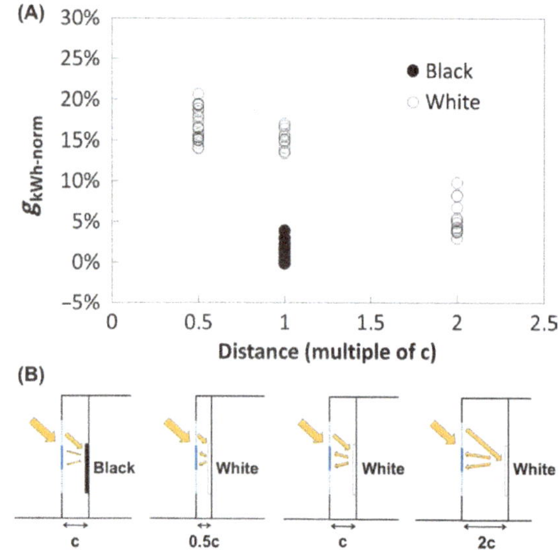

Figure 12. Daily bifacial electrical energy gain compared to monofacial (powers normalized to the front STC for each module) for different opto-geometrical configurations. The module reflector distance is shown in terms of module height, c = 36 cm.

Bifacial power at real temperature

In order to get the electrical power at real operating temperatures, we use a linear model based on outdoor data obtained for the monofacial module. Figure 11 shows the temperature difference T_{module}–$T_{ambient}$ as a function of irradiance G_{front} over the whole year in minute data. The linear approximation lead to a coefficient κ_{mono} = 0.035°C/(W/m²) in Equation 2. Therefore, the module temperature can be evaluated using the ambient temperature and irradiance measured on site, and the real powers can be calculated. These front and back powers are summed to give the bifacial power. We ensured with double illumination indoor measurements that this hypothesis is valid for the level of total irradiance involved in our application (<1.04 sun).

$$T_{module} = T_{ambient} + \kappa \cdot G_{front} \quad (2)$$

Influence of Opto-geometrical Parameters

Experimental data

Between November 2012 and March 2013, several opto-geometrical configurations have been tested one after each other varying the module–reflector distance (0.5*c*, *c* and 2*c*; with the module height *c* = 36 cm) and the inner reflector (black or white). In this test, we used a module of standard architecture (glass–glass structure, full-size cells in series) with the configuration of the test bench shown on Figure 7A. Figure 12 shows the daily electrical energy gain brought by the use of a bifacial module instead of a monofacial one for four configurations. The variability in the data points is due to the period of test and the meteorological conditions, which were more or less cloudy.

At a distance *c*, the use of a white reflector (weighted reflectivity with respect to the solar spectrum, *R* ~80.4%) gives an energy gain compared to monofacial module ($g_{kWh\text{-}norm}$) 13% higher than the one obtained with a black reflector (*R* ~4.9%). At a smaller distance (0.5*c*), maximal gains appear to be comprised between 15% and 20%.

Additionally, the temperature difference between bifacial and monofacial modules is shown in Figure 13 for the four configurations. At low irradiances, a measurement artefact due to the combination of data measured with different IV tracers is visible (temperatures calculated with Equation 1). For higher irradiances, all the configurations follow the same linear tendency, which reaches −2°C at a front irradiance of 1000 W/m². The main reason is that the bifacial module has twice as much surface compared to the monofacial one to release heat by thermal radiation. Using these data, we define a new coefficient κ_{bi} = 0.033°C/(W/m²) for the bifacial module, which will be used for simulating its thermal advantage over the year. It must be noticed that both the temperatures measured with thermocouples and calculated with V_{oc} gave the same linear trend (except at low irradiances) after removing thermal conduction contributions, which are different for a thermocouple glued on a 0.17 mm backsheet compared to a 3 mm backside glass [15].

Validation of the model

A sunny and a cloudy day have been used to compare the simulation and experimental results for winter period as well as for summer period. The model sources (see Fig. 9) are defined using direct irradiance measured with a sun tracker on our site and diffuse irradiance on the module plane calculated using global irradiance measured with the front pyranometer (filled orange circle on Fig. 7). The backside irradiance is simulated and converted in electrical power data as explained in section Optical to electrical conversion at 25°C. These data are temperature corrected using experimental temperature monitoring. Figure 14 shows the comparison for these 4 days. We use the determination coefficient R^2 and the Student test (see Equation 3 and Equation 4 – N being the number of data points over the day) to compare the experimental and simulated temporal series (x^i_{exp} and x^i_{sim}). Each R^2 is above 95.1% and the Student test allows us to say that simulations estimate measurements with a confidence interval of 99.9% [16].

$$R^2 = 1 - \frac{\sum_{i=1}^{N} \left(x^i_{sim} - x^i_{exp}\right)^2}{\sum_{i=1}^{N} \left(x^i_{exp} - x^{ave}_{exp}\right)^2} \quad (3)$$

$$t_s = \sqrt{\frac{(N-2)\cdot R^2}{1-R^2}} \quad (4)$$

Annual performance

In the previous sections, we obtained experimental data in different configurations of our test bench at specific periods of the year. The model being validated, we can now extrapolate our results over the whole year. Figure 15 shows the bifacial module energy production simulated in watt-hour for each typical day as well as over the year (cumulating each typical day) according to the distance between module and reflector in the white reflector case. As in the monofacial case, we can see that the energy production is maximum near spring and autumn solstices independently of the distance (see Fig. 8). The annual energy production (black curve) clearly shows a maximum between distance 0.5*c* and 0.75*c*. Above these distances, the decrease is due to side effects of our test bench, namely the black side walls between the module and the reflector, which absorb the light diffused by the white reflector and prevent the ambient diffuse light from reaching the backside of the module. Below these distances, the decrease is due to current limitation by the least irradiated cell in the module. Indeed, Figure 16 shows the non-uniformity of backside irradiance (daily average of hourly data points calculated with Equation 5) according to the distance for each typical day as well as over the year. The black curve on the chart shows a strong increase in non-uniformity at short distances. This is mainly due to the module shade as well as the black side walls

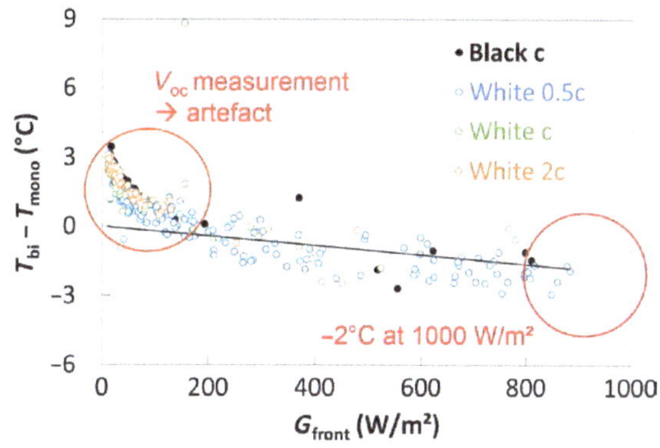

Figure 13. Temperature difference between bifacial and monofacial module as a function of front irradiance for the four configurations shown in Figure 12 (hourly data points). The linear tendency without taking into account the low irradiance artefact concerns the four configurations all together.

shade both projected onto the reflector (depending on the position of the sun).

$$NU = \frac{G_{max} - G_{min}}{G_{max} + G_{min}} \quad (5)$$

Influence of Module Architecture

Experimental data

Between April 2013 and December 2013, innovative module architectures have been tested one after each other with the configuration of the test bench shown on Figure 7B. We saw before that bifacial modules suffer from resistive losses due to higher currents, and the backside irradiance is often non-uniform and diffuse. In order to address these issues, we tested two different architectures. On the one hand, a module has been made with two series of half-cut cells connected in parallel in order to reduce the resistive losses and to be less sensitive to non-uniform irradiance (Fig. 17A). We ensured that this architecture has the same $I_{sc\text{-}front\text{-}STC}$ and $V_{oc\text{-}front\text{-}STC}$ as a standard architecture (four full size cells in series). On the other hand, we used a textured glass on both sides for collecting high incidence angles of diffuse radiation (Fig. 17B). The following subsections describe the experimental tests (indoor and outdoor) performed for finding empirical models, which will allow us later to identify the potential of these novel module architectures at the annual scale.

Influence of using half-cells

In order to separate the influence of half-cells from the influence of parallel interconnection, we had to measure

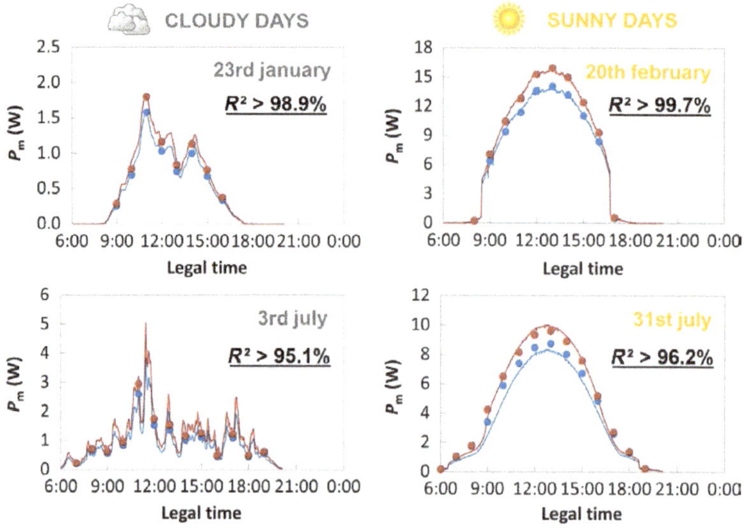

Figure 14. Comparison of simulated (dots) and experimental (curves) data for bifacial (red) and monofacial (blue) modules against legal time, with associated R^2. Cloudy days on the left, sunny days on the right, winter days on the top, and summer days on the bottom.

the module under uniform illumination. This was possible with our indoor double illumination setup (Fig. 18A) and with our outdoor test bench masking the backside of the module (Fig. 18B). Both charts show the gain brought by the novel architecture compared to the standard one for parameters I_m, V_m, and P_m. The indoor and outdoor measurements are in agreement showing a linear increase in P_m and V_m with total irradiance on the module: we observe a negative gain below 0.25 sun and ~3% at 1 sun. The indoor bifacial characterization shows up to 7% gain at 2 sun. Indeed, as the total irradiance increases on the module, the resistive losses increase on the standard architecture and the use of half-cells brings more and more advantage (higher power due to higher voltage). Note that the negative gain at low irradiances might be due to shunt currents induced by laser cutting of our N-type cells.

From the indoor experimental data, we used a linear relationship (green line shown on Fig. 18A) between the power gain g_{Pm} and the total irradiance incident on the module $I_{sc\text{-}bi}/I_{sc\text{-}front\text{-}STC}$ shown in Equation 6 for simulating the contribution of half-cells at the annual scale.

$$g_{Pm} = 4.6\% \cdot I_{sc-bi}/I_{sc-front-STC} - 1.4\% \quad (6)$$

Influence of parallel interconnection

The bifacial gain is critically dependent on the uniformity of irradiance on the module backside (examples in [4, 17]). For vertical facades, the section Annual performance showed that the backside irradiance can be strongly non-uniform if the module–reflector distance is very small. In this case, the upper and lower cells do not receive the same amount of radiation due to the module shade projection onto the back reflector. Therefore, we investigated the advantage of using two strings of cells connected in parallel in the module architecture in order to be less sensitive to non-uniform irradiance.

First, we artificially created strong non-uniformities on the backside of the module placed in the double illumination setup. Table 2 shows the gain between the novel architecture and the standard one for parameters I_m, V_m, and P_m in three different backside configurations: no mask, a $c \times 0.25c$ mask, and a $c \times 0.5c$ mask (no mask on the

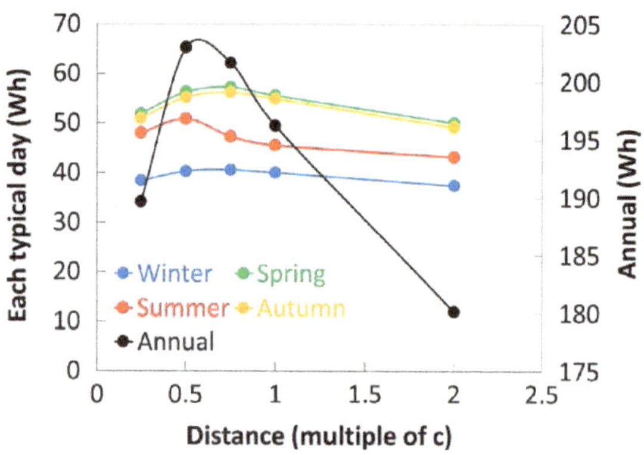

Figure 15. Simulated daily energy production for each typical day (winter in blue, spring in green, summer in red, and autumn in yellow) and over the year (black) as a function of the distance between module and reflector.

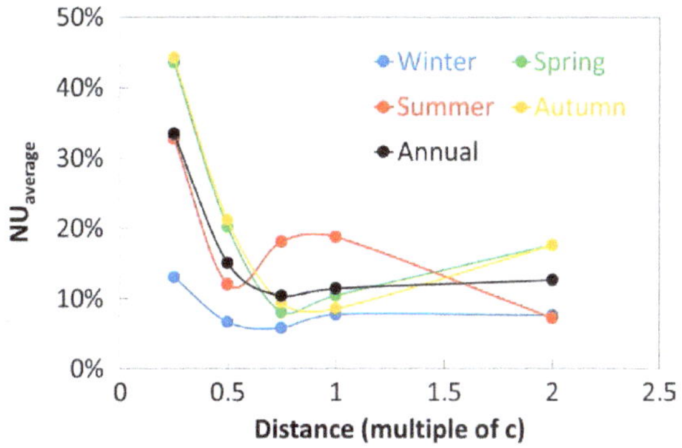

Figure 16. Simulated daily averaged non-uniformity of backside irradiance for each typical day (winter in blue, spring in green, summer in red, and autumn in yellow) and over the year (black) as a function of the distance between module and reflector.

frontside). For comparison, the gain taking into account resistive losses only (calculation with eq. 6) is written between brackets. The configuration $c \times 0.25c$ shows a strong current gain (12%) due to the interconnection in parallel since the shaded string does not limit the unshaded string in the novel architecture. In terms of voltage, the shaded string slightly limits the unshaded string: a 6% gain due to resistive losses becomes only a 1.4% gain with the parallel interconnection in addition. This still results in a positive effect on the power (13.5% gain). If one of the two strings is fully masked (configuration $c \times 0.5c$), this phenomenon is increased with a higher current gain (+41%) and a voltage loss (−2.2%).

Then, we compared two different configurations of our outdoor test bench by varying the distance between module and reflector: a more uniform irradiance and a less uniform irradiance on the backside of the module as shown on Figure 16. Figure 19 shows that the power gain is about 0.5% higher for the less uniform configuration. This confirms the slight advantage of using a parallel interconnection in outdoor conditions for our specific case. This is the result of about 1% gain for I_m (no current limitation by the least irradiated string) and about 0.5% loss for V_m (voltage limitation by the least irradiated string). Note that both curves have a similar slope due to the resistive loss effect as explained previously.

In order to be able to simulate the maximal contribution brought by the parallel interconnection on the annual scale, we expressed the I_{sc} gain using the spatial distribution of irradiance on each side. The following equations have been used to get Equation 7:

1. $I_{sc\text{-}bi} = I_{sc\text{-}front} + I_{sc\text{-}back}$ for both modules. This is verified at −0.5% in average, while Ohtsuka had +0.4% [7] and Ezquer −1.8% [8].
2. $I_{sc\text{-}front} = I_{sc\text{-}front\text{-}STC} \times G_{front}$ for both modules (uniform irradiance on their frontside). G is the global irradiance incident on one face in terms of number of sun.
3. $I_{sc\text{-}back} = \min(I_{sc\text{-}back\text{-}string1}, I_{sc\text{-}back\text{-}string2}) = I_{sc\text{-}back\text{-}STC} \times \min(G_{back\text{-}string1}, G_{back\text{-}string2})$ for the standard architecture.
4. $I_{sc\text{-}back} = I_{sc\text{-}back\text{-}string1} + I_{sc\text{-}back\text{-}string2} = 0.5 \times I_{sc\text{-}back\text{-}STC} \times (G_{back\text{-}string1} + G_{back\text{-}string2})$ for the novel architecture.
5. $I_{sc\text{-}back\text{-}STC} = r \times I_{sc\text{-}front\text{-}STC}$ for both modules, which have the same I_{sc} in STC conditions. r is the bifacial ratio.
6. the numbers of sun G_{front}, $G_{back\text{-}string1}$ and $G_{back\text{-}string2}$ are linked through the non-uniformity of irradiance NU

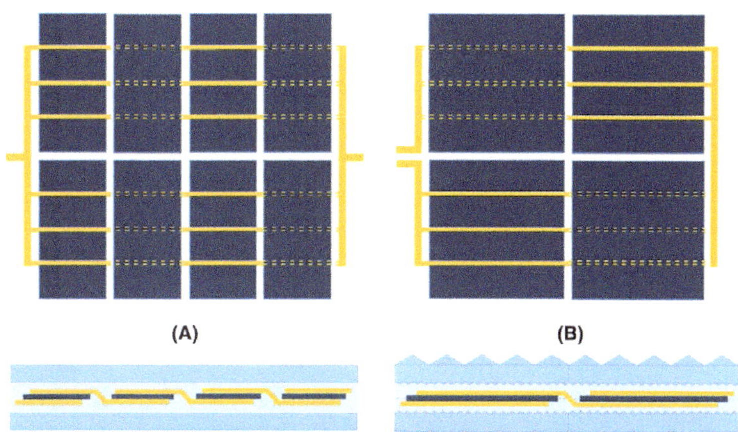

Figure 17. Schematics of the two different architectures tested: half-cells and parallel strings with flat glasses (A), and four full-size cells in series with textured glasses (B). The modules have been made with cells from the same batch and will be compared to a reference module made of four full-size cells in series encapsulated between flat glasses.

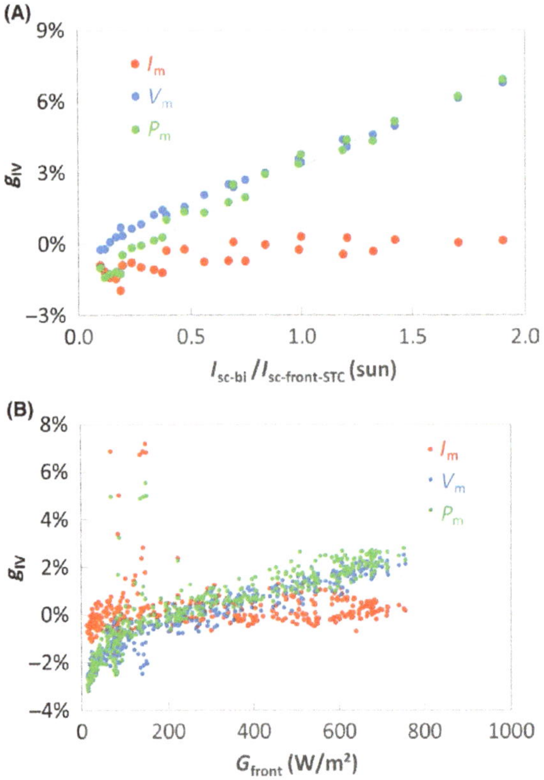

Figure 18. Gain of the novel architecture compared to a standard module for the parameters I_m, V_m, and P_m obtained with indoor measurements (A – bifacial illumination characterization) and outdoor measurement (B – front characterization only in hourly data).

between the upper string and the lower string of the backside (see eq. 5), and through the effective albedos $A_{min} = \min(G_{back\text{-}string1}, G_{back\text{-}string2})/G_{front}$ and $A_{ave} = 0.5 \times (G_{back\text{-}string1} + G_{back\text{-}string2})/G_{front}$.

$$g_{Isc} = NU \cdot \frac{r \cdot A_{ave}}{1 + r \cdot A_{min}} \tag{7}$$

Table 3 shows the comparison between calculation and measurements with our indoor setup. Our calculations are validated experimentally in this case of strong non-uniformities artificially created (masks $c \times 0.25c$ and $c \times 0.5c$), so we applied this model using outdoor I_{sc} data. We observed that the measurement values were lower than the calculated values for unknown reasons (chart not represented). Therefore, we decided to add an empirical coefficient of 0.4 in Equation 7 in order to fit outdoor observations. Finally, we associate the resulting I_{sc} gain to the upper limit of the P_m gain in Equation 8, since the shaded string can suffer from reverse currents. Equation 8 will therefore allow us to evaluate the maximum power gain due to non-uniform irradiance even though the real power gain will be always lower.

$$g_{Pm-max} = 0.4 \cdot NU \cdot \frac{r \cdot A_{ave}}{1 + r \cdot A_{min}} \tag{8}$$

Influence of linear textured glasses

Diffuse radiation plays an important role in bifacial applications particularly on the backside of the module. We compared a bifacial module with a unidirectional textured glass from the company AGC on both sides, to a bifacial module with standard flat glasses. Such a unidirectional texturation can have the advantage of self-cleaning with rain. In order to identify the effect of the glass only, we measured the gain between a monofacial module with the front linear textured glass and a monofacial module with the front flat glass as a function of the incidence angle of light θ (see Fig. 10C). Figure 20 shows the results for two orientations of the texturation: parallel and orthogonal to the rotation axis. For angles below 40°, the gain is about 2–3% due to a light trapping effect. Above 60°, the orthogonal orientation shows a gain increase due

Table 2. Gains of the novel architecture compared to standard one for parameters I_m, V_m and P_m for different shading configurations on the backside of the modules placed in our double illumination setup (see the image on the left corner). The measured gains (in bold) are compared with calculated gains taking into account resistive losses only (between brackets).

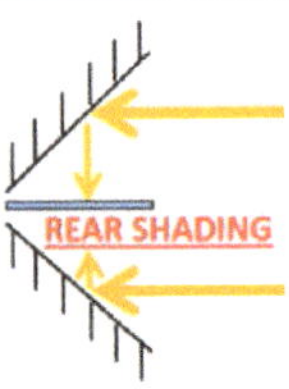	No mask 	Mask c × 0.25c 	Mask c × 0.5c 
g_{Im}	**−0.3%** (0.3%)	**12%** (0%)	**41%** (−0.4%)
g_{Vm}	**6%** (6.2%)	**1.4%** (4.6%)	**−2.2%** (3.1%)
g_{Pm}	**5.7%** (6.5%)	**13.5%** (4.6%)	**37.9%** (2.7%)

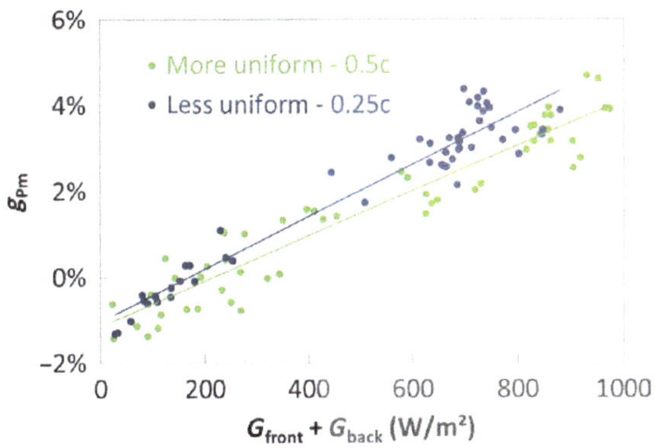

Figure 19. Hourly data point of the power gain g_{Pm} of the novel architecture compared to standard one plotted against the total irradiance on the module $G_{front} + G_{back}$ for two different configurations of the test bench: smaller distance (less uniform irradiance on the backside of the module) and larger distance (more uniform irradiance).

to the divergence of the PASAN source as well as parasitic reflections on the black curtain behind the module. For this range of angles, a clear increase in the gain confirms the positive effect of textured glass for the parallel orientation.

In order to be able to simulate the influence of textured glasses at the annual scale on our application, the angular distribution of light on each face of the module has been simulated hourly for each typical day (see Fig. 10D as an example). For each side of the module, the optical flux is integrated over ±45° around the horizontal direction on the one hand, and around the vertical direction on the other hand. We associated this simulated data (normalized with the total flux incident on the surface) with the indoor characterization data shown in Figure 20 in order to obtain the power gain brought by the use of a textured glass for each hourly data point and each side of the module in two different orientations.

Validation of the model

The empirical correlations obtained in section Experimental data have been used to simulate the evolution of P_m over two summer days. We chose a sunny day giving non-uniform irradiance on the backside of the module (NU comprised between 31% and 60%) for the half-cells and parallel interconnection architecture, and a cloudy day giving diffuse irradiance on both sides (diffuse to global ratio above 99.7%) for the linear textured glasses architecture. The comparison of temporal series (simulated and measured) gives R^2 above 98.8% for the half-cells and parallel interconnection architecture and R^2 above 94.3% for textured glasses architecture (see eq. 3). Knowing the number of hourly data points for each day, the Student test allows us to say that the model estimate the reality with a confidence interval of 99.9% (see eq. 4 and [16]).

Table 3. Comparison between calculated and measured g_{Isc} in the case of indoor artificially created non-uniformities on the backside of both modules (masks c × 0.25c and c × 0.5c).

Mask on backside	NU	A_{min}	A_{ave}	g_{Isc} calc.	g_{Isc} meas.
c × 0.25c	33.3%	0.5	0.75	15.2%	13.1%
c × 0.5c	100%	0	0.5	43.6%	44.3%

Figure 21 shows the resulting gain from experimental and simulated data for these two days. The upper chart (sunny day) shows a strong loss for the novel architecture compared to standard one (four full size cells in series) around 9:00 am and a strong gain around 4:30 pm. Indeed, in this configuration, the standard module is on the west side meaning that its backside receive more radiation when the sun rises, and the novel module is on the east side meaning that its backside receives more radiation when the sun goes down. By contrast, the lower chart (cloudy day) does not show this asymmetrical production as the optical flux is completely diffuse on the backside of the module.

Annual performance

Half-cells and parallel interconnection architecture

Figure 22 shows the annual simulation of the gain brought by the novel architecture compared to standard one with the influence of half-cells only and with the influence of the parallel interconnection in addition. We observe that the gain due to less resistive losses (half-cells influence) is limited independently of the distance (below 0.4%) as the total irradiance on our south-oriented vertical bifacial module is not very high (1.04 sun maximum, 0.32 sun in average). This type of architecture could be suitable in applications where the module receives more radiation, which implies more resistive losses. We also observe that the maximal gain due to non-uniform irradiance (parallel interconnection influence) is obviously higher for short distances when the non-uniformity of backside irradiance increases. Therefore, this architecture could be suitable for applications which suffer from high non-uniform irradiance on the module like in compact bifacial solar plants for example. Note that the real annual gain

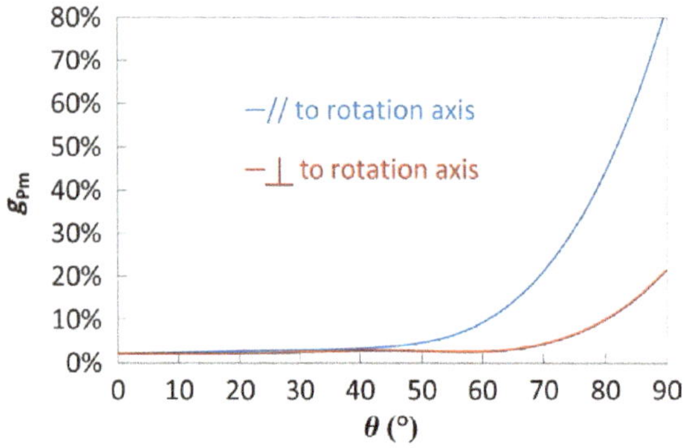

Figure 20. Power gain for a monofacial module with a unidirectional textured glass compared to flat glass as a function of the incidence angle of light (0° corresponds to normal incidence) for two different orientations of the textured glass.

considering both effects would be comprised between the dotted and plain curve since the plain red curve describes the maximal g_{Pm} (see eq. 8). However, we cannot quantify this with our simple model.

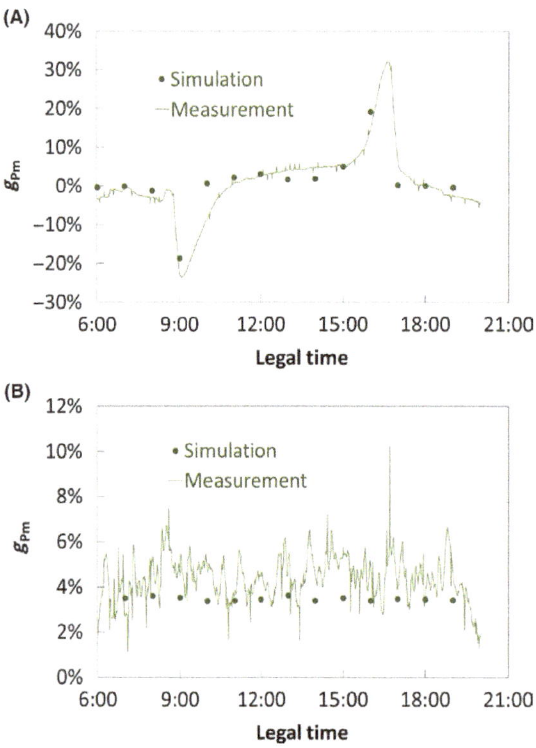

Figure 21. (A) Simulated and measured power gain of the half-cells and parallel interconnection architecture compared to standard one on the 31/07/13. (B) Simulated and measured power gain of the textured glasses architecture compared to standard one on the 03/07/13.

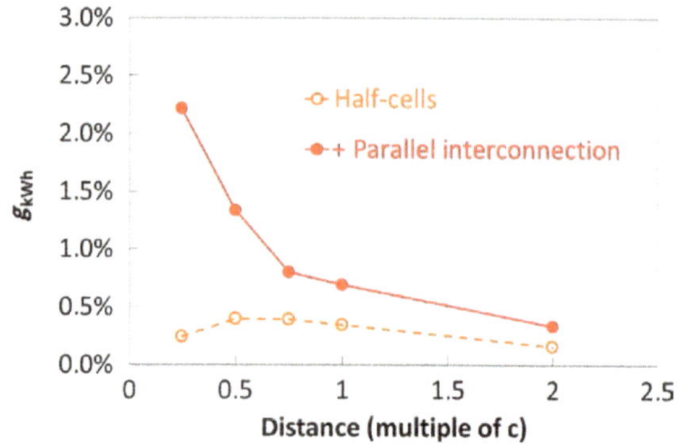

Figure 22. Simulation of the annual gain of the novel architecture compared to standard one varying the distance between module and reflector. The red plain curve shows the parallel interconnection influence in addition to the half-cells influence shown by the orange dotted curve.

Textured glasses architecture

Figure 23 shows the annual simulation of the gain brought by the novel architecture compared to standard one in different orientation of the linear textured glass and for each side of the module. We observe that the backside gain does not depend a lot on the texturation orientation as the white backside reflector is quasi-lambertian, and it increases for shorter distances as the proportion of high incidence angles increases. In a real facade application, the backside gain would be certainly higher since the black inner side walls of our test bench absorb a part of radiations. We also observe that the frontside gain is independent of the distance and lower for the horizontal orientation. This gain would be certainly higher if the ground albedo had been taken into account in our model since more radiation would be incident from the ground.

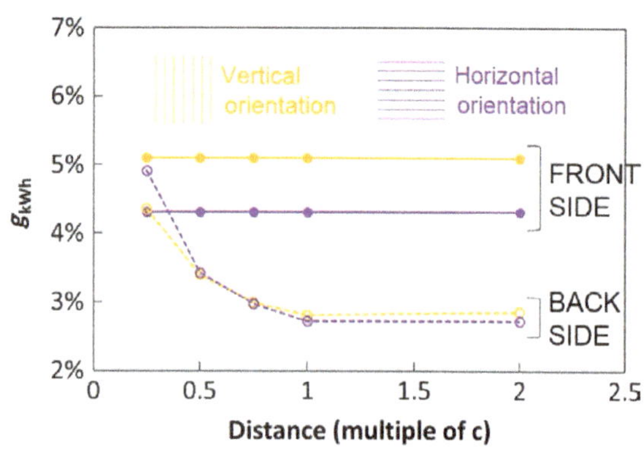

Figure 23. Simulation of the annual gain of the novel architecture compared to standard one varying the distance between module and reflector.

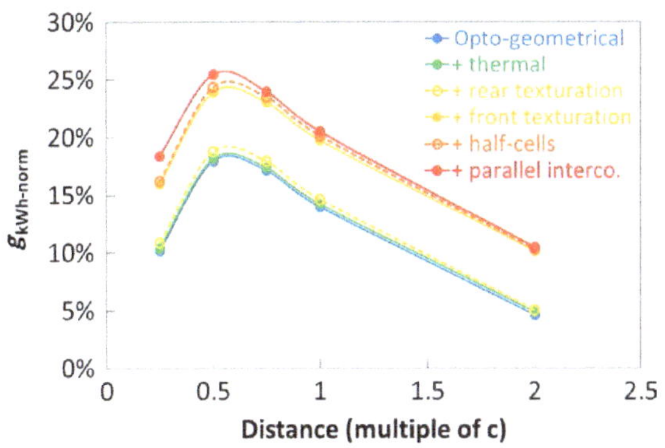

Figure 24. Simulation of the annual gain of a bifacial module compared to a monofacial one as a function of the distance between module and reflector. From the bottom to the top, each curve shows an additional contribution to the gain.

Bifacial gain considering all contributions

Finally, Figure 24 shows the annual gain of a bifacial module compared to a monofacial one in our reduced scale vertical facade. First, the largest part of the gain is due to the opto-geometrical environment (18.1% for the optimal distance – see section Annual performance). Then, a slight gain is due to the thermal advantage of the bifacial module (0.3% maximum – see section Experimental data). The use of textured glasses in their vertical orientation adds a large gain (up to 5.8%) mainly due to the front glass (up to 5.1%). In comparison, the gain brought by the half-cells and parallel interconnection architecture appears to be limited in our vertical facade application.

Perspective at Real Scale

Full-size module architecture

The electrical and optical architecture of bifacial modules need to be optimized regarding the issues of resistive losses, non-uniform and diffuse backside irradiance. Based on the work described here, we propose the module architectures shown in Figure 25.

On the electrical part, the module has two parallel blocks of 60 half-cut cells (Fig. 25A), three parallel blocks of 60 cells cut in three (Fig. 25B), or four parallel blocks of 60 cells cut in four (Fig. 25C). Possible positions of bypass diodes are shown on the images. The two blocks module seems to be the most suitable in practice, since the three blocks module requires electrical insulation between the yellow and brown busbars, and the four blocks module could imply complex soldering process due to the small size of the cells. However, with three or four parallel blocks (smaller cells), the robustness to non-uniform irradiance would be higher and the resistive losses would be lower compared with the two blocks module. Finally, the two and four parallel blocks modules seem to be suitable for vertical facade applications as their positive and negative outputs are positioned on both sides of the module. Therefore, the neighboring modules can be more easily connected in series and resistive losses are minimized at the system level.

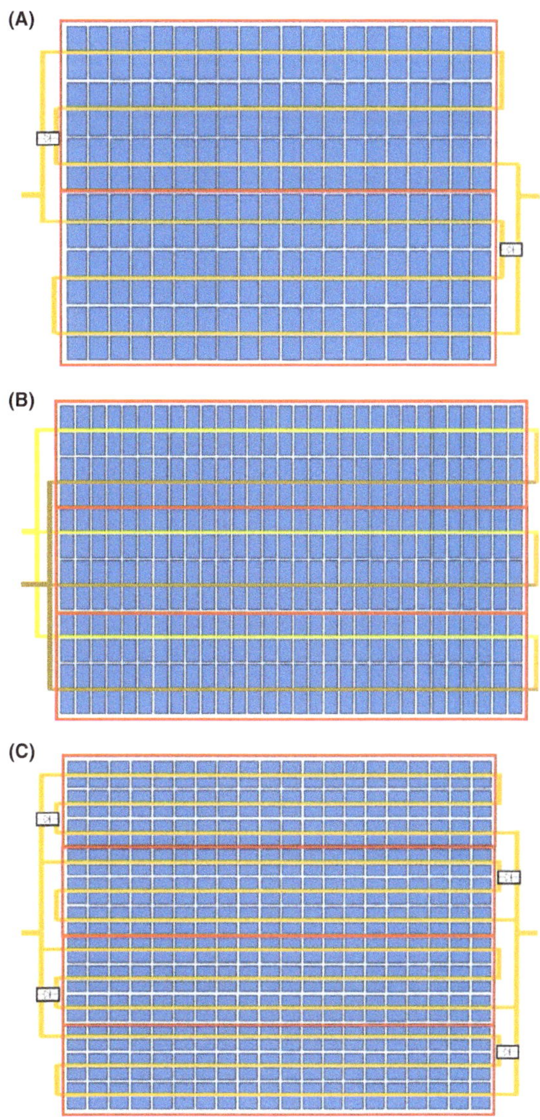

Figure 25. Module electrical architectures equivalent to a 6 × 10 full-size cells module. The division factor of a cell is equal to the number of blocks in parallel shown with red rectangles: factor two (A), factor three (B), and factor four (C).

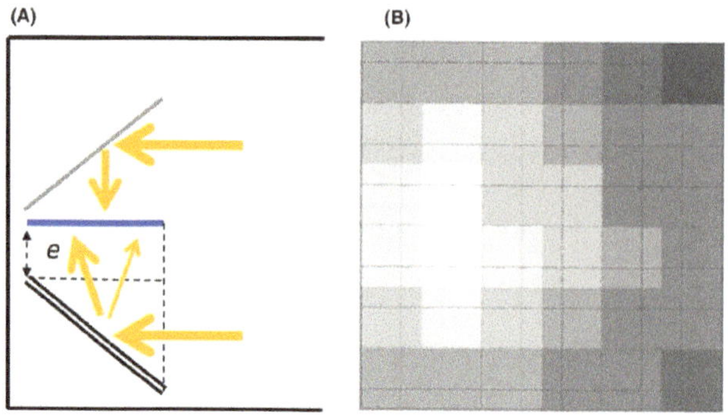

Figure 26. (A) Schematics of the adapted double illumination setup (aluminum mirror on the front, white outdoor reflector on the back). (B) Ray-tracing simulation of the distribution of irradiance on the backside of a 6 × 6 cells module (instead of a 6 × 10 cells in portrait orientation) with the adapted double illumination setup (white pixels show the most irradiated cells).

On the optical part, the textured glasses can be used on both sides knowing that most part of the gain will be due to the front glass. Ideally, the texturation of glasses would be optimized for several types of bifacial applications: tilted face to the equator or vertical face to east–west on a horizontal white ground or white roof, and vertically integrated onto a white facade.

Further experimental tests

The potential of the module architectures proposed in the previous section must be evaluated experimentally. This must be performed with both an outdoor and an indoor setup. On the one hand, a real size adjustable facade would allow evaluating the optimal module–reflector distance in order to compare it to the 0.5*c*–0.75*c* found on our reduced scale test bench. On the other hand, a real size double illumination setup could be designed in order to simulate real outdoor environment conditions. Figure 26A shows that a white reflector (the one used outdoor on a flat roof or vertical facade for example) can be used instead of the back mirror in the double illumination setup. In this case, the radiation incident on the backside of the module would be nonuniform (Fig. 26B), diffuse, and would overall take into account the spectral reflectivity of the reflector as in a real outdoor environment.

Finally, we designed an optimal real-size standard double illumination setup (mirrors on both sides of the module) with our ray-tracing model of the PASAN solar simulator. We found that a portrait orientation of the module is suitable otherwise the mirrors length would not fit inside the PASAN. We also found an optimal angle between mirror and module of 43° ± 1°. That leads to a similar uniformity of irradiance, a similar global irradiance and a normal incidence in average for our setup and for a standard STC test (considering a perfect mirror and for a distance e = 5 cm – see parameter e on Fig. 26A). For this angle, increasing parameter e does not improve the uniformity of irradiance. Note that an experimental validation of this setup is required since the use of real mirrors (with a part of diffuse reflection, dust, micro-scratches, and less clean areas) could make the nonuniformity of irradiance too high for the characterization standards (NU < 2% for class A [18]). The protocol described in section Double illumination characterization setup must be used and the uncertainties must be evaluated again in the real size case. Such a study could provide additional information to the actual discussions about defining a standard for indoor characterization of bifacial modules in solar simulators.

Conclusion

In this article, we developed a methodology for evaluating the annual performance of bifacial modules integrated on a vertical facade at a reduced scale. First, a protocol using a double illumination setup in a solar simulator has been used. We found that the use of aluminum mirrors does not impact the spectral class of measurement as long as the PASAN source intensity is adjusted. Also, the uncertainty of measurement has been quantified and does not exceed ±3%. Additionally, we performed outdoor tests in real conditions, which allowed us to validate our ray-tracing simulations for both sunny and cloudy days in winter as well as in summer: we found R^2 > 94.3% with Student test giving a high confidence interval. With these three tools, we were able to quantify the resistive losses which are due to high total irradiance on both sides of the module, the non-uniformity of irradiance on the backside which is due to the module shade projected onto the back reflector, and its diffuse character which is due to the use of quasi-lambertian reflectors.

Regarding the issues above mentioned and using our methodology, we varied opto-geometrical parameters of the application (particularly the distance between module and reflector and the type of reflector) as well as electrical and optical aspects of the module architecture (particularly the cell interconnection and the glasses) in order to maximize the annual performance. The annual electrical energy gain of a bifacial module compared to monofacial could reach 25% with all the cumulated contributions for a module–reflector distance ranging between 0.5*c* and 0.75*c*, *c* being the module height. The highest contribution comes from the opto-geometrical environment of the module namely the irradiance incident on its back, which gives a 18% maximum gain. Another large contribution comes from the use of a linear textured glass vertically oriented on the frontside, which gives a 5% maximum gain.

Based on our tests and simulations for a reduced scale application, we proposed a full-size module architecture which could be more robust to non-uniform irradiance on the backside with less resistive losses and more harnessing of high incidence angles. The next research step will be to evaluate the optimal module–reflector distance in the case of a real-size vertical facade, and to quantify the advantage of the proposed module architecture in several kinds of bifacial applications. This could be achieved with more general simulation tools taking into account the complex opto-geometrical environment of the bifacial module. Finally, the optimized applications will have to be long term tested at the real size power plant scale in order to give confidence to investors.

Acknowledgments

We thank AGC Glass Europe for providing the textured glass samples and for technical discussions. Bruno Soria's work has been supported by a PhD grant from the Commissariat à l'Energie Atomique et aux Energies Alternatives (CEA). The resulting thesis is available online in French [19]. This article is dedicated to the memory of Yves Delesse who actively participated to the instrumentation of the test bench applied in this study.

Conflict of Interest

None declared.

References

1. Cuevas, A. 2005. The early history of bifacial solar cells. Pp. 801–805 *in* Proceedings 20th European PV solar energy conference, Barcelona.
2. Luque, A., and S. Hegedus. 2003. Handbook of photovoltaic science and engineering. Wiley & Sons, UK.
3. Obara, S., D. Konno, Y. Utsugi, and J. Morel. 2014. Analysis of output power and capacity reduction in electrical storage facilities by peak shift control of PV system with bifacial modules. Appl. Energy 128:35–48.
4. Yusufoglu, U. A., T. M. Pletzer, L. J. Koduvelikulathu, C. Comparotto, R. Kopecek, and H. Kurz. 2015. Analysis of the annual performance of bifacial modules and optimization methods. IEEE J. Photovolt. 5:320–328.
5. Lo, C. K., Y. S. Lim, and F. A. Rahman. 2015. New integrated simulation tool for the optimum design of bifacial solar panel with reflectors on a specific site. Renew. Energy 81:293–307.
6. Cabal, R., Y. Veschetti, V. Sanzone, S. Manuel, S. Gall, F. Barbier, et al. 2013. Industrial process leading to 19.8% on N-Type Cz silicon. Energy Procedia. 33:11–17.
7. Ohtsuka, H., M. Sakamoto, M. Koyama, K. Tsutsui, T. Uematsu, and Y. Yazawa. 2001. Characteristics of bifacial solar cells under bifacial illumination with various intensity levels. Prog. Photovolt. Res. Appl. 9:1–13.
8. Ezquer, M., I. Petrina, J. M. Cuadra, A. R. Lagunas, and F. Cener-Ciemat. 2009. Design of a special set-up for the IV characterization of bifacial photovoltaic solar cells. Presented at 24th European Photovoltaic Solar Energy Conference and Exhibition, Hamburg.
9. Edler, A. 2012. Flasher setup for bifacial measurements. Presented at BIFIPV workshop, Konstanz.
10. Lambda Research Corporation. TracePro: Software for designing optical systems. Available at http://www.lambdares.com/.
11. King, D. L., J. A. Kratochvil, and W. E. Boyson. 1997. Temperature coefficients for PV modules and arrays: measurement methods, difficulties, and results. Photovoltaic Specialists Conference, 1997., Conference Record of the Twenty-Sixth IEEE:1183–1186.
12. Reich, N. H., W. G. J. H. M. van Sark, W. C. Turkenburg, and W. C. Sinke. 2010. Using CAD software to simulate PV energy yield – The case of product integrated photovoltaic operated under indoor solar irradiation. Sol. Energy 84:1526–1537.
13. Sprenger, W. 2013. Electricity yield simulation of complex BIPV systems. Delft University of Technology, Fraunhofer-Verlag. ISBN 978-3-8396-0606-3, 158 pages.
14. PVsyst. Software for designing PV plants. Available at www.pvsyst.com.
15. Soria, B., and E. Gerritsen. 2013. Vertical Facade Integration of Bifacial PV Modules: Outdoor Testing and Optical Modelling. Presented at 28th European Photovoltaic Solar Energy Conference and Exhibition, Paris.

16. Muneer, T. 2004. Solar radiation and daylight models: (with software available from companion web site), 2nd ed. Elsevier Butterworth Heinemann, Oxford; Burlington, MA.
17. Kreinin, L., N. Bordin, A. Karsenty, A. Drori, D. Grobgeld, and N. Eisenberg. 2010. PV module power gain due to bifacial design. Photovoltaic Specialists Conference (PVSC), 2010 35th IEEE:002171–002175.
18. International Electrotechnical Commission. 2007. Norme IEC 60904-9 Ed 2.0 Photovoltaic devices - Part 9: Solar simulator performance requirements. October, 2007.
19. Soria, B. 2014. Etude des performances électriques annuelles de modules photovoltaïques bifaces. Cas particulier? Modules bifaces intégrés en façade verticale. Université de Grenoble, https://tel.archives-ouvertes.fr/tel-01126959/document.

6

Systematic study of aqueous monoethanolamine-based CO_2 capture process: model development and process improvement

Kangkang Li[1,2], Ashleigh Cousins[3], Hai Yu[1], Paul Feron[1], Moses Tade[2], Weiliang Luo[4] & Jian Chen[4]

[1]CSIRO Energy, 10 Murray Dwyer Circuit, Mayfield West, New South Wales 2304, Australia
[2]Department of Chemical Engineering, Curtin University of Technology Australia, GPO Box U1987, Perth, Western Australia 6845, Australia
[3]CSIRO Energy, P.O. Box 883, Kenmore, Queensland 4069, Australia
[4]Department of Chemical Engineering, Tsinghua University, Beijing 100084, China

Keywords
CO_2 capture, MEA, model validation, process improvements, rate-based model

Correspondence
Paul Feron, CSIRO Energy, 10 Murray Dwyer Circuit, Mayfield West, NSW 2304, Australia.

E-mail: paul.feron@csiro.au

Funding Information
Kangkang Li thanks the Australian IPRS-APA scholarship and CSIRO Top-up scholarship to support his research. Weiliang Luo and Jian Chen thank support from the National Natural Science Foundation of China (key project no. 51134017).

Abstract

In this paper, we present improvements to postcombustion capture (PCC) processes based on aqueous monoethanolamine (MEA). First, a rigorous, rate-based model of the carbon dioxide (CO_2) capture process from flue gas by aqueous MEA was developed using Aspen Plus, and validated against results from the PCC pilot plant trials located at the coal-fired Tarong power station in Queensland, Australia. The model satisfactorily predicted the comprehensive experimental results from CO_2 absorption and CO_2 stripping process. The model was then employed to guide the systematic study of the MEA-based CO_2 capture process for the reduction in regeneration energy penalty through parameter optimization and process modification. Important process parameters such as MEA concentration, lean CO_2 loading, lean temperature, and stripper pressure were optimized. The process modifications were investigated, which included the absorber intercooling, rich-split, and stripper interheating processes. The minimum regeneration energy obtained from the combined parameter optimization and process modification was 3.1 MJ/kg CO_2. This study suggests that the combination of a validated rate-based model and process simulation can be used as an effective tool to guide sophisticated process plant, equipment design and process improvement.

Introduction

Global climate change caused by the increasing atmospheric concentration of greenhouse gases such as carbon dioxide (CO_2), has led to great interest in the development of CO_2 capture and storage (CCS) technologies [1–5]. Among them, the amine-based postcombustion CO_2 capture technique is likely to be the first technology applied on a large scale for CO_2 capture from fossil fuel combustion and energy-related processes [6]. One of the major challenges of the amine-based postcombustion capture (PCC) technology in commercial use is the substantial energy penalty involved in the CO_2 capture process, especially the regeneration energy that accounts for more than 50% total energy consumption [7, 8]. While significant efforts have been devoted to the development of novel solvents [9, 10] that have outstanding performance in terms of high CO_2 absorption rates or capacities and low energy consumption of solvent regeneration, the commercial application of these advanced solvents is still immature and requires more efforts to test their technical and economic feasibility on a large scale. In comparison, aqueous monoethanolamine (MEA), a simple and cheap amine, is still widely recognized as a first choice or at least the benchmark solvent for PCC technology, due to the fast CO_2 absorption rate and mature technology commercially applied in the gas processing industry [11].

Considerable efforts around the world have been made toward the commercialization of the MEA-based CO_2 capture technology in coal-fired power stations. Experimentally, plenty of academic research work including laboratory- and pilot-scale experiments have been carried out to deeply understand the chemical reaction mechanism and to evaluate the technical, economic, and environmental feasibility of this technology [12, 13]. Parallel to the experimental activities, several process simulators such as Aspen Plus [14–16], Aspen HYSYS [17], gProms [18], Fortran [19, 20] etc. were used to thermodynamically or kinetically simulate the absorption/stripping process in order to guide process optimization and development for energy efficiency. The research findings, however, suggest that commercial application for large-scale CO_2 reduction using aqueous MEA will still require significant technology advancements with respect to (1) the energy consumption, (2) solvent degradation, (3) absorption capacity, and (4) equipment corrosion. Among them, the large energy penalty associated with solvent regeneration constitutes a major obstacle for this technology in commercial use. The specific heat requirement of solvent regeneration in the MEA process is generally around 3.6–3.8 MJ/kg CO_2 [17, 21–23], which will lead to a significant drop of net power plant efficiency. For example in Australia, the energy consumption for solvent regeneration and operation of the PCC plant will lead to typically 10%-points decrease in net power generation efficiency if a PCC plant with 90% capture efficiency is integrated into a new black coal-fired power station, consequently resulting in more than doubling the cost of electricity generation across the board [24]. Reducing the energy requirement of solvent regeneration is therefore imperative in order to push forward MEA-based capture technology.

The optimization of the PCC plant including parameter optimization and process modification seems to be an effective method to reduce the energy penalty involved in the CO_2 capture process. Salkuyeh et al. [25] and Abu-Zahra et al. [26] carried out a parametric study and showed that the process operating conditions have a great impact on the stripper reboiler duty. These studies highlighted that the parameters of the CO_2 capture process should be optimized to determine the best conditions that minimize the energy demand. Cousins et al. [27, 28] evaluated sixteen different process configurations aiming at reducing the energy consumption of amine processes, while Karimi et al. [29] studied five different stripper configurations with respect to savings in capital cost and energy consumption. Their results show that the advanced process configurations such as the absorber intercooling, rich solvent split, stripper interheating, etc. play an important role in the energy saving of stripper reboiler duty. Moreover, Freguia et al. [30] and Leonard et al. [31] investigated both the parameter optimization and process modification, and indicated that the energy consumption had a great potential to be significantly reduced by the combination of these two process improvements.

Since large-scale PCC plants are very expensive to be built for research purposes, the rigorous process and equipment modeling is an efficient and economic tool for evaluating the performance of these process improvements. A rigorous rate-based model enables an accurate description and characterization of CO_2 absorption/stripping processes taking place along the packed column, such as mass and heat transfer across the gas and liquid phases, chemical reactions, material and energy balance, and hydraulic properties, etc. [32]. Aspen Plus®, a commercial software, has been widely used as an effective simulator to study the CO_2 capture process and evaluate the energy demands involved in PCC, specifically the stripper reboiler heat requirement [14–16, 33]. However, process modeling cannot always ensure sufficient confidence in the process performance in a commercial sense and sometimes even generates unpractical results due to a lack of necessary experimental validation from pilot plant results. Therefore, the combined work of pilot plant trials and process simulation would be the most appropriate way to develop the PCC plant economically and effectively.

In 2010, CSIRO in collaboration with Stanwell Corporation Limited based in Queensland, Australia, constructed a PCC pilot plant with a designed CO_2 capture rate of ~100 kg/h using aqueous MEA and real flue gas containing 11.0–13.5% CO_2 from Tarong power station [27, 34]. Using the pilot plant results, a rigorous, rate-based model was developed in Aspen Plus® V7.3 and used to evaluate the MEA-based CO_2 capture process. The model was validated against the experimental data from Tarong pilot plant trials in terms of the key parameters of both the absorber column and stripper column. Systematic studies including process parameter optimization and flow sheet modifications were investigated to significantly reduce the regeneration energy consumption.

Rate-based Model Development

Model description

The commercially available Aspen Plus® software was used to simulate the MEA-based CO_2 capture process. The process model consists of a thermodynamic model, a transport model, and a rate-based model. To simplify the process modeling, the absorber simulation and stripper simulation were conducted independently. In the stripping modeling, the inlet stream was copied from the outlet stream from the CO_2 absorber, and vice versa. It should be noted that although this simple method would slightly deviate the material balance of the absorption-desorption system, the marginal deviation has little effect on the technical and energy performance of the MEA process.

Physical and chemical properties

Rigorous physical and chemical properties are fundamentally essential for the model to accurately evaluate the performance and characteristics of the CO_2 capture process by aqueous MEA. In this simulation, the electrolyte non-random two-liquid (NRTL) method and the RK (Redlich–Kwong) equation of state were used to compute liquid properties (activity coefficient, Gibbs energy, enthalpy, and entropy) and vapor properties (fugacity coefficients) of the model MEA-CO_2-H_2O system, respectively. This electrolyte NRTL model has been validated to accurately predict the vapor–liquid equilibrium, aqueous speciation, heat capacity, and CO_2 absorption enthalpy of the MEA–H_2O–CO_2 system with a wide application range: MEA concentration up to 40wt.%, CO_2 loading up to 1.33, temperature up to 443 K and pressure up to 20 MPa [35]. These conditions cover all the conditions used in the pilot plant and simulations studied. The gases CO_2, N_2 and O_2 were selected as Henry-components to which Henry's law was applied. The Henry's constants, transport and thermal properties of the MEA-CO_2-H_2O system were retrieved from the Aspen Plus databanks, which have been proved to accurately describe the physical and transport characteristics based on experimental data [35].

The electrolyte solution chemistry of the MEA-CO_2-H_2O system was modeled taking into account the equilibrium and kinetic reactions shown in Table 1. The equilibrium constants for reactions (1)–(4) were calculated from the standard Gibbs-free energy change and the rate constants for reactions (5)–(8) were taken from the work of Pinsent et al. [36] and Hikita et al. [37]. The equilibrium and kinetic parameters have been built and updated in Aspen Plus databases and are described in more details elsewhere [14–16].

Table 1. Chemical reactions in the MEA-CO_2-H_2O system.

No.	Type	Reactions
1	Equilibrium	$2H_2O \leftrightarrow H_3O^+ + OH^-$
2	Equilibrium	$CO_2 + 2H_2O \leftrightarrow H_3O^+ + HCO_3^-$
3	Equilibrium	$HCO_3^- + H_2O \leftrightarrow CO_3^{2-} + H_3O^+$
4	Equilibrium	$MEAH^+ + H_2O \leftrightarrow MEA + H_3O^+$
5	Kinetic	$CO_2 + OH^- \rightarrow HCO_3^-$
6	Kinetic	$HCO_3^- \rightarrow CO_2 + OH^-$
7	Kinetic	$MEA + CO_2 + H_2O \rightarrow MEACOO^- + H_3O^+$
8	Kinetic	$MEACOO^- + H_3O^+ \rightarrow MEA + CO_2 + H_2O$

Rate-based modeling

The rate-based model validation of the CO_2 capture process using aqueous MEA was carried out based on the Tarong pilot plant configuration as shown in Figure 1. The RateSep simulator embedded in Aspen Plus was used to simulate the aqueous MEA-based CO_2 capture process. This simulator allows the user to divide the tray column or packed column into different stages and provides more accurate and detailed description of CO_2 absorption behavior at each stage based on the material and energy balance. In order to reflect the actual pilot MEA process, the rate-based model used the same column parameters as the pilot plant, such as packing material, column diameters, and packed heights. Table 2 lists the column parameters of both the CO_2 absorber and CO_2 stripper together with the primary correlations and settings of the rate-based absorber/stripper model. The interfacial area factor was varied from 1.0 to 2.0 to provide a good agreement between experimental and simulation results. The value 1.8 was chosen due to the excellent agreement between the experimental and simulation results. This was shown by the average relative error deviations between experimental and simulation results of 2.7% for CO_2-rich loading, 5.8% for CO_2 absorption rate, 1.3% for reboiler temperature, 0.3% for CO_2 purity, and 4.0% for the regeneration energy. Given the 10% uncertainty associated with the comparison between predicted and pilot results [14], the proposed rate-based model enabled a reasonable prediction of the CO_2 absorption and desorption process. Details are discussed further in the next section.

Model validation against Tarong pilot plant results

The CO_2 capture process using aqueous MEA in the Tarong PCC pilot plant consisted of two major parts: the absorption and desorption processes. Accordingly, the rate-based modeling for the MEA process was carried out through the validation of the packed absorber column and stripper column, respectively. Table 3 summarizes the operating conditions and pilot results of both the CO_2 absorber and CO_2 stripper, together with the simulation results based on the conditions of the 22 pilot plant trials.

Performance of CO_2 absorber

The CO_2 absorption rate is considered one of the most significant indicators for developing a reliable rate-based model, as it closely represents the reaction properties such as equilibrium and kinetic constants. Figure 2A shows the excellent match between model results and pilot plant

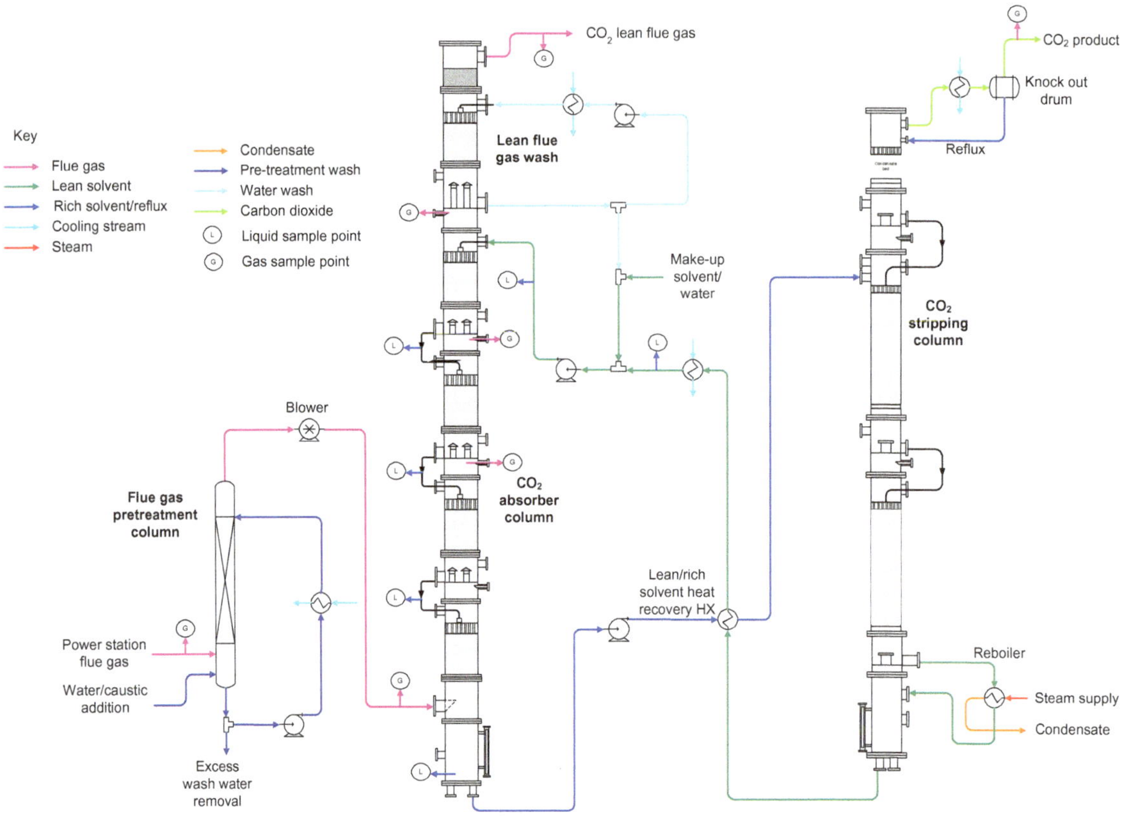

Figure 1. Process flow-sheet of Tarong CO_2 capture pilot plant cited from Cousins et al. [34].

data for a wide range of CO_2 absorption rates 40–100 kg/h. The average relative error deviation for the 22 tests was 5.6%.

Figure 2B shows the parity plot of CO_2 loading (mole ratios of CO_2/MEA) in the rich solvent after absorption between experimental and pilot results. It can be seen that the model gave a 2.7% overestimation on the rich CO_2 loading compared to the experimental data. This overestimation is likely caused by the samples being analyzed offline. A small portion of the absorbed CO_2 was most likely lost during sample collection and measurement due to the high CO_2 partial pressure of the rich solvent

Table 2. Summary of model parameters and column settings used in the rate-based model.

Model and column properties	Absorber	Desorber
Number of stages	20	20
Packing material	Mellapak M250X	Mellapak M350X
Total packed height	7.136 m (4 × 1.784 m)	7.168 m (2 × 3.584 m)
Column diameter	350 mm	250 mm
Flow model	Mixed model	Mixed model
Interfacial area factor	1.8	1.8
Initial liquid holdup	0.03 L	0.03 L
Film resistance	Discrxn for liquid; Film for vapor	Discrxn for liquid; Film for vapor
Discretization points for liquid film	5	5
Mass transfer correlation method	Bravo et al. [38]	Bravo et al. [38]
Heat transfer correlation method	Chilton-Colburn	Chilton-Colburn
Interfacial area method	Bravo et al. [38]	Bravo et al. [38]
Liquid holdup correlation method	Bravo et al. [39]	Bravo et al. [39]

Table 3. Comparison between pilot plant trials and rate-based model simulation results under a variety of experimental conditions.

Test date (2011)	1 Feb	3 Feb	7 Feb	11 Feb	22 Mar	24 Mar	25 Mar	30 Mar	4 Apr	5 Apr	13 Apr
Test conditions of CO_2 absorption											
Lean temp. °C	31.7	31.4	31.3	35.5	33.9	34.8	35.3	33.5	32.4	37.5	35.6
Lean flow rate, L/min	31.7	32.0	27.0	31.3	26.9	33.0	35.8	32.0	24.0	24.4	32.0
Lean MEA conc., wt.%	25.1	24	27.9	25.5	31.6	29.6	29.2	30.3	33.5	34.0	29.3
Lean CO_2 loading, mol/mol	0.279	0.294	0.284	0.280	0.314	0.333	0.347	0.316	0.254	0.278	0.414
Inlet flue gas temp. °C	51.7	50.5	48.1	51.3	53.6	54.2	55.8	65.5	66.1	63.9	49.8
Inlet gas pressure, kPa-a	106.1	105.6	106.3	106.6	107.5	107.3	107.3	107.3	108.7	108.3	107.2
Inlet flue gas flow rate, kg/h	489.6	491.1	488.6	489.4	482.8	483.9	484.8	483.2	646.1	644.4	487.5
Inlet flue gas CO_2 vol%	11.9	11.8	11.8	12.1	13.4	13.5	13.5	12.8	12.9	13.2	12.7
Inlet flue gas H_2O vol%	4.2	5.0	3.8	3.9	4.3	4.0	3.9	3.6	4.3	4.1	3.8
Test conditions of CO_2 desorption											
Condenser Temp. °C	30.2	29	26.2	29.6	32.0	29.9	29.4	25.4	26.9	24.8	28.8
Stripper top pressure, kPa-a	181.9	180.4	177.5	189.9	189.2	195.0	195.5	191.5	223.3	222.1	151.8
Temp. difference[1], K	16.7	16.9	18.5	17.4	21.0	20.0	19.8	19.0	23.8	25.6	17.6
CO_2 desorption rate, kg/h	71.6	71.0	70.6	73.6	72.6	76.8	76.7	76.0	93.9	94.2	49.3
Results comparison of pilot plant and simulation											
Expt. rich CO_2 loading, mol/mol	0.466 ± 0.007	0.480 ± 0.005	0.469 ± 0.004	0.471 ± 0.001	0.481 ± 0.001	0.486 ± 0.004	0.488 ± 0.002	0.494 ± 0.007	0.494 ± 0.001	0.489 ± 0.003	0.525 ± 0.003
Simu. rich CO_2 loading, mol/mol	0.491	0.497	0.493	0.493	0.509	0.499	0.496	0.490	0.504	0.505	0.502
Expt. CO_2 abs rate, kg/h[2]	73.5 ± 1.4	74.2 ± 1.6	72.3 ± 1.6	74.4 ± 1.6	74.2 ± 1.7	77.8 ± 1.8	77.4 ± 1.4	75.5 ± 1.5	96.9 ± 1.8	94.1 ± 1.4	49.2 ± 1.9
Simu. CO_2 abs rate, kg/h	80.6	75.3	80.2	81.0	68.8	80.8	80.4	83.5	96.3	91.1	45.2
Expt. reboiler temp., °C	116.9 ± 0.3	116.4 ± 0.2	117.0 ± 0.2	117.6 ± 0.2	117.2 ± 0.4	117.1 ± 0.2	115.9 ± 0.2	117.5 ± 0.2	125.3 ± 0.3	125.8 ± 0.4	104.7 ± 0.6
Simu. reboiler temp., °C	118.6	117.9	118.4	119.1	119.2	119.6	119.3	119.5	126.7	126.5	110.0
Expt. regen. energy, MJ/kg CO_2	4.48	4.45	4.35	4.50	4.33	4.51	4.60	4.49	4.01	4.04	4.55
Simu. regen. energy, MJ/kg CO_2	4.60	4.56	4.57	4.65	4.66	4.68	4.83	4.66	4.01	4.11	5.20
Expt. CO_2 product purity, vol%	97.7	97.7	97.8	97.7	97.6	97.8	97.8	98.0	98.9	99.0	98.1
Simu. CO_2 product purity, vol%	97.6	97.7	98.0	97.7	97.4	97.8	97.8	98.2	98.3	98.5	97.3

Test date	19 Apr	20 Apr	11 May	12 May	13 May	18 May	20 May	24 May	25 May	26 May	27 May
Test conditions of CO_2 absorption											
Lean temp. °C	38.7	38.7	39.4	39.3	39.2	39.7	40.5	39.6	39.4	39.6	39.5
Lean flow rate, L/min	21.1	21.1	27.1	27.0	27.0	32.1	27.0	27.0	26.9	27.2	27.0
Lean MEA conc., wt.%	32.3	32.8	29.2	29.2	28.9	26.2	27.4	28.1	28.3	28.5	28.5
Lean CO_2 loading, mol/mol	0.285	0.291	0.285	0.285	0.280	0.314	0.288	0.295	0.297	0.283	0.283
Inlet flue gas temp. °C	48.3	48.8	58.3	59.4	56.7	61.8	60.7	60.0	56.1	58.2	57.9
Inlet gas pressure, kPa-a	106.7	106.6	108.9	108.7	108.8	108.3	108.5	109.0	108.8	108.9	109.0
Inlet flue gas flow rate, kg/h	485.5	487.5	598.7	598.4	597.2	598.0	597.2	596.8	596.0	598.5	597.8
Inlet flue gas CO_2, vol%	12.2	12.2	11.1	11.2	11.2	11.1	11.0	11.6	11.6	11.6	11.1
Inlet flue gas H_2O, vol%	3.9	3.9	5.2	5.3	5.0	5.5	5.4	5.5	5.1	5.2	5.2

(Continued)

Table 3. (Continued)

Test date	19 Apr	20 Apr	11 May	12 May	13 May	18 May	20 May	24 May	25 May	26 May	27 May
Test conditions of CO_2 desorption											
Condenser Temp. °C	24.7	23.9	19.2	20.4	18.2	20.7	19.8	19.4	17.6	18.6	17.0
Stripper top pressure, kPa-a	186.9	182.0	202.7	200.0	202.2	200.5	202.2	203.1	202.0	203.3	202.2
Temp. difference[1], K	18.6	18.5	18.3	18.8	18.8	17.4	19.6	18.9	19.4	19.2	18.7
CO_2 desorption rate, kg/h	73.3	71.1	84.3	83.6	84.0	82.4	84.2	84.6	84.1	84.7	84.2
Results comparison of pilot plant and simulation											
Expt. rich CO_2 loading, mol/mol	0.486 ± 0.001	0.500 ± 0.001	0.488 ± 0.001	0.488 ± 0.001	0.491 ± 0.002	0.499 ± 0.005	0.503 ± 0.001	0.511 ± 0.006	0.512 ± 0.001	0.492 ± 0.001	0.492 ± 0.006
Simu. rich CO_2 loading, mol/mol	0.506	0.506	0.500	0.500	0.501	0.499	0.501	0.503	0.504	0.502	0.503
Expt. CO_2 abs rate, kg/h[2]	76.0 ± 2.1	72.0 ± 1.7	85.3 ± 1.5	83.3 ± 1.6	83.4 ± 2.0	82.8 ± 1.7	84.3 ± 1.6	83.9 ± 2.0	83.8 ± 1.8	85.7 ± 1.5	85.0 ± 1.4
Simu. CO_2 abs rate, kg/h	72.9	72.3	81.8	81.7	82.5	75.6	75.6	76.0	75.8	81.4	80.9
Expt. reboiler temp., °C	119.9 ± 0.3	118.7 ± 0.3	121.5 ± 0.2	121.1 ± 0.4	121.2 ± 0.3	119.9 ± 0.1	121.7 ± 0.3	121.3 ± 0.3	121.1 ± 0.2	121.4 ± 0.3	121.6 ± 0.2
Simu. reboiler temp., °C	120.6	119.7	122.4	122.0	122.3	121.0	122.3	122.4	122.2	121.8	122.3
Expt. Regeneration[3], MJ/kg CO_2	3.86	3.88	4.10	4.12	4.15	4.28	4.18	4.10	4.11	4.12	4.11
Simu. regeneration, MJ/kg CO_2	4.01	4.02	4.21	4.23	4.23	4.46	4.29	4.23	4.24	4.19	4.22
Expt. CO_2 product purity, vol%	98.5	98.8	99.4	98.7	99.4	99.2	99.4	99.0	99.1	99.3	99.5
Simu. CO_2 product purity, vol%	98.3	98.3	98.8	98.7	98.9	98.7	98.8	98.8	99.0	98.9	99.0

[1]Temperature difference of solvent in and out of stripper.
[2]The difference of CO_2 mass flow rate across the CO_2 absorber.
[3]Experiment regeneration was calculated by sum of three heat components.

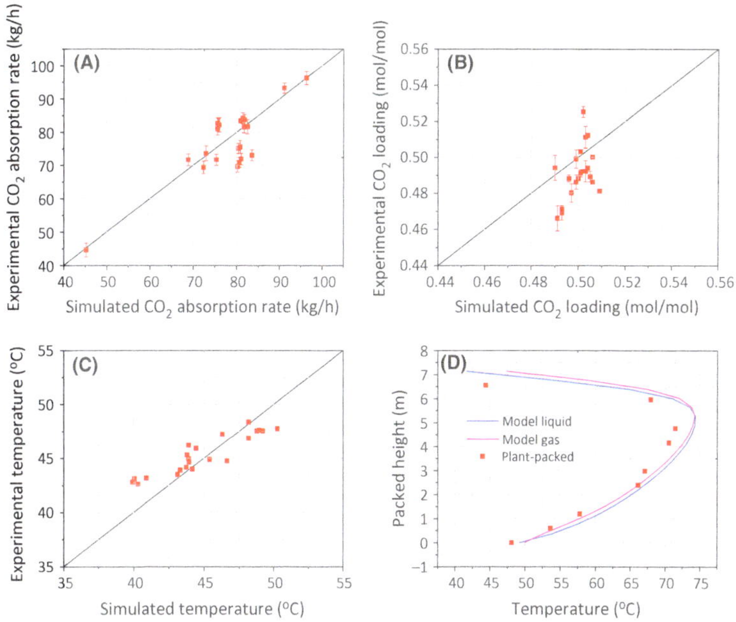

Figure 2. Results of comparison between simulation and Tarong pilot plant measurements: (A) CO_2 absorption rate in absorber; (B) CO_2 loading of rich solvent leaving absorber; (C) temperature of rich solvent leaving absorber and (D) temperature profiles along absorber column (01 Feb).

samples. Overall, the predictions of CO_2 loading were considered satisfactory.

Figure 2C and D suggest that the experimental temperature has a good agreement with the simulation results, implying that the rate-based model is able to predict the temperature profile along the absorber column. However, it should be noted that the experimental temperatures along the column in Figure 2D were always lower than the model data and the deviation was even greater at the high temperature sections at a packed height of 4–6 m. This is most likely due to heat loss along the column wall during the CO_2 absorption process. Another possibility is that the solvent lost heat through the uninsulated pipe, whilst the solvent was removed from the column between the packed sections.

Performance of CO_2 stripper

Figure 3A shows the parity plot of stripper reboiler temperature between simulation results and pilot plant data. It is found that the experimental values were consistently lower than those of the simulation because of the drastic heat loss in the high temperature stripper (90–130°C). This is demonstrated by the temperature profiles indicated in Figure 3B. Heat loss was always occurring along the stripper column, which is reflected by the temperature deviation from the model results. It is worthwhile to mention that the temperature deviation at the packed height 0–3.584 m (bottom section) was much greater than that at the height 3.584–7.168 m (top section), and that some temperatures in the bottom stage were surprisingly lower than that in the top stage. Two possible reasons can account for this phenomenon. One is the higher temperature in the bottom stage resulted in higher heat loss. The second is that heat loss took place from the solvent when being transported through pipelines between packed sections which are installed outside the stripper column (Fig. 1). This resulted in greater heat loss to the environment.

Due to water vaporization at high temperatures in the stripper, the CO_2 stream generated in the stripper may require further purification. Condensation was considered as the the effective approach to separate most of the water vapor from the CO_2 stream. In the pilot plant trials, the condenser temperature was controlled in the range of 17–30°C. This ensured a high purity of CO_2 product ranging from 97.5 to 99.5 vol.%, which meets the requirement for CO_2 compression. The simulation results in Figure 3C and 3D are in excellent agreement with the experimental results in terms of H_2O content and CO_2 concentration in the CO_2 product, which proves that the rate-based stripper model has the capability to predict the condensation process.

The experimental regeneration energy was calculated by summing three key components: heat for stripping

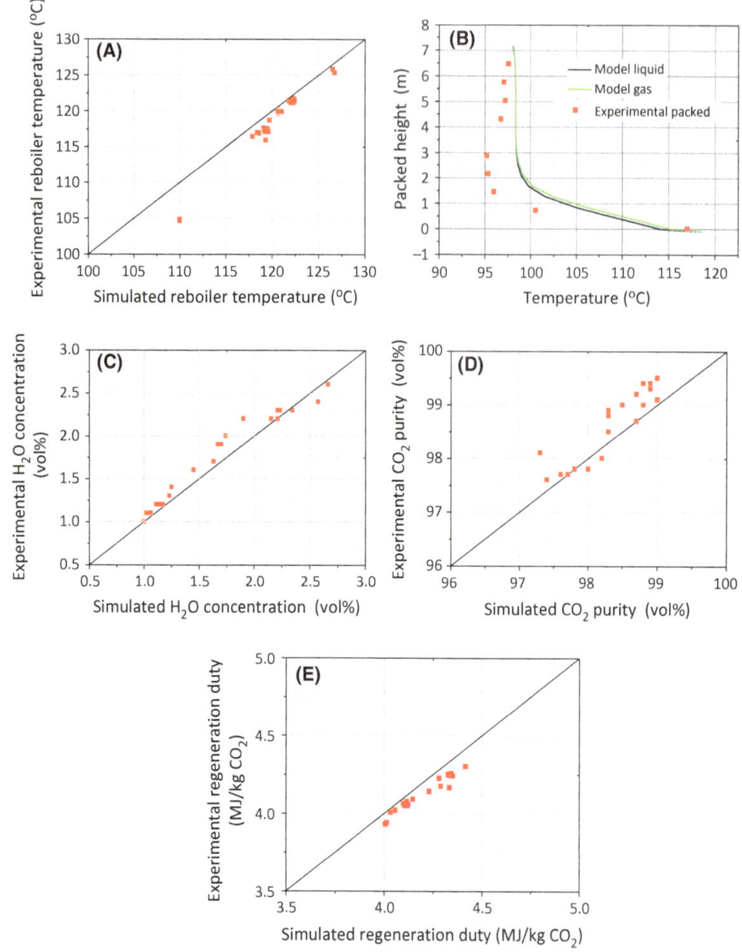

Figure 3. Results of comparison between simulation and Tarong pilot trials: (A) reboiler temperatures; (B) temperature profiles in packed desorber (Test 01 Feb); (C) CO_2 purity in the CO_2 product; (D) H_2O concentration in the CO_2 product; (E) solvent regeneration duty.

CO_2 from the solvent, sensible heat for heating up solvent to the required desorption temperature, and the water vapor leaving the stripper in the overhead gas stream. Due to the heat loss along the stripping column, the measured reboiler temperature in pilot trials would be always lower than the actual temperature (Fig. 3A), which resulted in the underestimation of sensible heat and the subsequent regeneration duty. While the model does not take the heat loss into account and take the high modeling reboiler temperature to calculate the regeneration duty. As a result, the simulated results of solvent regeneration duty had an average 4.0% overestimation over the pilot plant results as shown in Figure 3E.

In conclusion, the good agreement between the experimental results and the modeling results for both CO_2 absorber and CO_2 stripper suggests that the established rate-based model can satisfactorily predict the CO_2 capture process by aqueous MEA.

Process Improvement of MEA-Based CO_2 Capture Process

The process improvement was proposed to reduce the energy requirement of the MEA process by using the validated rate-based model to investigate parameter optimization and process modifications. The typical flue gas with 900 kg/h flow rate in the Tarong pilot trials was used and it contained 11.9% CO_2, 7.3% O_2, 4.3% H_2O, and 76.5% N_2. The gas pressure and temperature (at inlet to absorber) were 106 kPa (absolute pressure) and 50°C, respectively. The CO_2 removal efficiency of the MEA process was designed at 85%.

Heat requirement of a typical Tarong pilot case

During the process optimization, the regeneration duty was considered as the most important factor to optimize and improve the MEA process, as it is the main contributor to the total energy required for the CO_2 capture process. In order to understand how the three heat requirements are distributed, the representative Tarong pilot trial (Test 01 Feb) with a regeneration duty of 4596 KJ/kg CO_2 was investigated as shown in Figure 4. Analyzing the individual heat requirement was based on the following equations.

1. $Q_{des,CO_2} = n_{CO_2} H_{CO_2}$, where n_{CO_2} is the mole of regenerated CO_2, mol; H_{CO_2} the enthalpy per mole CO_2 desorbed from the solution and the calculation method is taken from Que [40], kJ/mol. The heat of CO_2 desorption required to break the chemical bond between MEA and CO_2, accounts for the largest energy consumption. The higher the MEA concentration and rich CO_2 loading, the lower the heat of CO_2 desorption.
2. $Q_{sens} = \overline{m}_{solv} \overline{c}_p (T_{in} - T_{out})$, where m_{solv} is the mass flow rate of the solvent flowing through the stripper, kg/h; C_p specific heat capacity of the solvent, kJ/kg·K; $T_{in} - T_{out}$ solvent temperature difference in and out of the stripper, K. Narrowing the temperature difference and lowering the solvent mass flow rate are the primary approaches to reducing the sensible heat.
3. $Q_{vap,\, H_2O} = n_{vap,\, H_2O} H_{vap,\, H_2P}$, where $n_{vap,\, H_2O}$ is the moles of excess steam leaving the stripping column, mol; $H_{vap,\, H_2O}$ latent heat of steam generation. An amount of stripping vapor is needed to maintain the driving force for CO_2 desorption in the stripper. However, if the amount of stripping vapor is high, large amounts of water vapor will leave the stripper and the energy is lost in the condenser. The heat of water vaporization is dependent on the temperature at the top of the stripper before the vapor enters the condenser. The best scenario is to make the temperature of the stripper exit as low as possible, whilst the CO_2 desorption process is maintained by a certain amount of water vapor leaving the stripper.

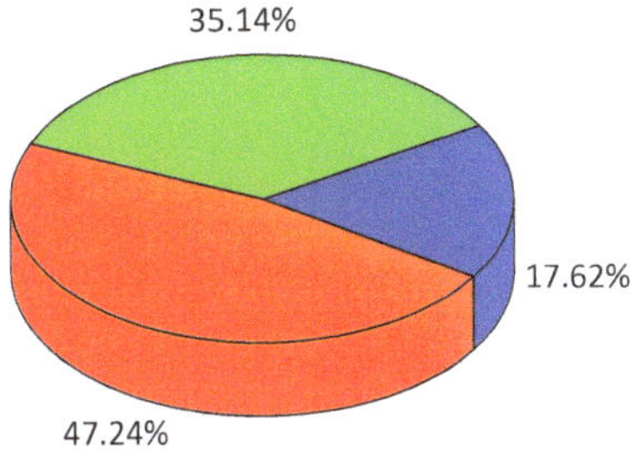

Figure 4. The distribution of three heat requirements: heat of CO_2 desorption, sensible heat, heat of water vaporization.

Parameter optimization

Important process parameters were studied, including MEA concentration, lean CO_2 loading, lean solvent temperature, and stripper pressure. The temperature difference of lean/rich cross heat exchangers was set as 15°C on the hot side.

MEA concentration

As shown in Figure 5A, the energy consumption for solvent regeneration decreased substantially with increasing MEA concentration. This is because high MEA concentration allowed for better CO_2 absorption performance including the improvement of CO_2 reaction rate and the CO_2 absorption capacity per kilogram solvent. The increasing MEA concentration also led to a decrease in the solution circulation rate which resulted in a decrease in sensible heat and a subsequent reduction in regeneration duty. Upon an increase in MEA concentration from 25 to 40 wt.%, the reboiler duty decreased by 14%. However, the use of high concentration MEA will considerably increase the degradation rate due to a higher O_2 mass transfer at higher MEA concentrations in the absorber [41]. Moreover, the higher solvent concentration causes an increase in viscosities and diffusion coefficient, thus increasing the operational difficulty in real practice. For balancing the energy saving and adverse effects, a MEA concentration of 35 wt.% was chosen in this study, which is a compromise concentration between 30% MEA used by Notz et al. [23] and 40% MEA by Abu-Zahra et al. [26]

Lean CO_2 loading

Figure 5B shows the influence of lean CO_2 loading on the regeneration duty and solvent flow rate. The lean CO_2 loading between 0.25 and 0.275 was the optimal with the energy ranging between 3.59 and 3.61 MJ/kg CO_2. At lean CO_2 loadings below 0.25, although the solution provided more free MEA for faster CO_2 absorption and a lower solvent flow rate, the regeneration duty increases. This is because an increasing amount of stripping steam is required to regenerate such a low loading solvent. At high lean CO_2 loadings, the solvent circulation flow

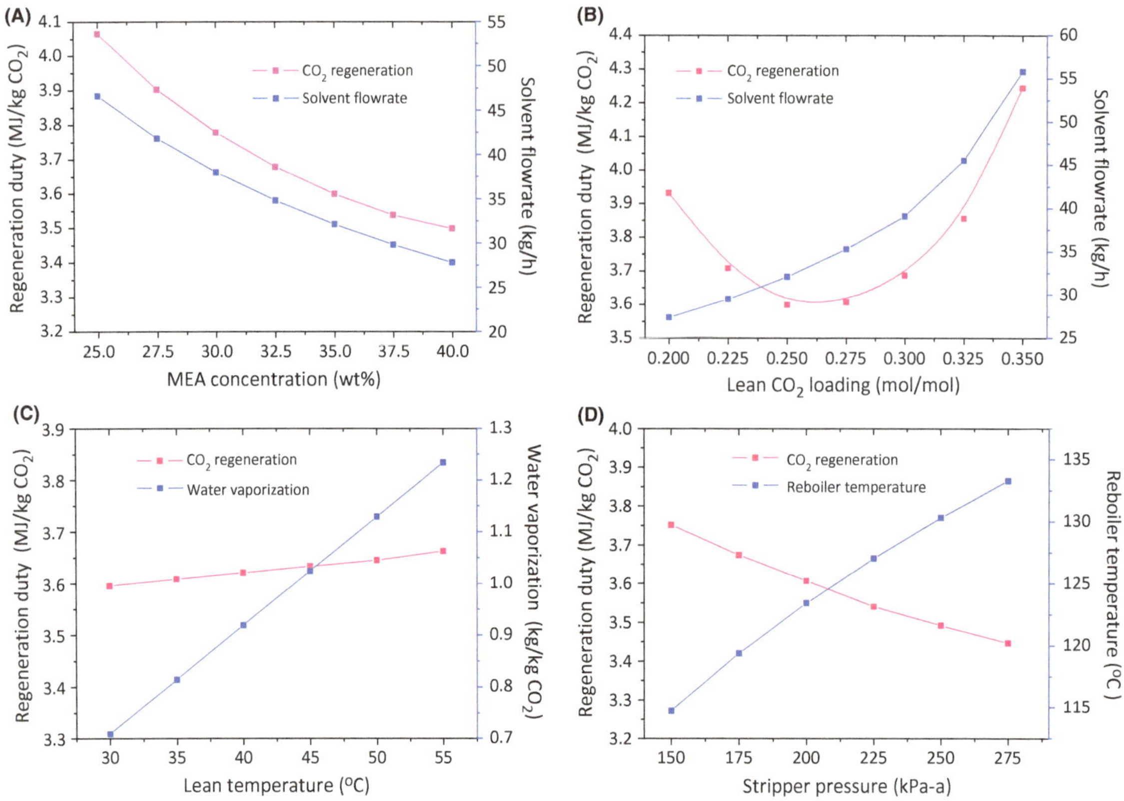

Figure 5. Effect of (A) MEA concentration, (B) lean CO_2 loading, (C) Lean solvent temperature, (D) stripper top pressure on the regeneration energy based on single factor analysis.

rate increased substantially resulting in an increase in sensible heat and risk of column flooding.

Lean solvent temperature

As shown in Figure 5C, increasing the lean solvent temperature from 30 to 55°C slightly increased the regeneration duty. This is because higher temperatures result in the increase of CO_2 equilibrium pressure and hence decreased the CO_2 absorption capacity of the MEA solvent. However, at high lean temperatures, less cooling duty is required for solvent cooling resulting in lower cooling water consumption. Moreover, from the viewpoint of practical operation, high lean solvent temperature led to the increase in the water vaporization rate from the absorption column, which is beneficial, in that the water condensing in the wash section would make it more effective for removing other trace constituents in the exiting flue gas. After water condensation, it is then periodically recycled back to the lean solvent to maintain the water balance of the system.

Stripper pressure

Elevating the stripper pressure has some benefits. One is to suppress the water vaporization and subsequently a reduction in heat of water vaporization; the second is to increase the stripper temperature to facilitate the CO_2 stripping process. As shown in Figure 5D, increasing stripper pressure from 150 to 275 kPa (absolute pressure) led to an 8.3% reduction in the heat requirement of solvent regeneration. However, the elevated pressure also brings the drawbacks: (1) the high temperature would enhance amine degradation rates and corrosion problems and subsequently increase the material and maintenance cost during operation [42]; (2) the high reboiler temperature would require extraction of higher quality steam which may results in a higher net efficiency penalty on the power station; (3) high pressure would place a burden on the capital cost of stripper design and construction. Thus, the suitable stripper pressure should be determined by considering the energy saving, material saving, construction cost, etc. In this simulation, a compromise stripper pressure of 200 kPa (absolute

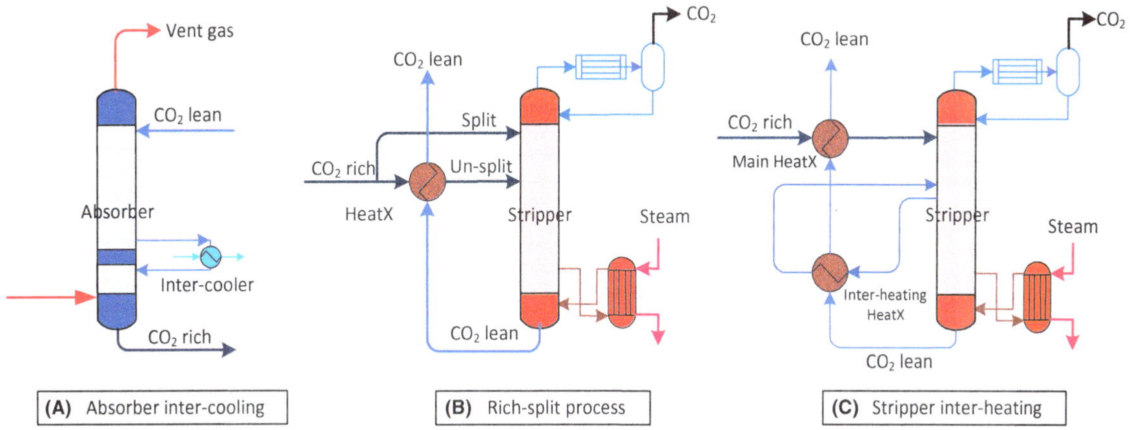

Figure 6. Process configurations of (A) absorber intercooling, (B) rich-split and (C) stripper interheating (HeatX represents heat-exchanger).

pressure) was chosen with condenser duty 1.2 MJ/kg CO_2, reboiler temperature 123°C, and regeneration duty 3.6 MJ/kg CO_2, in which heat of CO_2 desorption accounts for 59%, heat of water vaporization 22% and sensible heat 19%. The regeneration duty of 3.6 MJ/kg CO_2 agrees well with the numbers reported in the literature ranging 3.6–3.8 MJ/kg CO_2 [17, 21–23]. It is worth mentioning that the optimal operating conditions are likely to vary depending on flue gas composition, column size, and process configuration. A case by case study is recommended to obtain the best operating conditions for the individual process.

Process modification

Based on the optimized parameters, process modifications were proposed to further reduce the energy consumption of solvent regeneration via the absorber intercooling process, rich-split process and stripper interheating process. The corresponding process configurations are shown in Figure 6.

Intercooling process

The exothermic reactions of CO_2 absorption by MEA increase the solvent temperature along the absorber column. This, on the one hand, favors the reaction kinetics due to the improved mass transfer coefficients; on the other hand, limits the solvent absorption capacity due to the increased CO_2 partial pressure at a given loading. The intercooling process modification (Fig. 6A), however, is able to overcome the drawback of absorption capacity caused by the increasing temperature, because this process modification enables the increase of CO_2 loading in the rich solvent owing to the low equilibrium CO_2 partial pressure at low temperature. Knudsen et al. [21] evaluated absorber intercooling at pilot scale and revealed that intercooling enabled a higher CO_2 cyclic carrying capacity thereby reducing regeneration energy.

Table 4 lists the results of different cases with the intercooling process applied to the absorber. All the solvent went through the intercooler to maximize the benefits of

Table 4. Results of intercooling process at different scenarios with L/G ratio 2.3.

	Rich CO_2 loading, mol/mol	Regeneration energy, MJ/kg CO_2	Energy saving, %	Specification
Reference case	0.5007	3.600	−0.0	Total packed height 7.136 m
Intercooler position: height from bottom				
3.568 m	0.4989	3.626	+0.72	Cooling to 30°C
1.586 m	0.5026	3.567	−0.92	
0.792 m	0.5039	3.545	−1.53	
Cooling temperature				
40°C	0.5015	3.578	−0.61	Height 0.792 m
35°C	0.5027	3.561	−1.08	
25°C	0.5046	3.535	−1.80	
Total height of absorber column				
5.325 m	0.5013	3.585	−0.41	Cooling to 25°C; height 0.6 m
5.325 m	0.4962	3.660	+1.66	No intercooling

the intercooling process. It can be seen that if the intercooler was installed at a low position near the bottom of the absorber, the intercooling process was able to improve the CO_2 loading in the rich solvent and subsequently reduced the reboiler duty, a 1.8% energy saving at intercooling temperature 25°C. In addition to the energy reduction, the intercooling process also had the potential to reduce the column height while maintaining the CO_2 removal efficiency at 85% and similar solvent regeneration duty. When intercooling to 25°C was applied to the absorber, the packed height can be reduced from 7.136 m to 5.325 m. This is a 25% reduction in column height and allows for a great reduction in the capital cost of column design and construction. It also can be seen that the lower the temperature of the intercooling process, the better the performance in terms of regeneration duty. If lower temperature cooling water is available for the PCC plant, the regeneration energy and/or column height could be further reduced by the intercooling process. Furthermore, for scrubbing solvents that have a higher CO_2 absorption capacity than MEA, better benefits of energy decrease and column reduction could be obtained from the intercooling process.

Rich-split process

The rich-split process modification shown in Figure 6B is an efficient method to reduce the reboiler duty via the recovery of the steam generated in the stripper. The cold rich stream was split to recover the energy contained in the upcoming high temperature water vapor; meanwhile the rich solvent was heated to release part of the CO_2. This process has proven to be effective in process simulations [43–45] as well as in the Tarong pilot plant trials [34]. In this study, the unsplit stream was introduced at the stage 5 (20 stages in total) after crossing the heat exchanger while the split stream was fed to the top of the stripper. Figure 7A shows the effect of split fraction (the ratio of split stream to the total rich solvent) on the regeneration duty. It can be seen that the rich split has a notable reduction on the reboiler energy consumption when the split fraction was up to 0.45. The temperature difference of the heat exchanger on the hot side decreased but was limited to 10°C with increasing split ratio. A minimum was achieved when 25% of the cold rich solvent was split to the top of the column. This minimum regeneration duty was 3.31 MJ/kg CO_2 together with 0.67 MJ/kg CO_2 condenser duty. This is an 8.06% reduction in reboiler duty and 45.1% reduction in condenser duty compared with the reference case without rich-split process.

In order to figure out why this saving occurs, the temperature profiles and H_2O/CO_2 vapor pressure along the stripper column were investigated. As shown in Figure 7B, the split cold rich solvent decreased the temperature at the top stage of the stripper (stage 2) in the rich-split process, which was beneficial because it could recover the steam and subsequently reduce the reboiler duty and condenser duty. Meanwhile the temperatures along the column after stage 3 were elevated. This temperature lift came from the heat exchanger where the unsplit rich solvent was heated to a higher temperature as a result of the decreasing solvent flow rate. So the effectiveness of this rich-split modification will be significantly influenced by the efficiency of the lean/rich heat exchanger. The increased temperature along the stripper column resulted in a higher H_2O vapor pressure and subsequently a lower CO_2 vapor pressure after stage 3 (Fig. 7C). This indicated that the driving force of the CO_2 desorption process was enhanced, which has the potential to accelerate the CO_2 desorption rate and reduce the energy consumption of solvent regeneration in practice.

To further specify how the energy was saved by the rich-split process, the distribution of the three heat requirements – heat of CO_2 desorption, heat of water vaporization, sensible heat – was determined. The heat of CO_2 desorption as an inherent property of solvent, was considered to be unchanged since the same rich solvent entering the stripper and the same CO_2 desorption rate were used for all the different split cases. As shown in the Figure 7D, the rich-split configuration led to a significant decrease in the heat of water vaporization with increasing split fraction, which was favorable for lowering condenser duty and the subsequent regeneration duty. However, at higher split fractions, the cold split stream started to cool down the stripper and more sensible heat was required to heat up the split solvent to the required temperature, which led to an increase in reboiler duty. So the heat of water vaporization and sensible heat in the rich-split process were competing with each other, resulting in an appropriate split fraction in which the saving of reboiler duty was maximized. It is worth mentioning that the optimal split fraction was also affected by the temperature approach of heat exchanger on the hot side. The smaller the temperature approach means the higher the temperature of the unsplit rich solvent, which then required more split solvent to recover the stripper steam.

Interheating process

The interheating process shown in Figure 6C exchanges heat between the hot lean stream leaving from the bottom of the stripper and the semi-lean solvent extracted from the middle of the stripper (interheating HeatX) before the hot lean stream goes to the main

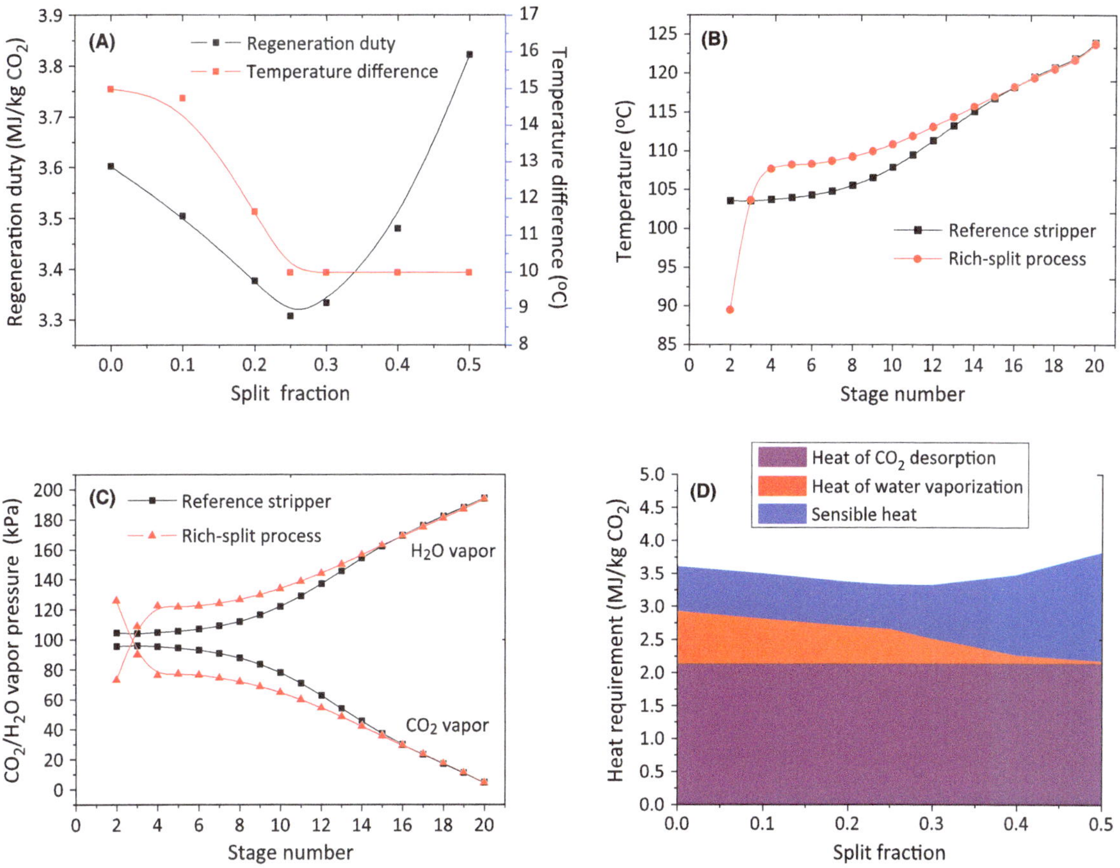

Figure 7. (A) Effect of rich split fraction on the total solvent regeneration duty and temperature difference of heat exchanger on the hot side; (B) temperature profile and (C) H_2O/CO_2 vapor pressure profile along the stripper column between reference stripper and rich-split stripper with 0.25 split ratio (condenser is at stage1; reboiler at stage 20); and (D) the distribution of three heat components: heat of CO_2 desorption, heat of water vaporization, sensible heat at different split fractions.

cross-exchanger (HeatX), thus making better use of the heat in the hot lean stream. This concept has been proposed by Leites et al. [46]. Paul et al. [47] patented this process and suggested using a heat-integrated stripping column to reduce the energy penalty associated with regenerating amine solutions and the modeling results revealed that the heat requirement of solvent regeneration can be reduced with the interheating process [48, 49]. This process design was aimed at reducing reboiler duty and condenser duty by means of (1) recycling the high-quality and high-temperature heat in the hot lean stream, which elevated the overall temperature along the stripper column; (2) reducing the energy loss associated with steam generation by lowering the temperature of the rich solvent entering the top of the stripper column. As shown in Figure 8A and B, like the rich-split process, the interheating process lowers the CO_2 partial pressure by increasing the temperature profiles and H_2O vapor pressure along the stripper column. Meanwhile the interheating process reallocated the heat distribution which elevated the temperature profiles along the bottom of the column whilst decreasing the temperature in the top section of the stripper. This accordingly reduced the energy requirement of sensible heat and heat of water vaporization (Fig. 8C). The regeneration duty with interheating process was reduced to 3.38 MJ/kg CO_2, a 6.1% reduction compared to the reference case.

It is interesting to find in Table 5 that the flow rate of the interheating stream extracted from the stripper had only a slight effect on the energy saving of the reboiler and condenser. Low flow rates mean that less solvent participated in the heat exchange during the interheating process, but this enabled a higher

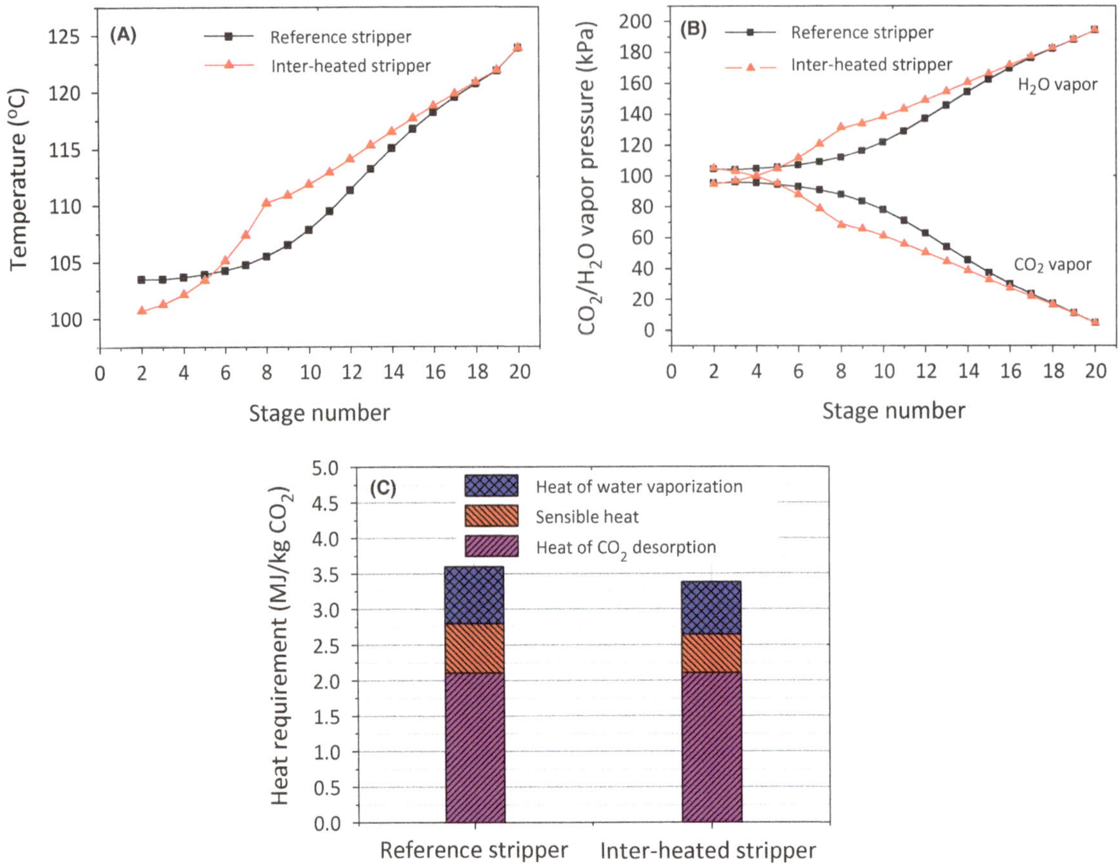

Figure 8. (A) temperature profiles, (B) H_2O/CO_2 partial pressure profiles along the stripper column and (C) heat requirement between reference stripper and interheated stripper with interheating at stage 8 (condenser is at stage1; reboiler at stage 20).

temperature and higher quality interheated stream entering the middle of the stripper. In contrast, high flow rates mean that more solvent can directly benefit from the interheating process, but results in a relatively lower temperature of the interheated stream. As a result, the extracted solvent obtained similar energy from the interheating process, leading to the results of unchanged energy saving. This can be accounted for by the heat duties of the two heat exchangers, the total of which remained the same with increasing solvent flow rate. This phenomenon suggests that the interheating process would be very flexible in practical application with respect to the amount of solvent extracted from the middle of the stripper. Low interheated solvent flow rates would have the potential of reducing the size of the interheating heat exchanger and energy saving of solvent pumping. It is also found that the total heat duties in the cross exchangers were very close between the reference stripper (135.7 kW) and interheated stripper (135.6–135.8 kW). This indicates that the role of interheating process was to make better utilization of heat contained in the hot lean stream, that is, extracting the high quality and high temperature stream to heat up the stripper column.

Combined intercooling, rich-split, and interheating process

The combined process modifications were proposed to make better use of the advantages of the intercooling process, rich-split process, and interheating process. Table 6 summarizes the main design specifications and simulation results of this combined process modification. It should be highlighted that the temperature difference of the main heat exchanger on the cold side was set as 8.8°C to be consistent with the reference case and make a reasonable comparison. Although the combined process increased the cooling duty by 0.14 MJ/kg CO_2

Table 5. Effect of interheated solvent flow rate on the duty of two heat exchangers and condenser/reboiler.

Interheating solvent flow rate[1], kg/h	Reference	500	1000	1500	1900
Temperature profiles in the interheating heat exchanger, °C					
Cold inlet	–	110.2	110.2	110.2	110.3
Cold outlet	–	119.6	118.1	117.4	117.1
Hot inlet	–	123.9	123.9	123.9	123.9
Hot outlet	–	116.1	116.1	116.1	116.1
Temperature profiles in the main heat exchanger, °C					
Cold inlet	39.2	39.2	39.2	39.2	39.2
Cold outlet	108.9	105.3	105.3	105.3	105.3
Hot inlet	123.9	116.1	116.1	116.1	116.1
Hot outlet	48.0	48.0	48.0	48.0	48.0
Heat duties of heat exchangers, kW					
Interheating HeatX	–	14.2	14.1	14.1	14.2
Main HeatX	135.7	121.6	121.6	121.5	121.5
Total	135.7	135.8	135.7	135.6	135.7
Energy requirement, MJ/kg CO_2					
Condenser duty	1.22	0.99	0.99	0.99	0.99
Reboiler duty	3.60	3.38	3.38	3.38	3.38

[1]The total flow rate of lean solvent was 1961.3 kg/h.

Table 6. Simulation conditions of combined absorber intercooling, rich split, stripper interheating.

Simulation conditions	Reference	Combined process
Intercooling temperature, °C	–	30
Intercooling stage (20 stages in total)	–	17
Feed stage of split stream (Split fraction 0.25)	–	2
Feed stage of unsplit stream	–	6
Interheated solvent flow rate, kg/h	–	500
Interheating stage (20 stages in total)	–	8
Temperature difference of main heat exchanger on cold side	8.8	8.8
Condenser temperature, °C	30	30
CO_2 desorption rate, kg/h	134.5	134.5
Results		
CO_2 mass purity, %	99.1	99.1
Intercooling duty, MJ/kg CO_2	0	0.61
Condenser duty, MJ/kg CO_2	1.20	0.73
Reboiler temperature, °C	123.7	123.7
Reboiler duty, MJ/kg CO_2	3.6	3.1

due to the intercooling process, the 25% reduction in absorber size can significantly reduce the capital investment of absorber column and thus compensate the increased cooling duty. More importantly, upon combining the benefits of the three process modifications, the regeneration duty was reduced to 3.11 MJ/kg CO_2, which is a 13.6% reduction in reboiler duty compared to the reference case.

Conclusion

In this study, the combination of pilot-plant trials and process modeling has demonstrated that a validated rate-based model was an effective and reliable tool to evaluate and improve the MEA-based CO_2 capture process. The rate-based model for the absorber and stripper has been successfully validated against the Tarong pilot plant results. The simulation results of the absorber model were in excellent agreement with the experimental results in terms of CO_2 loading in the rich solvent, temperature profiles along the column and CO_2 absorption rate, while the stripper model provided a very good predication of stripper parameters including temperature profiles along the stripper column, CO_2 product composition, and solvent regeneration duty. The rate-based model can be used as a good guide for modeling the MEA-based CO_2 capture process and used as a starting point for more sophisticated models for process development, debottlenecking, plant, and equipment design.

Process improvements including parameter optimization and process modification were carried out to reduce the energy consumption involved in solvent regeneration. After a sensitivity analysis, the optimal operating conditions selected were 35% MEA, 0.25 lean CO_2 loading, 40°C lean solvent temperature and 200 kPa stripper pressure which resulted in a regeneration duty of 3.6 MJ/kg CO_2. The intercooling process alone on the absorber was able to reduce the regeneration duty by 0.6–1.8% or reduce the column height by 25%. The rich-split process alone can reduce the energy by 8.5% with the reboiler duty reduced to 3.3 MJ/kg CO_2. The interheating process alone can also reduce the regeneration duty to 3.3 MJ/kg CO_2. The combination of the three process modifications showed the best energy reduction with the regeneration duty of 3.1 MJ/kg CO_2, which is a 13.6% reduction in reboiler duty compared to the reference case.

Acknowledgments

Kangkang Li thanks the Australian IPRS-APA scholarship and CSIRO Top-up scholarship that supported his research. Weiliang Luo and Jian Chen thank the support from the National Natural Science Foundation of China (key project no. 51134017).

Conflict of Interest

None declared.

Reference

1. International Energy Agency. 2013. 21st Century coal – advanced technology and global energy solution. IEA, Paris, France.
2. Li, K. K., H. Yu, M. Tade, and P. Feron. 2014. Theoretical and experimental study of NH_3 suppression by addition of Me(II) ions (Ni, Cu and Zn) in an ammonia-based CO_2 capture process. Int. J. Greenh. Gas Con. 24:54–63.
3. Li, H., J. Yan, J. Yan, and M. Anheden. 2009. Impurity impacts on the purification process in oxy-fuel combustion based CO_2 capture and storage system. Appl. Energy 86:202–213.
4. Kunze, C., and H. Spliethoff. 2012. Assessment of oxy-fuel, pre- and post-combustion based carbon capture for future IGCC plants. Appl. Energy 94:109–116.
5. Li, K., H. Yu, G. Qi, P. Feron, M. Tade, J. Yu, and S. Wang. 2015. Rate-based modelling of combined SO_2 removal and NH_3 recycling integrated with an aqueous NH_3-based CO_2 capture process. Appl. Energy 148:66–77.
6. Rochelle, G. T. 2009. Amine scrubbing for CO_2 capture. Science 325:1652–1654.
7. Herzog, H., J. Meldon, and A. Hatton. 2009. Advanced post-combustion CO_2 capture. Prepared for Clean Air Task Force. https://mitei.mit.edu/system/files/herzog-meldon-hatton.pdf. Accessed on May 2014.
8. Aaron, D., and C. Tsouris. 2005. Separation of CO_2 from flue gas: a review. Sep. Sci. Technol. 40:321–348.
9. Puxty, G., R. Rowland, A. Allport, Q. Yang, M. Bown, R. Burns, et al. 2009. Carbon dioxide postcombustion capture: a novel screening study of the carbon dioxide absorption performance of 76 amines. Environ. Sci. Technol. 43:6427–6433.
10. Porcheron, F., A. Gibert, P. Mougin, and A. Wender. 2011. High throughput screening of CO_2 solubility in aqueous monoamine solutions. Environ. Sci. Technol. 45:2486–2492.
11. DECC ScottishPower CCS Consortium. 2011. UK Carbon capture and storage demonstration competition. SP-SP 6.0–RT015 FEED Close Out Report.
12. Reynolds, A. J., T. V. Verheyen, S. B. Adeloju, E. Meuleman, A. Chaffee, A. J. Cottrell, et al. 2013. Chemical characterization of MEA degradation in PCC pilot plants operating in Australia. Energy Procedia 37:877–882.
13. Manzolini, G., E. S. Fernandez, S. Rezvani, E. Macchi, E. L. V. Goetheer, and T. J. H. Vlugt. 2015. Economic assessment of novel amine based CO_2 capture technologies integrated in power plants based on European Benchmarking Task Force methodology. Appl. Energy 138:546–558.
14. Zhang, Y., H. Chen, C. C. Chen, J. M. Plaza, R. Dugas, and G. T. Rochelle. 2009. Rate-based processXmodeling study of CO_2 capture with aqueous monoethanolamine solution. Ind. Eng. Chem. Res. 48:9233–9246.
15. Zhang, Y., and C. C. Chen. 2013. Modeling CO_2 absorption and desorption by aqueous monoethanolamine solution with Aspen rate-based model. Energy Procedia 37:1584–1596.
16. Zhang, Y., H. Que, and C. C. Chen. 2011. Thermodynamic modeling for CO_2 absorption in aqueous MEA solution with electrolyte NRTL model. Fluid Phase Equilibr. 311:67–75.
17. Øi, L. 2007. Aspen HYSYS simulation of CO_2 removal by amine absorption from a gas-based power plant. In: SIMS Conference, Gøteborg.
18. Lawal, A., M. Wang, P. Stephenson, and H. Yeung. 2009. Dynamic modelling of CO_2 absorption for post combustion capture in coal-fired power plants. Fuel 88:2455–2462.
19. Tobiesen, F. A., O. Juliussen, and H. F. Svendsen. 2008. Experimental validation of a rigorous desorber model for CO_2 post-combustion capture. Chem. Eng. Sci. 63:2641–2656.
20. Tobiesen, F. A., H. F. Svendsen, and O. Juliussen. 2007. Experimental validation of a rigorous absorber model for CO_2 postcombustion capture. AIChE J. 53:846–865.
21. Knudsen, J. N., J. Andersen, J. N. Jensen, and O. Biede. 2011. Evaluation of process upgrades and novel solvents for the post combustion CO_2 capture process in pilot-scale. Energy Procedia 4:1558–1565.
22. Knudsen, J. N., J. N. Jensen, P.-J. Vilhelmsen, and O. Biede. 2009. Experience with CO_2 capture from coal flue gas in pilot-scale: testing of different amine solvents. Energy Procedia 1:783–790.
23. Notz, R., H. P. Mangalapally, and H. Hasse. 2012. Post combustion CO_2 capture by reactive absorption: pilot

plant description and results of systematic studies with MEA. Int. J. Greenh. Gas Con. 6:84–112.
24. Dave, N., T. Do, D. Palfreyman, P. H. M. Feron, S. Xu, S. Gao, et al. 2011. Post-combustion capture of CO_2 from coal-fired power plants in China and Australia: an experience based cost comparison. Energy Procedia 4:1869–1877.
25. Salkuyeh, Y. K., and M. Mofarahi. 2012. Reduction of CO_2 capture plant energy requirement by selecting a suitable solvent and analyzing the operating parameters. Int. J. Energy Res. 37:973–981.
26. Abu-Zahra, M. R. M., L. H. J. Schneiders, J. P. M. Niederer, P. H. M. Feron, and G. F. Versteeg. 2007. CO_2 capture from power plants part I. A parametric study of the technical performance based on monoethanolamine. Int. J. Greenh. Gas Con. 1: 37–46.
27. Cousins, A., L. T. Wardhaugh, and P. H. M. Feron. 2011. Preliminary analysis of process flow sheet modifications for energy efficient CO_2 capture from flue gases using chemical absorption. Chem. Eng. Res. Des. 89:1237–1251.
28. Cousins, A., L. T. Wardhaugh, and P. H. M. Feron. 2011. A survey of process flow sheet modifications for energy efficient CO_2 capture from flue gases using chemical absorption. Int. J. Greenh. Gas Con. 5:605–619.
29. Karimi, M., M. Hillestad, and H. F. Svendsen. 2011. Capital costs and energy considerations of different alternative stripper configurations for post combustion CO_2 capture. Chem. Eng. Res. Des. 89:1229–1236.
30. Freguia, S., and G. Rochelle. 2003. Modeling of CO_2 capture by aqueous monoethanolamine. AIChE J. 49:1676–1686.
31. Leonard, G., and G. Heyen. 2011. Modelling post-combustion CO_2 capture with amine solvents. Comput. Aided Chem. Eng. 29:1768–1772.
32. Li, K., H. Yu, M. Tade, P. Feron, J. Yu, and S. Wang. 2014. Process modeling of an advanced NH_3 abatementand recycling technology in the ammonia-based CO_2 capture process. Environ. Sci. Technol. 48:7179–7186.
33. Lee, A. S., J. C. Eslick, D. C. Miller, and J. R. Kitchin. 2013. Comparisons of amine solvents for post-combustion CO_2 capture: a multi-objective analysis approach. Int. J. Greenh. Gas Con. 18:68–74.
34. Cousins, A., A. Cottrell, A. Lawson, S. Huang, and P. Feron. 2012. Model verification and evaluation of the rich-split process modification at an Australian-based post combustion CO_2 capture pilot plant. Greenh Gas: Sci. Technol. 2:329–345.
35. Aspen Technology. 2011. Rate-Based Model of the CO2 Capture Process by MEA using Aspen Plus. Aspen Technology, Burlington, MA.
36. Pinsent, B. R., L. Pearson, and F. J. W. Roughton. 1956. The kinetics of combination of carbon dioxide with hydroxide ions. Trans. Faraday Soc. 52:1512–1520.
37. Hikita, H., S. Asai, H. Ishikawa, and M. Honda. 1977. The kinetics of reactions of carbon dioxide with monoethanolamine, diethanolamine, and triethanolamine by a rapid mixing method. Chem. Eng. J. 13:7–12.
38. Bravo, J. L., J. A. Rocha, and J. R. Fair. 1985. Mass transfer in gauze packings. Hydrocaborn Process 64:91–95.
39. Bravo, J. L., J. A. Rocha, and J. R. Fair. 1993. A comprehensive model for the performance of columns containing structured packings. I. ChemE. Symp. Ser. 32:641–651.
40. Que, H., and C. Chen. 2011. Thermodynamic modeling of the NH_3-CO_2-H_2O system with electrolyte NRTL model. Ind. Eng. Chem. Res. 50:11406–11421.
41. Léonard, G., C. Crosset, D. Toye, and G. Heyen. 2015. Influence of process operating conditions on solvent thermal and oxidative degradation in post-combustion CO_2 capture. Comput. Chem. Eng. J. 83:121–130.
42. Bailey, D., and P. Feron. 2005. Post-combustion decarbonisation processes. Oil. Gas. Sci. Technol. 60:461–474.
43. Van Wagener, D.H. 2011. Stripper modeling for CO_2 removal using monoethanolamine and piperazine solvents. Ph.D. Dissertation, The University of Texas at Austin, Austin, TX.
44. Lin, Y. J., T. Madan, and G. T. Rochelle. 2014. Regeneration with rich bypass of aqueous piperazine and monoethanolamine for CO_2 capture. Ind. Eng. Chem. Res. 53:4067–4074.
45. Li, K. K., H. Yu, P. Feron, M. Tade, and L. Wardhaugh. 2015. Technical and energy performance of an advanced, aqueous ammonia-based CO_2 capture technology for a 500-MW coal-fired power station. Environ. Sci. Technol. 49:10243–10252.
46. Leites, I. L., D. A. Sama, and N. Lior. 2003. The theory and practice of energy saving in the chemical industry: some methods for reducing thermodynamic irreversibility in chemical technology processes. Energy 28:55–97.
47. Paul, S., and K. Lance. 2002. Improved aqua-ammonia absorption system generator utilizing structured packing. World Intellectual Property, Patent WO 02/12803 A1, 2002.
48. Karimi, M., M. Hillestad, and H. Svendsen. 2012. Positive and negative effects on energy consumption by interheating of stripper in CO_2 capture plant. Energy Procedia 23:15–22.
49. Frailie, P., T. Madan, B. Sherman, and G. Rochelle. 2013. Energy performance of advanced stripper configurations. Energy Procedia 37:1696–1705.

7

High moisture corn stover pelleting in a flat die pellet mill fitted with a 6 mm die: physical properties and specific energy consumption

Jaya Shankar Tumuluru

Biofuels and Renewable Energy Technology Department, Idaho National Laboratory, 750 University Blvd, Idaho Falls, Idaho 83415-3570

Keywords
High moisture corn stover, pellet properties, pelleting, specific energy consumption

Correspondence
Jaya Shankar Tumuluru, Biofuels and Renewable Energy Technology Department, Idaho National Laboratory, 750 University Blvd, Idaho Falls, Idaho 83415-3570.

E-mail: JayaShankar.Tumuluru@inl.gov

Funding Information
This work was supported by the Department of Energy, Office of Energy Efficiency, and Renewable Energy under the Department of Energy Idaho Operations Office Contract DE-AC07-05ID14517.

Abstract

The quality and specific energy consumption (SEC) of the biomass pellets produced depend upon pelleting process conditions. The present study includes understanding the effect of feedstock moisture in the range of 28–38% (wet basis [w.b.]) and preheating in the range of 30–110°C at two die speeds of 40 and 60 Hz on the physical properties and SEC. A flat die pellet mill fitted with a 6 mm die was used in the present study. The physical properties of pellets such as moisture content, unit, bulk and tapped density, durability, and expansion ratio and SEC of the pelleting process are measured. The results indicate that the pellets produced have durability values in the range of 87–98%, and unit bulk and tapped density in the range of 670–1100, 375–575, and 420–620 kg/m^3. Increasing the feedstock moisture content from 33% to 38% (w.b) decreased the unit, bulk and tapped density by about 30–40%. Increasing feedstock moisture content increased the expansion ratio and decreased the density values. A higher feedstock moisture content of 38% (w.b.) and higher preheating temperature of 110°C resulted in lower density and a higher expansion ratio, which can be attributed to flash off of moisture as the material extrudes out of the die. The SEC was in the range of 75–275 kWh/ton. Higher feedstock moisture content of 38% (w.b.) and a lower die speed of 40 Hz increased the SEC, whereas lower to medium preheating temperature (30–70°C), medium feedstock moisture content of 33% (w.b.), and a higher die speed of 60 Hz minimized the SEC to <100 kWh/ton.

Introduction

Bulk density of the plant-based biomass is low (50–130 kg/m^3) and needs to be increased to 500–700 kg/m^3 to reduce transportation and logistics costs and increase the storage stability [1]. Some of the advantages of densification include improved bulk density by five to seven times compared to raw biomass, improved handling and conveyance efficiencies, uniform size and shape, meeting the specifications of supply systems, and improved biochemical conversion performance [1]. The most commonly used densification systems include pellet mill, briquette press, and screw extruder. The pellets made from pellet mill are the most widely produced densified product and are transported intercontinentally for bioenergy applications. For example, wood pellets made in Canada and the United States are mainly exported to Europe for cofiring applications [2]. The pellets, which are transported to other countries from Canada and the United States, have to meet the standards established by the European Committee for Standardization (CEN) [2]. In the United States, the Pellet Fuel Institute (PFI), has established standard to transport pellets within and outside the nation. Typical dimensions of pellets used for residential and commercial application are 6 mm in diameter and 11–17 mm in length [2].

Pellet mill process variables

The process variables (like die rotational speed, die geometry [length to diameter (L/D) ratio], preheating temperature), and feedstock variables (like moisture content and particle size) influence the pellet physical properties like density, durability, and specific energy consumption (SEC) [1]. Studies on the effect of the preheating temperature on the density of wheat straw briquettes indicated that temperatures up to 90°C increased the density, but further increasing to 110°C reduced it [3]. This effect depends on the biomass composition where a high temperature (>90°C) is favorable for biomass that is high in protein content [4]. The die dimensions, such as diameter and length, affect the amount of material that can be pelleted, the quality of the pelleted material, and the energy required for compression. In general, smaller die diameters and lower rotational speeds are desirable for fibrous material, whereas bigger dies and higher die speeds are desirable for a wet and sticky biomass. Increasing the length of the pellet die increases pelleting pressure, whereas increasing the diameter of the pellet die decreases pelleting pressure [5]. Tabil and Sokhansanj [6] found that a bigger pellet die accommodates higher feedstock moistures. Their results indicated that a 6.2 mm die could process alfalfa with a moisture range of 7.5–9.0%, where a 7.8 mm die could tolerate moisture up to 12%. Feedstock moisture content and particle size also have a significant impact on the binding properties and quality of the pellets produced. A number of studies have indicated that fine grinding of biomass up to 2 mm produces denser pellets and increases the throughput of the machine [1, 6, 7]. At the same time, smaller particles less than 2 mm can lead to jamming of the pellet die and affect production capacity. Tumuluru et al. [1] in their review on densification systems for producing commodity type products from biomass indicated that feedstock moisture content lowers glass-transition temperature, promotes solid-bridge formation, and increases contact surface area. Many researchers have worked on pelleting and briquetting of biomass by adjusting feedstock moisture content in the range of 7–23% (w.b.) [8–14]. Some of the recent research articles published in 2014 have reported on pelleting raw, chemically and thermally pretreated biomass at moisture content >23% (w.b.) [15–19]. Recent preliminary studies on pelleting ammonia fiber explosion (AFEX) pretreated corn stover and municipal solid waste (consisting of about 30% plastic and nearly 70% being paper and noncorrugated cardboard) at a moisture content of about 19% and 15–25% (w.b.) in the ring die pellet mill (Bliss Pioneer Pellet Mill [B35A-75]), Bliss Industries, LLC), indicated good quality pellets in terms of bulk density (672 and 544 kg/m^3) and durability (99.6% and 98.4%) can be produced [20].

Pellet physical properties and SEC

Knowledge of the biomass pellet physical properties and the process variables influencing them is essential for producing high quality fuel pellets, designing equipment for processing facilities, and optimizing the unit operations. Some key quality attributes of fuel pellets are moisture content, density, and durability [1]. Pellet moisture content is important for safe storage and transportation. A higher moisture content of >15% (w.b.) can lead to pellet breakage as well as spoilage due to microbial decomposition, whereas a low moisture content of <3% can result in physical disintegration and generate more fines [1, 2]. Both higher and lower pellet moisture content can result in revenue loss for the pellet manufacturers. Density is important from a storage and transportation standpoint. Higher density helps to transport higher volumes of material and also helps to reduce the storage footprint at the conversion facilities. The durability indicates the integrity of the pellets during storage and transportation [21]. This can also be defined as the ability of the pellets to withstand the abrasive, impact, and frictional resistances during handling, storage, and transportation. Expansion ratio is defined as the ratio of diameter of the pellet to the diameter of pellet die. Expansion ratio is inversely proportional to the density of the pellet, where lower expansion ratio results in higher density [22]. In the case of fuel pellets, the lower expansion ratio is more desirable as expanded products will have high porosity and can be more reactive to storage environments. SEC of pelleting varies with different processing parameters, such as temperature and pressure, feedstock constituents (moisture content, particle size, and distribution), and chemical composition (starch, protein, lignin fat, waxes, and other lignocellulosic components) [1, 23]. Pellet mill rollers exert a constant pressure contingent upon the design of the pellet mill. The L/D ratio of the die determines the compression pressure. This parameter greatly influences the pellet quality and SEC [1, 24]. In the process of pelleting, the loose biomass is initially precompressed and deformed. The precompressed biomass further enters the die and balances the die's frictional forces as it extrudes through it [25]. To have a steady state pellet production, the roller pressure should be higher than both the compression pressure and frictional resistances. High fictional resistances in the die result in unsteady state pellet production and can increase the SEC.

High moisture pelleting

Tumuluru [15] and Tumuluru et al. [26] have indicated that pelleting corn stover feedstock at a high moisture content and further drying of the high moisture pellets at lower temperature of about 70–80°C using a grain dryer can be cost and energy efficient. Yancey et al. [27] and Pirraglia et al. [28] analyzed the energy consumption of various unit operations in a pellet production process and concluded that biomass drying from an initial feedstock moisture content of 30% (w.b.) to final moisture content of 10% (w.b.) using a rotary dryer consumes about 65–70% of the energy in the pellet production process, whereas pelleting requires approximately 9–13% of the energy. Sakkampang and Wongwuttanasatian [29] indicated that drying of biomass is the major energy consumer of the briquetting process. Their studies indicated that drying of biomasses such as rice husk, sawdust, sugarcane bagasse, and sugarcane leaf took about 2.25 MJ/kg, whereas the energy consumption for briquetting process was in the range of 0.04–0.10 MJ/kg. Figure 1 indicates the energy consumption reported by Yancey et al. [27] for the various unit operations like stage-1 and stage-2 grinding, drying, and pelleting during a conventional pellet production process. Tumuluru et al. [26] conducted techno-economic analyses on conventional and high moisture pelleting and briquetting production processes and concluded that high moisture pelleting offers significant energy savings (>50%) compared to conventional pelleting and briquetting processes. The low drying temperature associated with the high moisture pellet production process results in reducing the fuel cost, and in turn, the greenhouse gas emissions.

Objective

A continuous pelleting system, like a flat die pellet mill is typically used to develop a process at a laboratory scale [1]. The data on the continuous pelleting system will be helpful in understanding the effect of process conditions on the quality attributes and energy consumption. Tumuluru et al. [1] in their review on different densification systems to produce commodity type product from biomass indicated that the operating principle of flat die and commercial scale ring die pellet mill are the same, which makes it easier to scale up the process to a commercial scale system. Not much literature is available on how herbaceous biomass, such as corn stover, behaves during pelleting at high feedstock moistures of >28% (w.b.) in a continuous pelleting system. The study conducted by Tumuluru [15] on pelleting high moisture corn stover in a laboratory scale, flat die pellet indicated that pellets with bulk density (>500 kg/m^3) and durability (>95%) can be produced using an 8 mm pellet die.

The diameter of the pellets used in Europe varies between 6 and 12 mm. The 6 mm diameter pellet is preferred by the residential and commercial market since it generates less dust [2] and fits to different boiler configurations for steam generation. Six millimeter pellets also have higher packing density compared to other bigger die pellets. A study conducted by Tumuluru [15] on high moisture corn stover was done using an 8 mm pellet die and did not include data on the SEC. In his studies, the data on expansion ratio, which is an important physical property when extruding high moisture feedstock through a narrow constricted pellet die hole, is not discussed. Many researchers have identified expansion ratio as a good index to understand the effects of high moisture, and preheating temperature on the density of the food and feed pellets [22, 30–33]. Tumuluru [15] study on pelleting of high moisture corn stover indicated that lower (40 Hz) and higher (60 Hz) die speed had significant impact on pellet properties like density and durability. The specific objective of the present research is to understand how feedstock moisture content in the range of 28–38% (w.b.) and preheating temperature of 30–110°C at two die rotational speeds (40 and 60 Hz) impact the quality attributes, such as pellet moisture content; unit, bulk and tapped density; durability; expansion ratio; and SEC of pellets produced in a flat die pellet mill fitted with a 6 mm die.

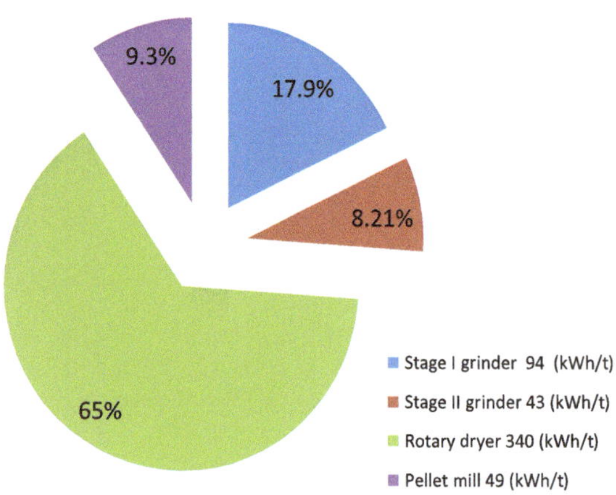

Figure 1. Energy consumption for the different unit operations for the pelleting of a lodgepole pine biomass.

Material and Methods

Feedstock

Harvested corn stover was procured from farms in Iowa in the form of bales and was initially ground to bigger particle sizes using a 50.8 mm screen with a Vermeer HG200 grinder (Vermeer Corporation-Agriculture, Pella, IA). The ground material was further reduced to smaller grinds (4.8 mm) using a hammer mill (Bliss Eliminator Hammer Mill, Model E-4424-TF) manufactured by Bliss Industries, Ponca City, OK [15, 27]. The ground material was evaluated for moisture content as well as bulk and tapped density using American Society of Agricultural and Biological Engineers (ASABE) standards [34]. The ground samples were further stored in airtight containers prior to pelleting tests.

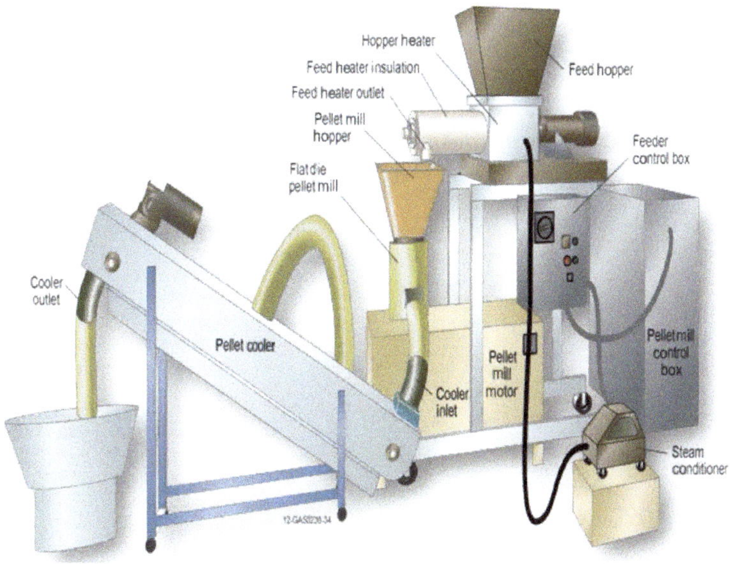

Figure 2. Flat-die pellet mill with other accessories (adapted from [15]).

Pelleting

A laboratory-scale ECO-10 (Colorado Mill Equipment, Canon City, CO, USA) flat-die pellet mill was used for the present studies (see Fig. 2). Tumuluru [15] has discussed in detail the pellet mill configuration. The pellet mill was equipped with a 10 HP, 460-V, three-phase motor. The hopper and screw feeder of the pellet mill were provided with flexible tape heaters, temperature controllers, and thermocouples to preheat the biomass to desired temperatures. A variable frequency drive (VFD) was provided to the pellet die to control the rotational speed and to the feeder to control the throughput rate. A horizontal pellet cooler was provided to cool the warm pellets coming out of the pellet die. Tumuluru [15] has discussed in detail the process for pelleting high moisture corn stover. About 2–3 kg of raw corn stover was mixed in a ribbon blender (RB 500, Colorado Mill Equipment, Canon City, CO, USA) with a calculated quantity of water to adjust moisture content to desired levels (28%, 33%, and 38%, w.b.). Experiments were conducted at different process conditions as indicated in Table 1. The moisture-adjusted corn stover was stored overnight in a refrigerator set at 4–5°C to equilibrate. This material was further loaded to the feed hopper of the pellet mill where it was preheated to different temperatures for short times (≈2–4 min) prior to pelleting. The feeding was carried out uniformly to assure there were no flow irregularities inside the pelletizer and to maintain a steady-state output. The pellets produced after cooling still had high moisture and were further dried in a mechanical oven at 70°C for about 3 h to reduce moisture content to <9% (w.b.). This makes pellets storable without any degradation. The dried pellets were further analyzed for physical properties like unit, bulk and tapped density; durability; and expansion ratio. The mill was provided with a power meter (Model no: APT-48T-MV-220-420; NK Technologies, San Jose, CA, USA). The percent motor power, pellet mill motor current, voltage, and die speed data were logged into a computer using LabVIEW software (Austin, TX, USA). Further, the percent motor power data collected for different experimental runs (Table 1) were used to calculate the power (kW) consumption.

Table 1. Experiments conducted at 40 and 60 Hz die speed at different feedstock moistures and preheating temperatures.

Experiment no.	Process variables	
	Feedstock moisture content (w.b.) (x_1)	Preheating temperature (°C) (x_2)
1	28	30
2	33	30
3	38	30
4	28	70
5	33	70
6	38	70
7	28	110
8	33	110
9	38	110

Physical properties

ASABE standards [34] were used to measure moisture content and bulk and tapped densities of moisture adjusted corn stover. The same standard was used to measure the pellets physical properties like unit, bulk, tapped density and durability. Brief descriptions of the methods followed are discussed below.

Moisture content

About 25–50 g of raw and pelleted corn stover samples were dried in a convective oven at 105°C for 24 h [34]. The samples were weighed before and after drying, and moisture content was calculated based on the change in weight. The values reported are in wet basis (w.b.) and are an average of three measurements.

Unit density

Unit density was measured to determine the density of individual pellets. The two ends of the pellets were made smooth with sandpaper. A vernier caliper was used to measure length and diameter of the pellet, and mass of the pellet was measured by using a balance with a 0.01 g precession [34]. Unit density was calculated by dividing the mass of an individual pellet by its volume. The reported values are an average of 10 pellets randomly selected from each experimental run.

Bulk and tapped density

The bulk density of the pellets produced was determined using a cylinder with a height to diameter (H/D) ratio based on ASABE Standard S269.4 [34]. Pellets were poured into the cylindrical container until they overflowed, and the surface of the cylinder was leveled by striking a straight edge across the top. The weight of the material and the cylinder volume were recorded. In case of tapped density, the loosely filled cylinder was tapped on a bench approximately five times, and the cylinder was again filled with pellets and leveled by striking the top. The reported values are an average of four measurements.

Durability

A pellet durability tester (Seedburo Equipment Co., Des Plaines, IL) with four compartments was used to measure the durability of the pellets. Fines were removed from pellets by sieving, and about 500 g of each sample was placed in each compartment of the tumbler. After tumbling at 50 rev/min for 10 min, the sample was removed and sieved using a 4.7 mm screen [34]. Pellet durability is defined as the ratio of weight of the tumbled sample after sieving to the total weight of the sample used for testing. The reported values are an average of four measurements.

Expansion ratio

The expansion ratio was calculated by using the pellet diameter data collected for the unit density measurement. The diametrical expansion ratio was calculated using the below equation [22].

$$ER = \frac{D^2}{d^2}, \tag{1}$$

where D is the diameter of the pellet extruded (mm) and d is the diameter of the die (mm). The reported values are an average of 10 measurements.

Specific energy consumption (SEC)

In the present study, the SEC for the pelleting process was calculated. For each pellet mill run, the percent motor data were logged using LabVIEW software. The percent motor power data were further converted to power (kW) based on Equation (2). This equation was provided by the power meter vendor. The size of the pellet mill motor is 10 HP, and approximate power loss due to VFD is about 114 W. Also, the vendor has indicated approximately 15% variability in the power (kW) consumption calculated using the percent motor power data.

$$\text{Power (kW)} = \frac{\left(\frac{\%\text{motor power}}{100}\right) \times (\text{Normal size of the motor horsepower}) \times (746\ \text{Watts/horsepower}) + (114)}{1000}. \tag{2}$$

The no-load power (kW) data of the pellet mill were recorded by running the pellet mill empty at 40 and 60 Hz die speed. This allowed determination of the net power required to pelletize the biomass material for each experimental run. SEC was calculated using the following formula: (eq. 3).

$$\text{SEC} = \frac{(\text{Full load Power (kW) - No load power (kW)}) \times \text{time (hr)}}{\text{weight of material (kg)}} \times 1000 = \frac{\text{kWh}}{\text{ton}}. \tag{3}$$

Results and Discussion

Initial raw material properties

The average moisture content of the raw materials was about 8.39% (w.b.), while the measured bulk and tapped densities were about 111 and 139 kg/m^3, respectively. Increasing the moisture of the corn stover to 28–38% (w.b.) brought a slight increase in bulk and tapped density by about 10–20 kg/m^3.

Pellet moisture content (%, w.b.)

Figure 3 indicates the effect of feedstock moisture content, die speed, and preheating on the pellet moisture content. Results indicated that moisture content of feedstock was reduced by about 5–10% (w.b.) after pelleting and cooling. The resulting pellets had moisture contents in the range of about 20–28% (w.b.). The present pellet moisture results have corroborated with earlier work on pelleting high moisture corn stover using an 8 mm pellet die [15]. At a feedstock moisture content of 28% (w.b.), die speed of 40 Hz, and preheating temperature of 110°C, the pellet moisture content observed was 21.29% (w.b.). By increasing the die speed to 60 Hz, the change in pellet moisture content was marginal (20.19% [w.b.]). At a higher feedstock moisture content of 38% (w.b.) and preheating temperature of 110°C, the pellet moisture observed was 27.97% (w.b.). The decrease in feedstock moisture content by about 10% (w.b.) when pelletized and cooled can be due to frictional heat developed in the pellet die, which can result in moisture flash off as the pellet is extruded. Further cooling of these pellets in a pellet cooler has also helped to remove some of the surface moisture. The present results have corroborated with our earlier research on pelleting of high moisture corn stover using an 8 mm pellet mill die [15], where feedstock moisture content influenced the final pellet moisture content of the pellets, and lower feedstock moisture content and higher preheating temperatures lowered the final pellet moisture content. A pilot scale pelleting study conducted on AFEX pretreated corn stover at about 19% (w.b.) moisture content using a ring die pellet mill (1 ton/h) indicated a decrease of about 9–10% (w.b.) feedstock moisture after pelleting [20]. The final moisture content of the AFEX pellets was found in-between 9% and 10% (w.b). Tumuluru et al. [1], in their review on biomass densification, indicated that moisture in the biomass feedstock helps in solid bridge formation by van der Waals forces and also lower the glass-transition temperature of some of the biomass components like lignin, protein and starch, which further takes an active part in the binding process. Furthermore, the presence of moisture, temperature and pressure in the pellet die enables some of the water soluble carbohydrates in the biomass to get solubilized and act as natural binder. The other biomass components which will be activated by moisture and temperature during the pelleting process are protein (which undergoes unfolding and denaturation) and starch

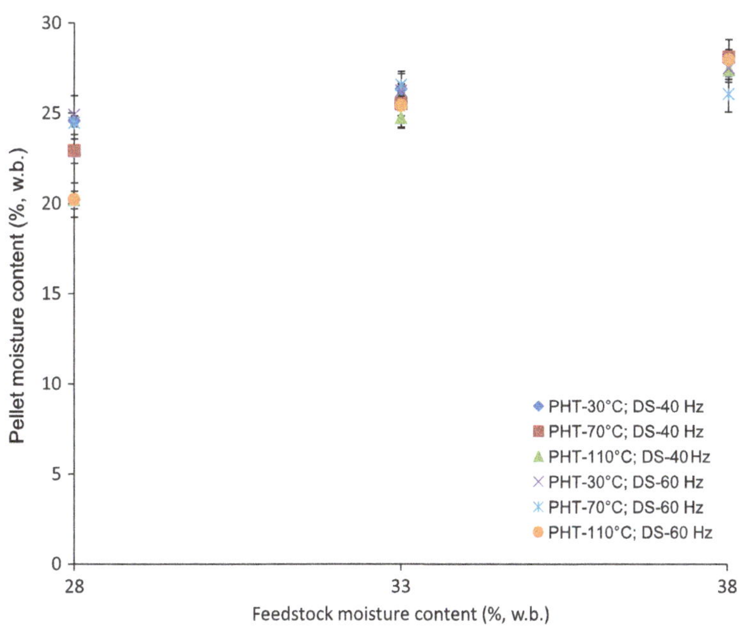

Figure 3. Effect of preheating, die speed, and feedstock moisture content on pellet moisture content.

(gelatinization) which result in formation of protein starch complexes.

Unit, bulk and tapped density (kg/m^3)

The highest unit density recorded was 1065 kg/m^3 at a lower moisture content of 28% (w.b.) with higher die speeds of 60 Hz, whereas bulk and tapped density values were 554 and 611 kg/m^3. Figures 4–6 indicate changes in unit, bulk and tapped densities with respect to preheating, die speed, and feedstock moisture content. Lower (28%, w.b.) and medium (33%, w.b.) feedstock moisture content and higher die rotational speed (60 Hz) maximized these properties. Typical unit densities of biomass pellets can be as high as 1100–1300 kg/m^3, and bulk densities are about 700 kg/m^3 [1, 2]. In their studies on pelleting of distillers dried grains with solubles, Tumuluru et al. [10] concluded that both unit and bulk densities of pellets made from the distillers dried grains are dependent on feed moisture and die temperature, where a maximum unit density of 1200 kg/m^3 and bulk density of 700 kg/m^3 is achievable at die temperature of about 100°C and feed moisture content of about >5%. Hall and Hall [35], in their studies of densification of Bermuda grass, indicated that moisture, pressure, and temperature influence density. They concluded the upper limits of the moisture content at which a certain pressure was able to produce a certain density also increased by preheating the biomass. Briquetting studies on *Caragana korshinskii* Kom biomass by Zhang and Guo [36] indicated that increasing the moisture content from 5–17% (w.b.) decreased the density. They found that moisture content is the most significant variable influencing the density. In the present study, it has been found that at a feedstock moisture content of 38% (w.b.) there is about 30–40% reduction in unit, bulk and tapped density compared to pellets made at 28% and 33% (w.b.) moisture content. Pelleting studies conducted by Mani et al. [37] on four biomass grinds such as wheat and barley straws, corn stover, and switchgrass at different moisture contents of 5, 10, and 15% (w.b.), indicated that pellets made at 5% (w.b.) had higher unit density as compared to pellets made at 15% (w.b.). This trend has corroborated with the present results where a lower feedstock moisture content of 28% (w.b.) resulted in higher unit, bulk and tapped density. Similar observations were reported by Rehkugar and Buchele [38] in their studies on forage wafers. They found that increasing the moisture content to 25% (w.b.) reduced the density. Pilot scale pelleting studies on AFEX pretreated corn stover at 19% (w.b.) moisture content resulted in bulk density of about 672 kg/m^3, whereas pelleting municipal solid waste (consisting of 30% plastic and about 70% paper and non-corrugated cardboard) at 15–25% (w.b.) moisture content produced pellets with density value of 544 kg/m^3 [20]. Both laboratory and pilot scale pelleting and briquetting indicated that the density of pellets or briquettes decreases with increase in feedstock moisture content. The possible explanation for lower bulk density at higher feedstock moisture content can be due to volumetric expansion of the pellet or briquettes due to moisture flash-off as it exits the die. Tumuluru et al. [13], in their studies on briquetting of wheat, oat, barley, and canola straws, indi-

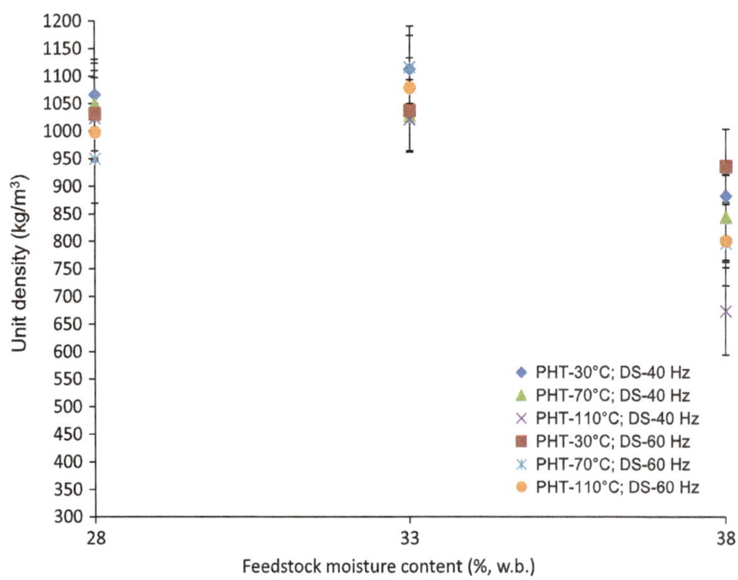

Figure 4. Effect of die speed, preheating, and feedstock moisture content on unit density.

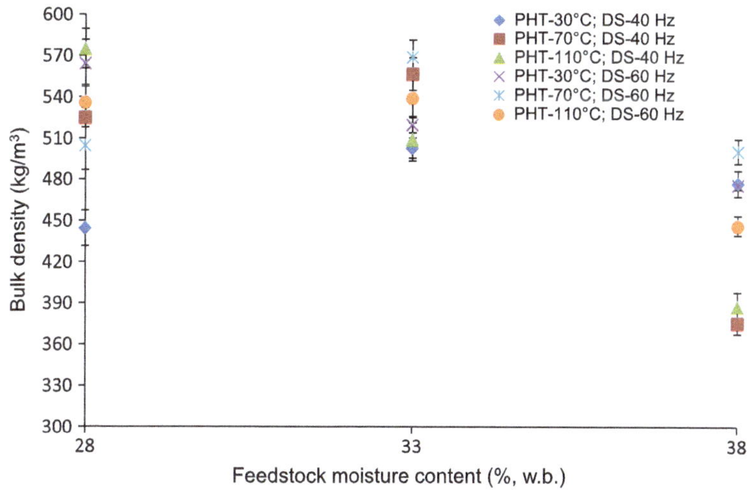

Figure 5. Effect of die speed, preheating, and feedstock moisture content on bulk density.

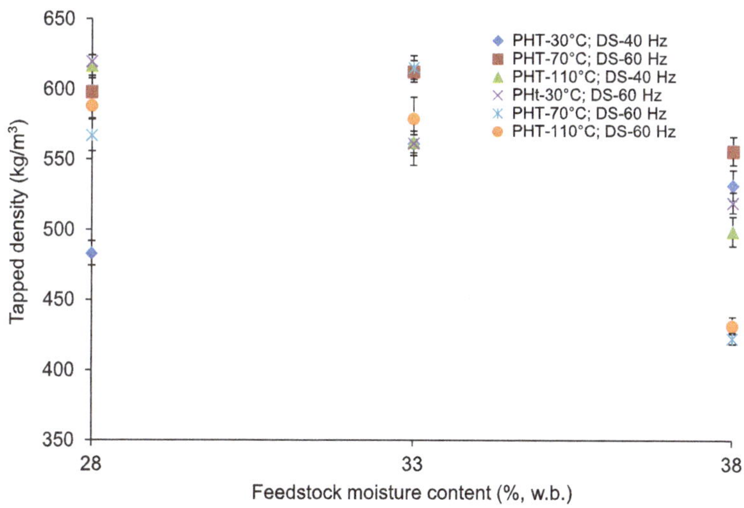

Figure 6. Effect of die speed, preheating, and feedstock moisture content on tapped density.

cated that increasing the moisture content to 15% (w.b.) and briquetting temperature to 130°C resulted in lower unit density. They reasoned that at higher feedstock moisture contents dimensional instability is higher for briquettes in both lateral and axial directions. In some cases, dimensional instability tends to be very marginal depending on the combination of the material and operating variables. A similar expansion trend was also reported by Mani et al. [39] on the compaction of corn stover and by Al-Widyan et al. [40] on the stability of olive cake briquettes.

Durability (%)

Durability values indicate the impact and shear resistances biomass pellets can withstand during transportation. Durability is one of the mandatory standards for pellets based on the Pellet Fuel Institute (PFI), USA and the European Committee for Standardization (CEN) [1, 2]. Pellets for international transport are expected to have durability values >97.5%, to retain the pellet integrity during long distance transportation. If pellets have lower durability values they can break and result in dust genera-

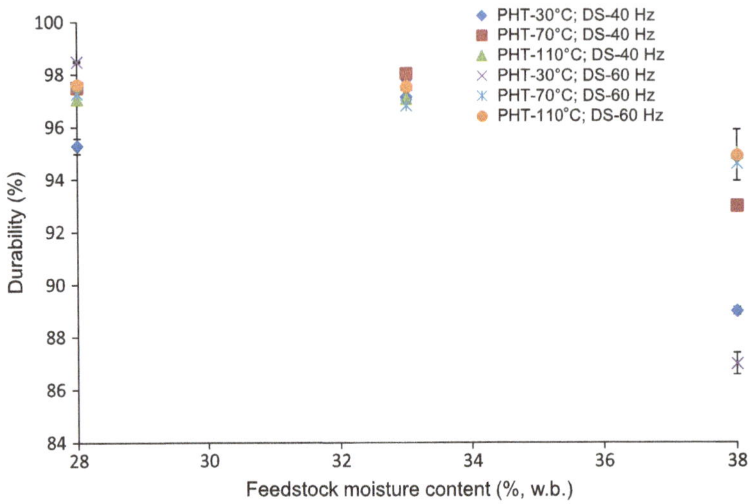

Figure 7. Effect of preheating, die speed, and feedstock moisture content on pellet durability.

tion which can lead to safety issues like spontaneous combustion, higher storage off-gas emissions, and revenue loss for the pellet producers. Figure 7 shows the plot drawn for durability values of pellets produced at different experimental conditions. From this figure, it is clear that feedstock moisture content of 28 and 33% (w.b.), and preheating temperatures of 70 and 110°C increased the durability values >95% (w.b.). Castellano et al. [41] observed that higher power and die temperatures increased durability values. In the present study, higher durability values of >95% observed can be due to the lowering of the glass-transition temperature of lignin and the increase in water soluble carbohydrates, which might have acted as natural binders during the pelleting process. Increasing the feedstock moisture content to 38% (w.b.) reduced the durability values to <90% at both 40 and 60 Hz die speeds. According to Koullas and Koukios [42], water in wheat straw helps in binding to make more durable briquettes. However, if the moisture content in the biomass feedstock crosses a threshold value, it acts more like a lubricant and results in less binding and more fines generation during the pellet or briquette production process. Kaliyan and Morey [21, 43] suggested that the binding of biomass during densification is strongly affected by the glass-transition temperature of lignin, which in turn is dependent on moisture content and preheating temperatures. Zhang and Guo [36], Tumuluru et al. [13], and Tumuluru [15] have found that the durability of *C. korshinskii* Kom and agricultural straw biomass briquettes and pellets made from high moisture corn stover has increased with an increase in the preheating temperature. This result has corroborated with present findings where higher preheating temperature of >70°C resulted in higher durability pellets.

Expansion ratio

The measurement of the expansion ratio helps to understand the diametrical expansion of the extruded biomass pellets. Figure 8 indicates the effects of feedstock moisture content and preheating temperature on the expansion ratio. Expansion ratio of the biomass pellets made from low moisture (10–12%, w.b.) feedstock using conventional pelleting process is typically ≤1.0. In the present study, increasing the feedstock moisture content to 38% (w.b.) and preheating temperatures to 110°C increased the expansion ratio of pellets to >1.26, whereas at a lower feedstock moisture content of 28% (w.b.), a minimum expansion ratio of <1.06 was observed. The results also indicated that the expansion ratio values did not change much by increasing the moisture content from 28% to 33% (w.b.). In general, lower expansion ratio values are preferred for fuel pellets as they result in higher densities. Poddar et al. [16] indicated that biomass pellets expanded after pelleting at higher feedstock moistures. They have reasoned that heat generated during pelleting, owing to friction between the biomass particles and prechannel of the die hole as well as the application of pressure, results in expelling of moisture as the pellet exits the die. Mohesenin and Zaske [44] and Faborode [45] indicated that high moisture in the feedstock makes briquettes expand, thereby resulting in higher residual stresses in the briquettes after ejection from the die, causing an elastic spring back or expansion. Kaliyan and Morey [43] indicated that an increase in feedstock moisture content subsequently increased the expansion of the briquettes at higher preheating temperatures. Figure 9 is drawn to understand the relationship between expansion ratio and

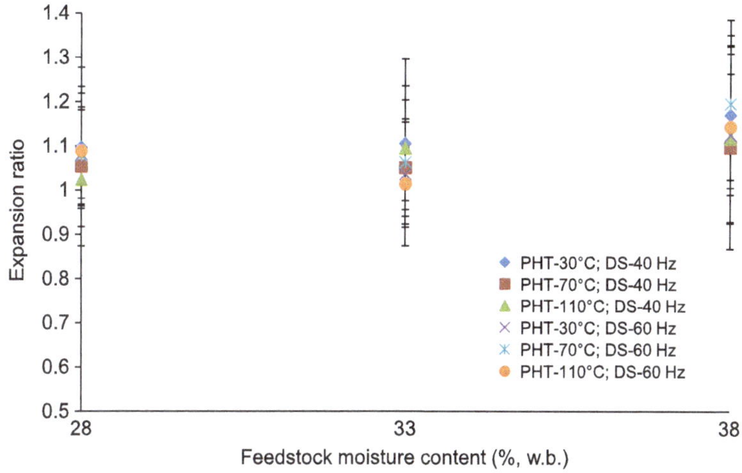

Figure 8. Effect of preheating, die speed, and feedstock moisture content on expansion ratio.

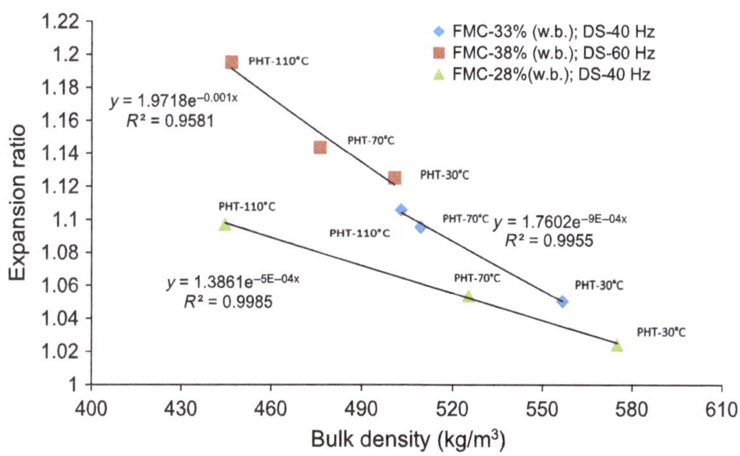

Figure 9. Effect of expansion ratio on bulk density at different die speed, preheating temperature and feedstock moisture content.

bulk density of the pellets produced. It is very clear from the figure that higher expansion ratio reduced the bulk density of the pellets. Higher feedstock moisture content led to a higher expansion ratio and, in turn, lowered the bulk density. The exponential equation fitted (Fig. 9) indicated that at lower feedstock moisture of 28% (w.b.), the expansion ratio is lower, whereas at a higher feedstock moisture content (38%, [w.b.]), expansion ratio is higher.

Specific energy consumption (SEC) (kWhr/ton)

In general, the SEC during pelleting is governed by feedstock moisture content, die dimensions, feedstock composition, feeding rate, and die rotational speed [1]. Higher die rotational speed can increase the throughput of the pellet mill and reduce the SEC. Figure 10 indicates the effect of preheating and feedstock moisture content on the SEC (kWh/ton). It is very clear from the plot that at feedstock moisture content of 33% (w.b.), and increasing the die speed to 60 Hz and preheating temperature to 70°C decreased the SEC to about 75 kWh/ton. At feedstock moisture content of 38% (w.b.), die speed of 40 Hz and preheating temperature of 70 and 110°C the SEC increased to about 275 kWh/ton. In the present study at higher feedstock moisture content of 38% (w.b.), slower feeding rates were used to produce more pellets and fewer fines. Lowering the feed rate at higher feedstock moisture has resulted in reduced throughput and increased the energy consumption. Also, die speed had a prominent effect on the throughput and in turn the SEC, where at

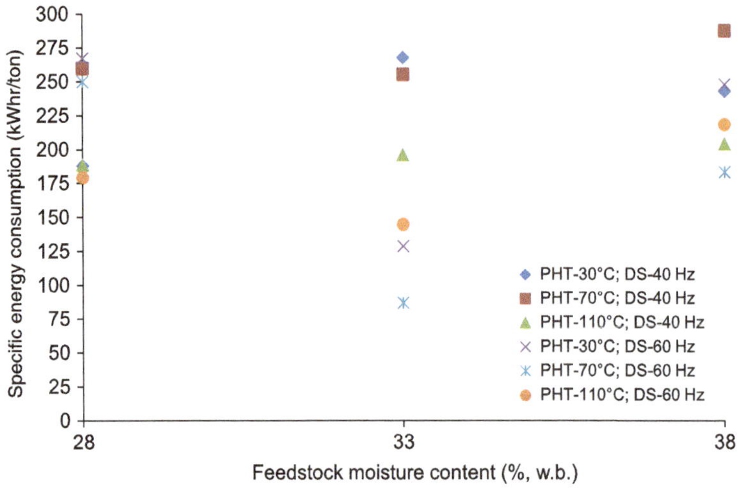

Figure 10. Effect of preheating, die speed, and feedstock moisture content on specific energy consumption.

40 Hz the observed SEC values were higher compared to 60 Hz. Reed et al. [23] indicated that increasing the preheating temperature reduces the SEC. According to Holm et al. [5], the length and diameter of the pellet die has a significant effect on the compression pressures. Increasing the length of the pellet die increases the pelleting pressure, whereas increasing the diameter decreases the pelleting pressure. So for the same L/D ratio, a bigger pellet die like 8 mm may need lower compression pressure, which might result in lowering the SEC compared to smaller die (≤6 mm).

Discussion

The current research results on density and durability for 6 mm pellets were compared with the 8 mm pellets produced by Tumuluru [16]. The densities of the pellets produced using a 6 mm die are slightly lower by about 20–30 kg/m^3 compared to an 8 mm die. This reduction in density can be due to a higher expansion of pellets as the high moisture biomass material exits through a smaller, constricted die size. High moisture biomass at temperatures >100°C, when passing through a constricted hole, results in moisture loss due to flash-off. Shankar and Bandyopadhyay [22] and other researchers have indicated that extruding high moisture (>30, w.b) food at a preheating temperature >100°C results in expansion of the extrudate due to flash-off of moisture as it leaves the die. The moisture flash-off from the extrudates can be higher when the die dimensions are smaller. In the case of durability for both 6 and 8 mm die corn stover pellets, pelleting at high moisture feedstock (28–38%, w.b.) resulted in durability values in the range of 85–96%. According to the Pellet Fuel Institute (PFI), USA, and the European Committee for Standardization (CEN), durability and bulk density are normative specification for grading of pellets [2]. Pellets with durability values >96.5% and bulk density >640 kg/m^3 are designated as super premium pellets based on PFI standards; whereas in the case of CEN the durability and bulk density values should be >97.5% and >700 kg/m^3 [2] for international transport. Both PFI and CEN have other density and durability standards for lower grade pellets [2]. In general to transport pellets for shorter distances (e.g., interstate), very high density and durability values may not be required. This gives an opportunity to customize the pellet production process to produce pellets with varying durability and density values to meet different transportation scenarios. The present study has indicated that pellets with different density and durability can be produced at different feedstock moistures, die speeds, and preheating temperatures and these process conditions have a significant impact on the pelleting energy. For both 6 and 8 mm die sizes, lower to medium moisture contents of 28–33% (w.b.) and preheating temperatures of about 70°C resulted in higher durability values of >90% and bulk density of >520 kg/m^3.

The present results indicated that pellets produced using high moisture corn stover (28–38%, w.b), has resulted in lower density (575–375 kg/m^3) compared to pellets produced using the same material at 10% (w.b.) moisture content using conventional or low moisture pelleting process (645 kg/m^3) [27]. The major reason for loss of pellet density at higher feedstock moisture content is due to diametrical expansion of the pellet as it exits the die. The expansion is a common phenomenon when extruding high-moisture feedstock. One way to reduce diametrical expansion of a pellet or briquette at higher feedstock moisture content is to add natural or commercial pellet binders. Tumuluru et al. [46] studies on extrusion of high-protein foods indicated that higher levels of proteins in the food mix can reduce the expansion and increase the hardness values of food pellets. The same authors indicated that the presence of higher amounts of starch in the biomass can increase the expansion ratio thereby reducing the density of the extrudates. Tumuluru et al. [13] in their studies found that canola straw which is high in protein content produced denser and more durable briquette compared to wheat, oat, and barley. They also found that the diametrical expansion or the spring back is lower in canola straws. Another important pellet mill parameter that has a great impact on the quality and energy consumption of the pelleting process is the L/D ratio of the pellet die. The L/D ratio controls the residence time of the material in the die and impact the quality attributes like density and durability. The L/D ratio also impacts the extrusion energy needed in the pelleting process. For the same die diameter, increasing the length of the pellet die increases the extrusion energy. Mewes [47] indicated that about 37–40% of the energy used during pelleting is needed to compress the biomass, whereas as the remaining energy of about 60–63% is needed to extrude the biomass out of the pellet die. Our further research is aimed at understanding the impact of binders and the L/D ratio of the pellet die on the pellet quality attributes (density and durability) and energy consumption of high moisture pelleting process.

Techno-economic analysis conducted by Tumuluru et al. [26] on conventional and high moisture pelleting indicated that there is a significant advantage in pelleting biomass at high moisture in terms of cost and energy. This process also enables the use of agricultural machinery like grain dryers which have lot of advantages compared to rotary dryers [15]. Low cost and energy-efficient dryers (such as grain or belt dryers) operate at low temperatures (60–80°C), and can be operated using low-quality heat. Low operating temperatures of the grain dryer helps to reduce the volatile organic compounds (VOC) emissions and can have a significant impact on the greenhouse gas emissions. The low drying temperatures associated with grain dryers will also reduce the risk associated with fire and explosion. Drying biomass pellets at a low temperature and for longer drying times also ensures that the product is dried more uniformly. Additionally, our studies are aimed at how the drying times impact the moisture distribution within the pellets and the quality of the pellets (density and durability).

Recently, chemical pretreatment methods like leaching, acid, alkali, and ammonia fiber explosion (AFEX) are gaining importance for both thermochemical and biochemical conversion applications [17, 48–51]. These pretreatments result in high moisture in the biomass feedstock, which need to be dried to a final moisture of <12% (w.b.) to increase their storage stability and for further pelleting or briquetting. Drying these chemically pretreated biomass materials, lowers the bulk density compared to their original state, which can be a major limitation to transport over longer distances. One way to reduce the preprocessing costs and overcome the low density limitation is by pelleting high moisture chemically pretreated biomass and further drying the high moisture pellets using a grain or belt dryer. This process helps to reduce the preprocessing costs, improve the storage stability, and increase the transportation efficiencies of chemically pretreated biomass. Our recent studies on pelleting the AFEX pretreated corn stover at 26% and 19% (w.b.) moisture content in a laboratory scale (flat die) and a pilot scale (ring die) pellet mill indicated that good quality pellets with high density (>600 kg/m^3) and durability (>98%) can be produced [17, 20].

Conclusions

The experimental data indicated that good quality pellets in terms of density and durability can be made using high moisture corn stover in a flat die pellet mill fitted with a 6 mm die. Different pelleting conditions resulted in the different quality of pellets as well as different SEC. Feedstock moisture content had a significant impact on the quality of the pellets. High moisture contents resulted in lower quality in terms of density and durability. Lower unit, bulk and tapped densities were observed at higher feedstock moisture contents, and the expansion ratios observed were higher. The present findings indicate that a higher expansion ratio results in lower bulk density. At 38% (w.b.) feedstock moisture content there was a decrease in bulk density by about 30–40% compared to pellets made at 28% and 33% (w.b.) feedstock moisture content. The high quality of pellets in terms of density and durability at lower SEC can be produced at feedstock moisture contents of 33% (w.b.), higher die speeds of 60 Hz, and low and medium preheating temperature of 30 and 70°C.

Acknowledgments

The author would like to acknowledge Kevin L. Kenney, Lead Bioenergy Program, Idaho National Laboratory, for reviewing the manuscript, Craig C. Conner for supporting experimental work, Matthew Anderson and Rod Shurtliff for instrumenting the pellet mill, Dave Pace for providing pilot scale pelleting data and Brendi Heath for Editorial Support. This work was supported by the Department of Energy, Office of Energy Efficiency, and Renewable

Energy under the Department of Energy Idaho Operations Office Contract DE-AC07-05ID14517. Accordingly, the publisher, by accepting the article for publication, acknowledges that the U.S. government retains a nonexclusive, paid-up, irrevocable, worldwide license to publish or reproduce the published form of this manuscript, or allow others to do so, for U.S. government purposes.

Conflict of Interest & Author disclosure statement

The author declare no conflict of interest exist. This information was prepared as an account of work sponsored by an agency of the U.S. government. Neither the U.S. government nor any agency thereof, nor any of their employees, makes any warranty, express or implied, or assumes any legal liability or responsibility for the accuracy, completeness, or usefulness of any information, apparatus, product, or process disclosed, or represents that its use would not infringe privately owned rights. References herein to any specific commercial product, process, or service by trade name, trademark, manufacturer, or otherwise, do not necessarily constitute or imply its endorsement, recommendation, or favouring by the U.S. government or any agency thereof. The views and opinions of the authors expressed herein do not necessarily state or reflect those of the U.S. government or any agency thereof.

References

1. Tumuluru, J. S., C. T. Wright, J. R. Hess, and K. L. Kenney. 2011. A review of biomass densification systems to develop uniform feedstock commodities for bioenergy application. Biofuel. Bioprod. Bior. 5:683–707.
2. Tumuluru, J. S., S. Sokhansanj, C. J. Lim, X. T. Bi, A. Lau, S. Melin, et al. 2010a. Quality of wood pellets produced in British Columbia for export. Appl. Eng. Agric. 26:1013–1020.
3. Smith, I. E., S. D. Probert, R. E. Stokes, and R. J. Hansford. 1977. The briquetting of wheat straw. J. Agr. Eng. Res. 22:105–111.
4. Tabil, L. G.. 1996. Binding and pelleting characteristics of alfalfa. Ph.D. diss., Department of Agricultural and Bioresource Engineering, University of Saskatchewan, Canada.
5. Holm, J. K., U. B. Henriksen, J. E. Hustad, and L. H. Sorensen. 2006. Toward an understanding of controlling parameters in softwood and hardwood pellets production. Energ. Fuels 20:2686–2694.
6. Tabil, L. G., and S. Sokhansanj, S. 1996. Process conditions affecting the physical quality of alfalfa pellets. Trans. ASAE. 12:345–350.
7. Dobie, J. B. 1959. Engineering appraisal of hay pelleting. Agr. Eng. 40:72–76.
8. Mani, S., L. G. Tabil, and S. Sokhansanj. 2006. Specific energy requirement for compacting corn stover. Bioresour. Technol. 97:1420–1426.
9. Li, Y., and H. Liu. 2000. High–pressure densification of wood residues to form an upgraded fuel. Biomass Bioenerg. 19:177–186.
10. Tumuluru, J. S., L. G. Tabil, A. Opoku, M. R. Mosqueda, and O. Fadeyi. 2010b. Effect of process variables on the quality characteristics of pelleted wheat distiller's dried grains with soluble. Biosyst. Eng. 105:466–475.
11. Tabil, L. G., and S. Sokhansanj. 1996. Compression and compaction behavior of alfalfa grinds, Part 1: compression behavior. Powder Handl. Proces. 8:17–23.
12. Ndiema, C. K. W., P. N. Manga, and C. R. Ruttoh. 2002. Influence of die pressure on relaxation characteristics of briquetted biomass. Energ. Convers. Manag. 43:2157–2161.
13. Tumuluru, J. S., L. G. Tabil, Y. Song, K. L. Iroba, and V. Meda. 2015. Impact of process conditions on the density and durability of wheat, oat, canola, and barley straw briquettes. Bioenerg. Res. 8:388–401.
14. Zafari, A., and M. H. Kianmehr. 2013. Factors affecting mechanical properties of biomass pellets from compost. Environ. Technol. 35:478–486.
15. Tumuluru, J. S. 2014. Effect of process variables on the density and durability of the pellets made from high moisture corn stover. Biosyst. Eng. 119:44–57.
16. Poddar, S., M. Kamruzzaman, S. M. A. Sujan, M. Hossain, M. S. Jamal, M. A. Gafur, et al. 2014. Effect of compression pressure on lignocellulosic biomass pellet to improve fuel properties: higher heating value. Fuel 131:43–48.
17. Hoover, A. N., J. S. Tumuluru, F. Teymouri, M. Moore, and G. Gresham. 2014. Effect of pelleting process variables on physical properties and sugar yields of ammonia fiber expansion (AFEX) pretreated corn stover. Bioresource Technol. 164:128–135.
18. Sarkar, M., A. Kumar, J. S. Tumuluru, K. N. Patil, and D. D. Bellmer. 2014a. Thermal devolatilization kinetics of switchgrass pretreated with torrefaction and densification. Trans. ASABE 57:1199–1210.
19. Sarkar, M., A. Kumar, J. S. Tumuluru, K. N. Patil, and D. D. Bellmer. 2014b. Gasification performance of switchgrass pretreated with torrefaction and densification. Appl. Energ. 127:194–201.
20. Pace, D. 2015. Pelleting of municipal solid wast and ammonia fiber explosion (AFEX) pretreated corn stover in a pilot scale ring die pellet mill. Biofuels Department, Chief Engineer, Biomass National User Facility, Idaho National Laboratory (Unpublished data).
21. Kaliyan, N., and R. V. Morey. 2009a. Factors affecting strength and durability of densified biomass products. Biomass Bioenerg. 33:337–359.

22. Shankar, T. J., and S. Bandyopadhyay. 2005. Process variables during single-screw extrusion of fish and rice-flour blends. J. Food Process. Pres. 29:151–164.
23. Reed, T. B., G. Trezek, and L. Diaz. 1980. Biomass densification energy requirements, Chapter 13. Pp. 169–177 *in* J. L. Jones, S. B. Radding, S. Takaoka, A. G. Buekens, M. Hiraoka, R. Overend, eds. Thermal conversion of solid wastes and biomass. American Chemical Society, Washington, DC. doi: 10.1021/bk-1980-0130
24. Stelte, W. 2011. Fuels pellets from biomass: processing, bonding, raw materials. Ph.D. thesis, National Laboratory for Sustainable Energy, Riso, Technical University of Denmark.
25. Winter, E. 1981. Fundamental considerations for preparing densified refuse derived fuel. U.S. Environmental Protection Agency, Technical Report EPA- 600/S2-81-180, Mun Env Res Lab, Cincinnati, OH.
26. Tumuluru, J. S., K. G. Cafferty, and K. L. Kenney. 2014. Techno-economic analysis of conventional, high moisture pelletization and briquetting process. American Society of Agricultural and Biological Engineer Annual Meeting, Paper No. 141911360, Montreal, Quebec Canada July 13 – 16, 2014, (doi: 10.13031/aim.20141911360).
27. Yancey, N. A., J. S. Tumuluru, and C. T. Wright. 2013. Drying, grinding and pelletization studies on raw and formulated biomass feedstock's for bioenergy applications. J. Biobased. Mater. Bioener. 7:549–558.
28. Pirraglia, A., R. Gonzalez, and D. Saloni. 2010. Technoeconomic analysis of wood pellets production from U.S. Manufactures. BioReosurces 5:2374–2390.
29. Sakkampang, C., and T. Wongwuttanasatian. 2014. Study ratio of energy consumption and gained energy during briquetting process for glycerin – biomass briquette fuel. Fuel 115:186–189.
30. Oke, M. O., S. O. Awonorin, and T. S. Workneh. 2013. Expansion ratio of extruded water yam (Dioscorea alata) starches using a single screw extruder. Afr. J. Agr. Res. 8:750–762.
31. Filli, K. B., I. Nkama, and V. A. Jideani. 2013. The effect of extrusion conditions on the physical and functional properties of millet-Bambara groundnut based fura. Am. J. Food Sci. Tech. 1:87–101.
32. Yuryey, V. P., D. V. Zasypkin, V. V. Alexeev, and A. N. Bogatyryev. 1995. Expansion ratio of extrudates prepared from potato starch-soybean protein mixtures. Carbohydr. Polym. 26:215–218.
33. Shankar, T. J., and S. Bandyopadhyay. 2004. Optimization of extrusion process variables using a genetic algorithm. Food Bioprod. Process. 82:143–150.
34. ASABE Standards. 2007. S269.4. Cubes, pellets, and crumbles – definitions and methods for determining density, durability, and moisture content. ASABE, St. Joseph, MO.
35. Hall, G. E., and C. W. Hall. 1968. Heated-die wafer formation of alfalfa and bermudagrass. Trans. ASAE 11:578–581.
36. Zhang, J., and Y. Guo. 2014. Physical properties of solid fuel briquettes made from *Caragana korshinskii* Kom. Powder Technol. 256:293–299.
37. Mani, S., L. G. Tabil and S. Sokhansanj. 2002. Compaction characteristics of some biomass grinds. AIC 2002 Meeting, CSAE/SCGR Program, Saskatoon, Saskatchewan, Canada, July 14–17, 2002.
38. Rehkugar, G. E., and W. F. Buchele. 1969. Bio-mechanics of forage wafering. Trans. ASAE 12:1–8.
39. Mani, S., L.G. Tabil, and S. Sokhansanj. 2004. Compaction of corn stover. ASAE Paper No 041160, presentation at the 2004 ASAE/CSAE Annual International Meeting.
40. Al-Widyan, M. I., H. F. Al-Jalil, M. M. Abu-Zreig, and N. H. Abu-Handeh. 2002. Physical durability and stability of olive cake briquettes. Can. Biosyst. Eng. 44:341–345.
41. Castellano, J. M., M. Gomez, M. Fernandez, L. S. Esteban, and J. E. Carrrasco. 2015. Study on the effects of raw materials composition and pelletization conditions on the quality and properties of pellets obtained from different woody and non woody biomasses. Fuel 139:629–636.
42. Koullas, D. P., and E. G. Koukios. 1987. Hot briquetting of wheat straw. Pp. 53–60 *in* P. Keller, ed. In Handling and processing of biomass for energy. Report and proceedings CNRE Bulletin (FAO), No. 18; European Cooperative Networks on Rural Energy. Workshop on Handling and Processing of Biomass for Energy, 1, Hamburg, Germany, F.R., 14–15 September 1987. FAO, Rome, Italy. Regional Office for Europe.
43. Kaliyan, N., and R. V. Morey. 2009b. Densification characteristics of corn stover and switchgrass. Trans. ASABE. 52:907–920.
44. Mohsenin, N., and J. Zaske. 1976. Stress relaxation and energy requirements in compaction of unconsolidated materials. J. Agr. Eng. Res. 11:193–205.
45. Faborode, M. O., and J. R. O. Callaghan. 1987. Optimizing the compression/briquetting of fibrous agricultural materials. J. Agr. Eng. Res. 38:245–262.
46. Tumuluru, J. S., S. Sokhansanj, S. Bandyopadhyay, and A. S. Bawa. 2013. Changes in moisture, protein and fat content of fish and rice flour coextrudates during single-screw extrusion cooking. Food Bioprocess Tech. 6:403–415.
47. Mewes, E. 1959. Berechung der druckverteilung an stroh- und heupressen [Calculation of the pressure distribution in straw and hay balers]. Landtechnische Forschung 9:160–170.
48. Balan, V., B. Bals, S. P. Chundawat, D. Marshall, and B. E. Dale. 2009. Lignocellulosic biomass pretreatment using AFEX. Methods Mol. Biol. 581:61–77.
49. Bals, B., C. Rogers, M. Jin, V. Balan, and B. E. Dale. 2010. Evaluation of ammonia fibre expansion (AFEX) pretreatment for enzymatic hydrolysis of switchgrass

harvested in different seasons and locations. Biotechnol. Biofuels 3:61–77.

50. Kumar, P., D. M. Barrett, M. J. Delwiche, and P. Stroeve. 2009. Methods for Pretreatment of Lignocellulosic Biomass for Efficient Hydrolysis and Biofuel Production. Ind. Eng. Chem. Res. 48:3713–3729.

51. Thompson, D. N., T. Campbell, B. Bals, T. Runge, F. Teymouri, and L. P. Ovard. 2013. Chemical preconversion: application of low-severity pretreatment chemistries for commoditization of lignocellulosic feedstock. Biofuels 4:323–340.

8

Amorphous single-junction cells for vertical BIPV application with high bifaciality

Nies Reininghaus, Clemens Feser, Benedikt Hanke, Martin Vehse & Carsten Agert

NEXT ENERGY EWE Research Centre for Energy Technology, University of Oldenburg, Carl-von-Ossietzky-Str. 15, 26129 Oldenburg, Germany

Keywords
a-Si, bifacial, multijunction solar cell, PECVD, thin film

Correspondence
Nies Reininghaus, NEXT ENERGY EWE Research Centre for Energy Technology, University of Oldenburg, Carl-von-Ossietzky-Str. 15, 26129 Oldenburg, Germany.
E-mail: nies.reininghaus@next-energy.de

Funding Information
No funding information provided.

Abstract

Solar cells used in building integration of photovoltaic cells (BIPV) are commonly made from crystalline wafer cells. This contribution investigates the challenges and benefits of using bifacial solar cells in vertical installations. We show that those cells get up to 13% more irradiance compared to optimum tilted south facing monofacial modules in Germany. The role of the n-layer in thin amorphous bifacial single-junction cells intended to be used as bifacial cells in BIPV applications is investigated. In contrast to the superstrate cell design, a transparent n-layer and back contact play a key role to achieve high bifaciality. We therefore increased the transparency of the n-layer by adding CO_2, increasing the PH_3 flow in the deposition gas and tested different thicknesses. With those measures, we reached a bifaciality of 98% for short-circuit current density and 99% for open-circuit voltage.

Introduction

The changing situation in the social awareness of climate change and the resulting need for renewable energy facilitates the intention to implement photovoltaic devices into already established products to generate additional benefit. Especially, the thin film silicon solar technology has some great advantages compared to conventional silicon wafer solar cells concerning the integration into existing products:

- Large area with homogenous optical appearance
- Nonvisible circuitry by laser interconnection
- Can be deposited on different substrates (glass, ceramic, plastic, metal, etc.)
- Small temperature coefficient
- Very short energy payback time
- Nontoxic in case of recycling

An area with huge potential which has seen almost no attention until now [1], is the integration of solar cells into the traffic infrastructure. This would have the benefit of no additional intrusion into nature because the traffic infrastructure already exists. Cost-cutting efforts in the area of infrastructure in the EU lead to a lack of investments that have to be redeemed in the future. Adding photovoltaic modules to building elements in traffic infrastructure has the benefit of return of those investment costs in the long term, making them viable investment projects.

Bifacial cells in vertical installation on highways and rail ways have the potential to generate 42 TWh/a [2] without changing the appearance of the landscape because the noise barrier walls already exist.

The combination of sustainable energy generation and noise barriers next to highways would increase the public acceptance of structural modifications as an additional benefit.

Integrating photovoltaic power generation into noise barrier walls has been tested before [3] by integrating silicon wafer modules. The bifacial concept leads to an increase in power generated per module [4]. It also pointed out some disadvantages of using silicon wafer technology: They cannot be integrated in every object and every position because of the restricted dimensions of those modules. Integrating semitransparent wafer modules has the optical disadvantage of casting a shadow like a chess board [4] that can be very distracting.

With wafer technology, it is therefore not possible to use an existing substrate.

As we demonstrated in a previous paper [5], there is more than one way to fabricate functional bifacial thin film silicon solar modules. As we have shown, multijunction cells are not feasible in bifacial applications. On the other hand, single-junction thin film solar cells do not suffer from this problem. Compared with μc- silicon thin film modules, amorphous silicon thin film modules have some advantages:

- Higher absorption coefficient makes very thin cells and short fabrication times possible
- Stable deposition regime

These are the reasons why we believe that amorphous silicon thin film modules are a promising technology for this application.

Irridiance and Bifaciality

Using a conventional wafer-based solar cell in vertical configuration has the disadvantage of using only one-half space and therefore only a reduced part of the incoming sunlight.

The efficiency characterization of solar modules is dependent on the solar spectrum. The solar spectrum can be divided into three parts: The direct, indirect (diffuse), and reflected solar irradiance. The direct sunlight reaches the module surface after crossing the atmosphere with no further changes. Indirect sunlight is scattered on clouds or particles and far away objects before reaching the solar modules. The reflected part of the irradiation spectra results from the reflection of the near environment and back scattered sunlight of the ground. For monofacial modules on slanted rooftops, the reflected sunlight of the near environment does not play a significant role. Figure 1 depicts the composition of those parts for three German cities. Even the southernmost city Munich has an indirect solar irradiance of over 50% in average over the year. This is true for the northern European area in general.

To get a deeper understanding, in what way the mounting orientation affects the module generation, we calculated the irradiance on a plane in relation to the angle of tilt and orientation. Data for this simulation was extracted from the meteonorm [6]. The results are depicted in Figure 2. To get the optimum yield in the yearly average for a conventional one-sided module in Oldenburg Germany, it should face to the south and slightly to the west. In this optimum tilt, the average irradiance would reach 1137 kWh/m²/a.

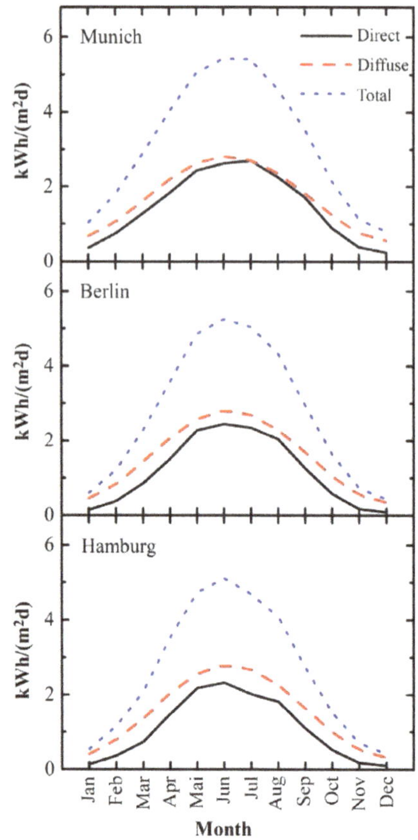

Figure 1. Composition of the solar spectrum in terms of direct and indirect irradiance for three German cities.

The same calculation can be done for a vertical plane facing east, south, west, and north. Because of the before mentioned diffuse irradiation, the average irradiance on those vertical planes per year is still very high. The results of the calculations can be seen in Figure 3. A vertical one-sided module facing south would generate the highest average yearly yield with an irradiance of 844 kWh/m²/a, which is 74% of the irradiance of the optimum tilted installation.

We propose to use bifacial modules for the vertical installation. Those modules are able to use both half spaces of the light hitting the back- and the front side of the module and are able to compensate the negative effect of partial shadowing to a certain extent. Furthermore, the vertical installation increases the naturally occurring cleaning effects and reduces the negative soiling problems caused by leaves, bird drops, and similar impurities. Thinking about noise barrier walls lining the motorways and railways going from north to south, we can use the data of Figure 3 to calculate the irradiance shining on a

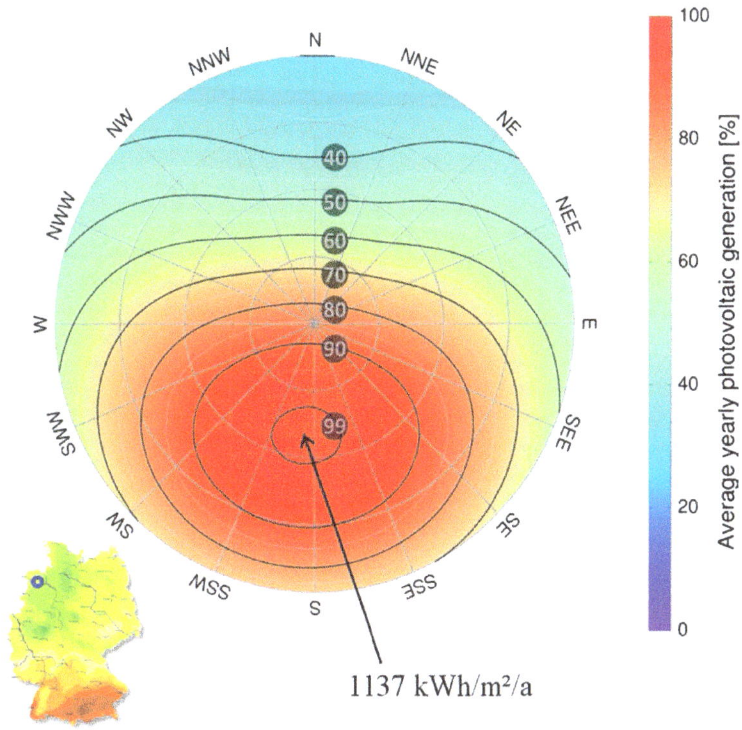

Figure 2. Tilt in relation to average yearly photovoltaic irradiance for Oldenburg, Germany. Black lines represent the isoline in percentage from the maximum irradiance.

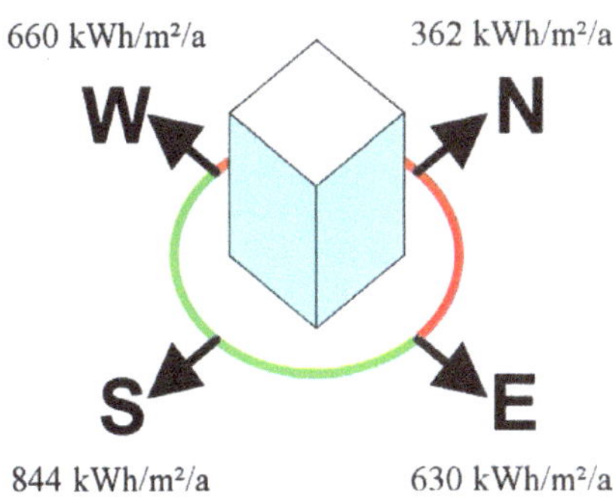

Figure 3. Photovoltaic yearly average irradiance for vertical planes facing north, east, south and west in Oldenburg with an albedo of 0.25.

bifacial module (100% bifaciality, vertical installation facing east and west) which it could convert to power. In yearly average it would face 1290 kWh/m²/a, which is 113% of a conventional module in the optimum tilt position would face. Even for transport infrastructure going from east to west the yearly average irradiance is still higher than for the modules in optimum tilt position (1206 kWh/m²/a vs. 1137 kWh/m²/a).

When using modules in vertical installation, the daily generation profile also changes. To give an example, we used measured irradiance data to simulate the irradiance on different planes of orientation and tilt on an average day in July. The results are depicted in Figure 4.

Using bifacial solar cells in vertical installation could shift the power generation peaks to the morning and evening hours. This would reduce the demand for power storage and increase the self-consumption rate since the demand profile for northern European regions is not dominated by air-conditioning.

The shift in the irradiance profile exists also on a yearly timescale. Using the same irradiance data, we calculated the same graph for those tilts and orientations on a monthly timescale. The results are shown in Figure 5.

Bifacial cells facing in north/south direction have the benefit of being able to generate more power in the winter month because of the sun being lower in the sky. In contrast, bifacial modules facing east and west could generate more power in the sunny months compared to conventional tilted modules.

Experimental

The ratio between rear- and front-side efficiency, and therefore the current generation density, is called bifaciality

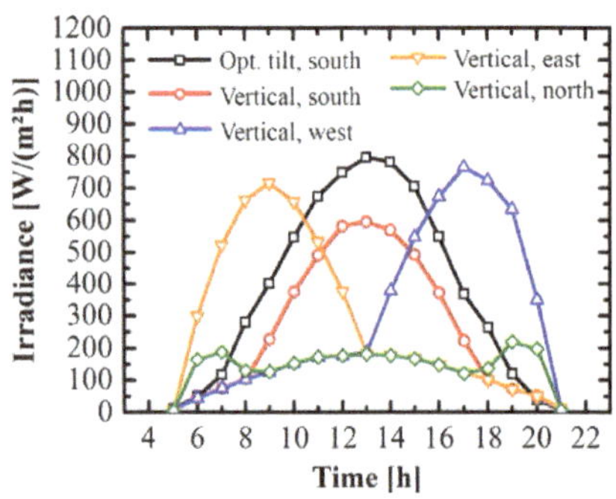

Figure 4. Simulated irradiance of an average day in July in Oldenburg for different tilts and orientations of the module plane.

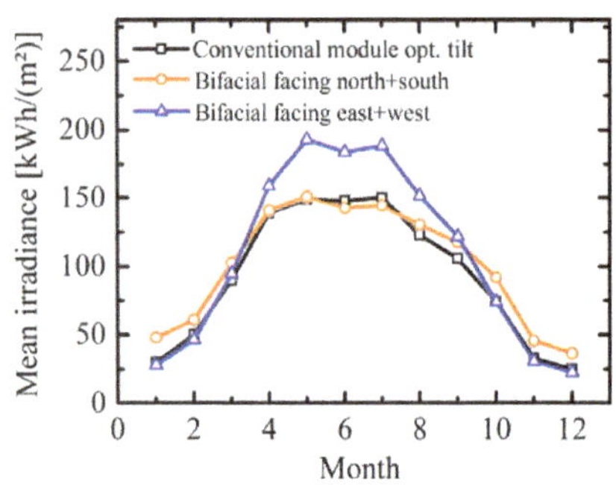

Figure 5. Monthly irradiance for three different planes in Oldenburg in different orientations.

[7]. In our previous work, we reached a bifaciality of 50% using a standard process [5] for the deposition of single a-Si thin-film solar cells in combination with a transparent back contact. The i-layer thickness was fixed at 250 nm and showed the expected light-induced degradation of 8% in average (1000 h, @STC). No further optimizations where done. To improve the bifaciality of our single-junction amorphous solar cells, we focused on the n-layer of the cell-stack. In superstrate pin configuration, amorphous silicon thin film solar cells see light only from the p-side. Therefore, there was no need for blue transparent n-layers. In the situation of a high bifaciality, the n-layer has to fulfill the same requirements as the p-layer. This means, a high transparency, high doping, less surface recombination, and good enough cross conductivity [8].

To improve the transmittance of the amorphous n-layer for low wavelengths, we varied:

- the CO_2 concentration
- the doping concentration
- the layer thickness

We decided to use an amorphous n-layer instead of the commonly used μc-$Si_{1-X}O_X$:H layer [9,10] because we expected a better long-term stability of highly doped amorphous n-layers compared to highly doped μc-$Si_{1-X}O_X$:H layers [work in prep.]. Furthermore, the process stability is better.

All presented samples were produced with a LEYBOLD Phoebus three chamber cluster PECVD-System (plasma enhanced chemical vapor deposition) at 13.56 MHz under a pressure of 4 mbar and substrate temperature between 190° to 220°. As substrates, glass with rough etched commercial sputtered ZnO:Al from the company Solayer were used. We also tested (results not shown) our cells on flat substrates and found no difference in terms of efficiency. In all samples the hydrogen flow (300 sccm), the silane flow (37.2 sccm), and the plasma power (70W) were the same. All samples were prepared with ITO-back contacts and the cell area of 1 × 1 cm² was defined by a laser isolation process.

IV-curves of solar cells under illumination (AM 1.5 g, 1000 W/cm², 25°C) were measured under a continuous light (DC) sun simulator of class A (WACOM WXS-155S-L2-AM1.5GMM). For determining the external quantum efficiency (EQE) of the solar cells, monochromatic probe beams using RERA System equipment with wavelength from 300 to 800 nm were used. Data for the irradiance graphs were extracted from the meteonorm [6].

Bifaciality Optimization

CO_2 variation

To increase the transparency of the n-layer, especially in the 300–500 nm wavelength regime, a common approach is to alloy the layer by adding CO_2 to the deposition gas mixture [9,10]. When adding too much CO_2 gas to the mixture, a decrease of the n- conductivity [8] can be observed. To prevent such an effect, we increased the CO_2 flow in small steps until we reached a gas flow of 10 sccm. As shown in Figure 6, the addition of CO_2 did not change the behavior of the cell when illuminated from the p-side. Therefore, we conclude that the used max gas flow of 10 sccm is small enough to neglect the influence on the n-layer conductivity. In contrast to the results of the p-side illumination, the short circuit current density increases slightly with the increasing gas flow. As expected, an increase in the transparency of the n-layer can be observed by increasing the CO_2 proportion at the n-layer. The maximum benefit of 0.6 mA/cm² is reached by adding a flow of 10 sccm CO_2 to the deposition gas mixture. To see where the additional current is generated, we measured the external quantum efficiency and extracted the results depicted in Figure 7. The reflectance (not shown) does not change measurably for all samples.

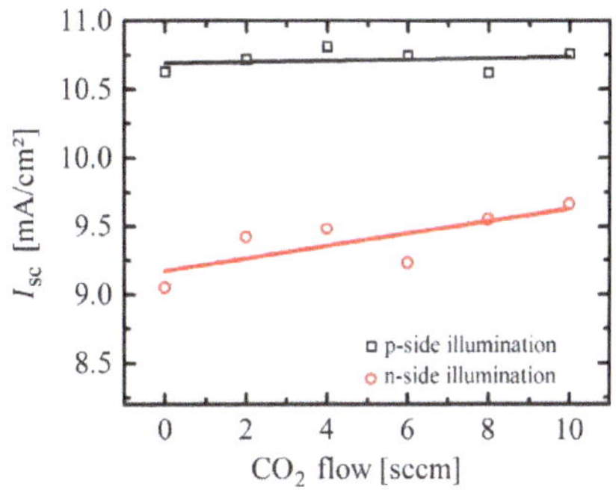

Figure 6. Short circuit current density of amorphous single-junction cells with oxygenated n-layers. Lines are fits of the measurement points.

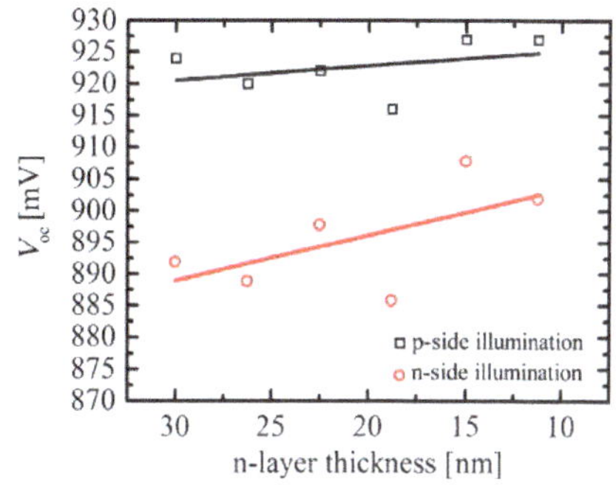

Figure 8. Open circuit voltage values of bifacial amorphous single-junction cells with decreased n-layer thickness. Lines are fits of the measurement points.

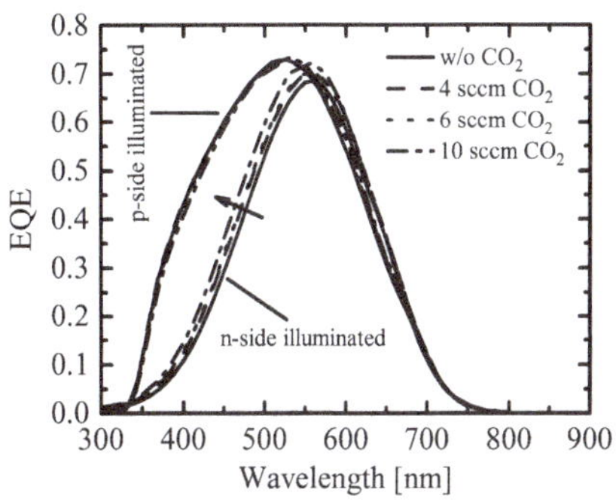

Figure 7. External quantum efficiency measurement of bifacial amorphous single-junction cells with CO_2 doping in the n-layer. The arrow indicates the tendency.

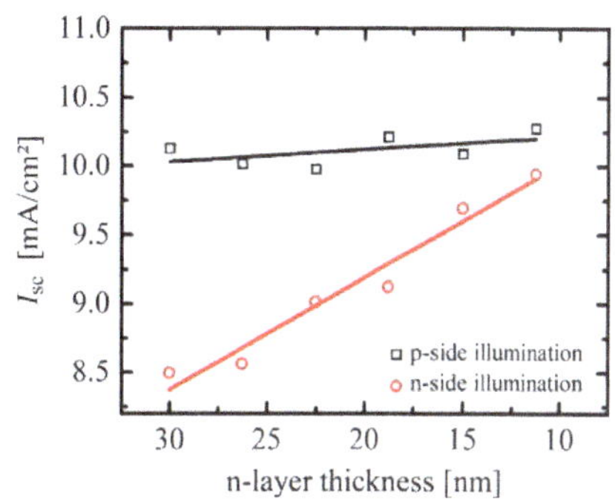

Figure 9. Short circuit current density of bifacial amorphous single-junction cells with decreased n-layer thickness. Lines are fits of the measurement points.

While the addition of CO_2 to the n-layer did not change the behavior, a distinct increase in conversion efficiency in the wavelength area between 400 and 600 nm can be observed (indicated by the arrow), when the cells are illuminated from the n-side. This can be attributed to the increased transparency due to the incorporation of oxygen and/or carbon to the layer. V_{oc} is not affected by the variation and remains nearly constant at an average of 920 mV for p-side illumination and 900 for n-side illumination.

Thickness variation

Besides mixing the n-layer deposition gas with CO_2 to increase its transparency, a thickness variation in the n-layer was performed to investigate the influence of the n-layer thickness on the absorption behavior, cell efficiency, and the bifaciality. The standard p-layer used in our cells is between 5 and 10 nm thick, whereas the default n-layer is 20–30 nm thick to ensure a strong built-in field for the improved carrier extraction and the homogeneity of the field. Figure 8 depicts the V_{oc} values of the series, proving that the used n-layer of approximately 10 nm is thick enough to conserve the strong built-in field.

As shown in Figure 9, the decreasing thickness of the n-layer does not influence the current generation density when the cell is illuminated from the p-side.

In case of the illumination from the n-side, an increase in short-circuit current density generation can be observed, reaching a short-circuit current generation bifaciality value of 96% for the cell with an n-layer of approximately 10 nm thickness.

As demonstrated in Figure 10, the additional current generated originates from the short wavelength regime. Compared to the CO_2 variation and the doping gas variation, the thickness variation shows the highest impact on the transparency for small wavelength.

The observed improvement in conversion efficiency for light from the high-energy regime is caused by a reduced absorption in the n-layer. Light absorbed in the n-layer does not contribute to the photocurrent. By reducing the

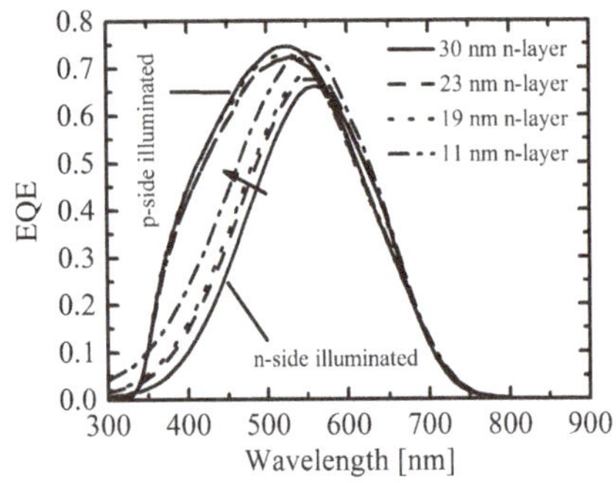

Figure 10. External quantum efficiency measurement of bifacial amorphous single-junction cells with reduced n-layer thickness. The arrow indicates the tendency.

thickness, more high energy light passes the n-layer to reach the i-layer where it is absorbed and converted into photocurrent.

Doping variation

Because a variation in the thickness of the n-layer might reduce the strength of the built-in field generated by the phosphorus doping, we deposited our bifacial a-Si:H cells with an increasing amount of PH_3 doping gas. For a better comparability to the other experiments, we used the same n-layer as a starting point as in the experiments before. The results of the doping variation are depicted in Figure 11.

With n-side illumination, an increase in the short-circuit current density of 0.4 mA/cm^2 can be seen when an additional gas flow of 5 sccm is added to the deposition gas mixture. The p-side formed by short-circuit current does not show such dependence.

The V_{oc} is not affected by the variation and remains constant at an average of 925 mV for p-side illumination and 900 for n-side illumination, respectively.

This asymmetric behavior is in good accordance to theory. The built-in field of amorphous silicon thin film cells largely originates from the boundary between i-layer and adjacent doped layer. Free charge carriers excited by absorption of light at the i-layer are forced toward the contacts by the built-in field (p for holes, n for electrons). Because of their lower mobility, free holes have a higher recombination probability in the i-layer compared to electrons [11]. This is why conventional a-Si:H modules are built in superstrat configuration, where the light enters the cell through the p-layer and the high density of exited holes on the first few nanometers can reach the contact before recombining.

The depicted n-doping increase in Figure 11 influences mainly the hole transport in the area of the n-i junction. This is why, only the short circuit current generation under n-side illumination increases. A faster excited charge transport has the additional benefit of removing the holes faster from the area where the highest concentration of free carriers is located in case of illumination from the n-side. This improved charge separation behavior resulted in a reduced recombination probability and consequently in an improved fill factor we also observed.

Combination of variations

The reference cell at the start of the investigations had a bifaciality in short-circuit current generation of only 77%. Due to a combination of the discussed improvements, we were able to fabricate a bifacial amorphous single-junction cell with improved bifaciality. This cell incorporated a 40% thinner n-layer an increased PH_3 and CO_2 deposition gas flow rate and was deposited on rough etched commercial ZnO:Al substrates. With that n-layer we managed to reach a bifaciality of 98% for short-circuit current density and 99% for open circuit voltage. The EQE in Figure 12 depicts the spectra of the optimized and the reference cell. The optimization leads to an increased

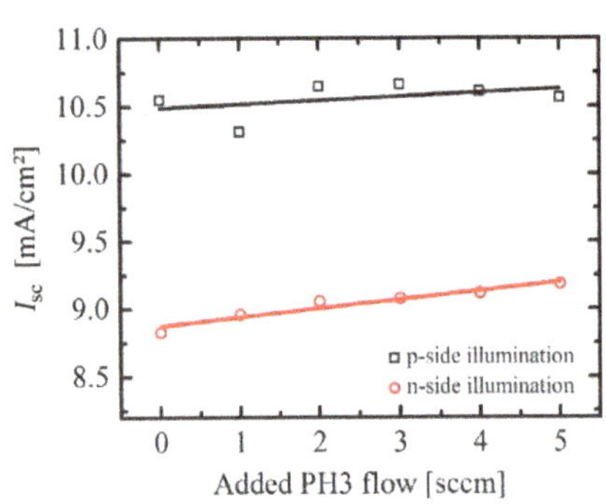

Figure 11. Short circuit current density of bifacial amorphous single-junction cells with increased doping gas flow. Lines are fits of the measurement points.

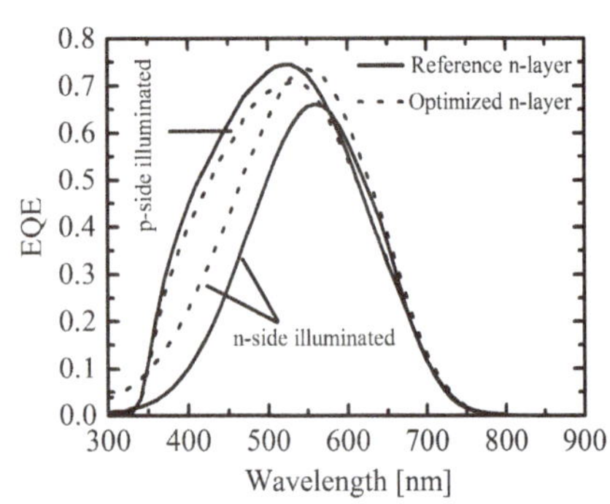

Figure 12. External quantum efficiency measurement of bifacial amorphous single-junction cells with reference and optimized n-layer.

conversion efficiency up to a wavelength of 600 nm when illuminated from the n-side while maintaining the performance when illuminated from the p-side.

The average power conversion efficiency of the monofacial reference cells was 7.0% while the optimized bifacial cells reached an efficiency of 7.4%.

Since the deposition time of the final n-layer is less than half of the deposition time of the reference layer and the only additional gas is cheap CO_2, the new n-layer would cost even less in industrial production than the reference n-layer.

Conclusion

Simulating the irradiance on different planes, we demonstrated that bifacial modules in freestanding vertical installation face more irradiance than monofacial modules in optimum tilt and orientation, regardless of the modules facing east-west or north-south.

We could show that a slight tuning of the n-layer of a standard amorphous single-junction cell in combination with the removal of the reflecting layer (Ag/white light reflector) can produce a cell with a bifaciality of 98%. Our improved bifacial amorphous single-junction solar cells show the same efficiency as our conventional cells under STC. Due to their bifacial nature this could lead to twice the output power generated under optimum conditions. We estimate that they could reach 40% to 70% improved power generation in a yearly average compared to monofacial solar cells under real-life conditions.

Outlook

To make measurements of bifacial modules comparable under STC, additional conditions like the reflectivity of the measurement block are being proposed [12]. To gage the performance of a plant with bifacial cells, additional measurement standards have to be defined [13]. To determine the plant yield, new factors like the albedo and the ratio between direct and indirect sunlight have to be taken into account.

Bifacial solar cell applications differ from classic rooftop or power plant installations demanding new solutions and different circuitry to compensate for shadowing by moving objects for example.

Current building material regulations hinder the incorporation of solar cells in building materials. To establish solar cells as part of building materials different hazards have to be evaluated:

- Danger of electrocution
- Fire hazards (ignition and toxicity)
- Blinding of vehicle drivers by reflections
- Breakage of glass

Since building material glass is a different type of glass compared to solar-grade glass, it has to be investigated how combination of thin film silicon solar cells and the building type glass used in noise barrier walls performs.

Conflict of Interest

None declared.

References

1. http://www.theguardian.com/environment/2014/nov/05/worlds-first-solar-cycle-lane-opening-in-the-netherlands (accessed 09.07.2015).
2. Quaschning, V. *Systemtechnik einer klimaverträglichen Elektrizitätsversorgung in Deutschland für das 21 Jahrhundert*. Fortschritts-Berichte, VDI Reihe 6, 2000; Nr. 437.
3. Pukrop, D. *Zur Modellierung großflächiger Photovoltaik-Generatoren* 1997; These; Carl von Ossietzky University Oldenburg.
4. Duran, C. *Bifacial Solar Cells: High Efficiency Design, Characterization, Modules and Applications* 2012; These; University Konstanz.
5. Reininghaus, N., C. Feser, and K. von Maydell. 2014. Investigation of bifacial thin-film silicon solar cells for building integration. Proc. EUPVSEC 29th:3837–3841. doi:10.4229/EUPVSEC20142014-6AV.5.39.
6. http://meteonorm.com (accessed 09.07.2015).
7. Kopecek, R., Y. Veschetti, E. Gerritsen, A. Schneider, C. Comparotto, V. D. Mihailetchi, et al. 2014. One small step for technology, one giant leap for kWh cost reduction. Photovolt. Int. 26:32–45.
8. Cuony, P. *Optical Layer for Thin-film Silcion* 2011, These, Ecole polytechnique federale de Lausanne.
9. Smirnov, V., A. Lambertz, and F. Finger. 2014. Bifacial microcrystalline silicon solar cells with improved performance due to μc-SiOx: H doped layers. Can. J. Phys. 92:913–916.
10. Veneri, P. D., L. V. Mercaldo, and I. Usatii. 2011. Improved micromorph solar cells by means of mixed-phase n-doped silicon oxide layers. Prog. Photovolt. Res. Appl. 21:148–155.
11. Shah, A. 2010. Thin-film silicon solar cells. EPFL Press, Switzerland.
12. Hohl-Ebinger, J., and W. Warta. 2010. Bifacial solar cells in STC measurement. Proc. EUPVSEC 25th:1358–1362. doi:10.4229/25thEUPVSEC2010-2CO.4.1.
13. Singh, J. P., A. G. Aberle, and T. M. Walsh. 2014. Electrical characterization method for bifacial photovoltaic modules. Sol. Energy Mater. Sol. Cells 127:136–142.

9

The great solar boom: a global perspective into the far reaching impact of an unexpected energy revolution

Francesco Meneguzzo[1], Rosaria Ciriminna[2], Lorenzo Albanese[1] & Mario Pagliaro[2]

[1]Institute of Biometeorology, CNR, via Caproni 8, Firenze 50145, Italy
[2]Institute for the Study of Nanostructured Materials, CNR, via U. La Malfa 153, Palermo 90146, Italy

Keywords
Photovoltaic, PV industry, PV manufacturers, solar boom, solar energy

Correspondence
Francesco Meneguzzo, Institute of Biometeorology, CNR, via Caproni 8, Firenze, 50145 FI, Italy.
E-mail: f.meneguzzo@ibimet.cnr.it
and
Mario Pagliaro, Institute for the Study of Nanostructured Materials, CNR, via U. La Malfa 153, Palermo, 90146 PA, Italy.

E-mail: mario.pagliaro@cnr.it

Funding Information
No funding information provided.

Abstract

This study offers a unified perspective into the unexpected solar energy photovoltaic revolution, and its far reaching impact onto both energy generation and electricity markets. Practically relevant aspects, such as those related to the value of solar PV electricity, land consumption, energy return on energy invested, reliability of the technology, the structure of the global PV industry, the cost of Li ion batteries and related market trends are clarified. We identify the main barriers to overcome for solar PV to expand beyond a niche market (say, <10% of a country's power generation), and the related societal benefits with electrification of energy end uses.

Introduction

Solar photovoltaic energy has been of crucial applicative relevance since its very inception, in the mid 1950s, when scientists in the United States built the first solar cells based on *p–n* junctions in single Si crystals with efficiencies of 5–6% [1]. As put it by Varadi, co-founder in 1973 of a pioneering solar company which 5 years later became the largest solar producer of silicon-based photovoltaic (PV) modules in the world (0.5 MW of nominal power, in 1978), "few people realize that without the invention of solar power many things we are using today such as cell phones, TV, internet, global weather service, the GPS system, and manned space stations would not be possible" [2].

Indeed, no communication satellite could function without solar cells as batteries in the space simply do not work; while perhaps even less people, to quote Varadi again, today are aware that the satellite powered by a nuclear reactor launched in 1964 in the United States broke apart during the launch, releasing all radioactive uranium that was scattered across the globe [2].

However, due to high cost and consequent limited demand, for decades "terrestrial" photovoltaic modules were mainly sold to customers in the oil, telecommunication and naval regulation sectors [1]. All changed starting in year 2000 with the introduction, in Germany, of the Feed-in Tariff (FiT) program conceived by Hermann Scheer [3].

This original incentive scheme, rewarding for 20 years with a high and flat tariff every kWh of a PV plant connected to the grid, overcame the financial barrier that for decades had limited solar power market penetration. Suddenly, the high-priced solar PV plants became "bankable" [4].

Demand surged and, with it, mass production and technological advances, stimulating investment and forcing competition among suppliers of different modules based on a wide spectrum of technologies. For example, the first CdTe modules commercialized by the leading supplier in 2002 had 7% nominal efficiency, while those sold as of early 2015 had 16.3% certified efficiency [5]. The owners of newly installed PV plants, furthermore, were forced to undertake active maintenance of their systems, as without real and prolonged production of clean energy, no financial incentive was going to be gained.

For about 4 years (2003–2007), following the deployment of the FiT incentives also in Spain and in Italy, the price of the solar modules remained high ($5–6/W). When, however, silicon prices dramatically decreased as a consequence of massive investment in polysilicon production plants by "vertically integrated" PV module manufacturers in China and in Taiwan the price of solar modules started to fall at unprecedented rate.

Unexpectedly, the sale price of photovoltaic modules, once thought to be "physically limited" by the cost of crystalline silicon (until recently a cost price of $1.00/W was referred to as a "tipping point" for the solar industry) [6], decreased to such a low level (<$0.5/W, as of May 2015) that, rather than "grid parity" [7], a true "generation parity" with the cheapest energy source, namely coal, is now being approached. We remind here that coal continues to be massively used to produce electricity in the world's largest economies (the United States, China, Germany, and India).

Huge PV plants are rapidly built in many countries across the world and connected to the grid, giving solar PV energy the status of a true utility-scale energy resource, with dramatic consequences on the electricity price formation in countries with significant market penetration [8]. In Germany, for example, where the largest PV power capacity in a single country is installed, on April 15, 2015 the PV power reached the record value of 27.7 GW pushing wholesale electricity price into negative territory. In detail, between 1 PM and 2 PM of that day German electricity was retailing for –€1/MWh, and that tariff further decreased to –€1.1/MWh between 2 PM and 3 PM [9].

Finally, and opposite to what had been observed in the past, growth in the installation of PV plants is going on at very high rate, despite the price of oil having halved from around $100/barrel to about $55/barrel in <1 year since July, 2014.

In 2015, reputed market analysts predict the deployment of further PV power by 61 GW, namely a 30% growth over the previous year, equivalent to one-third of the cumulated global PV power installed worldwide at the end of 2014 [10]. These facts may justify the use of a frequently abused word such as "revolution" in power generation.

Aimed to energy scholars as well as to policy makers, this perspective study provides a unified overview of the global solar photovoltaic boom, taking into consideration technical, economic and energy aspects. By doing so, this study aims to address a gap in current energy and scientific literature. After clarifying a few aspects concerning the PV technology for both decentralized and utility scale electricity generation, the study identifies the barriers to overcome towards massive, global adoption of solar PV power replacing electricity obtained by burning fossil fuels or through Uranium nuclear fission. Within two decades, we argument in the conclusions, massive adoption of solar PV energy across the world will have resulted in long awaited end-use energy electrification affording the economic and environmental benefits that will solve the current energy and environmental crises, opening the route to a true solar-based economy.

A Reliable, Convenient Technology

Photovoltaics is an inherently reliable industrial technology delivering a product, the solar module, that since the early days of its commercialization needed to reliably provide power in the demanding environment in which the first "terrestrial" PV modules were used by customers (telecommunication and oil companies, and naval authorities), namely remote power generating stations, remote pipelines, offshore platforms, and buoys [1].

Crucially important quality and testing programs were then financed by the United States and other European governments right around those years. For example, the first quality control program started by the Jet Propulsion Laboratory in Pasadena, lowered the failure rate of existing solar cells from 45 to 0.1%, namely achieving the quality levels of advanced industrial productions [2].

A PV system consists of the PV module and the balance of system (BoS) components encompassing support racks, the wiring, transformer to raise the voltage, inverter to convert direct current to alternating current, and switches for connecting to the electric grid. All are reliable and long lasting.

For example, the long-term reliability of the PV technology since its early days is exemplified by the 80 kW system array installed in the island of Vulcano (a volcanic island north of Sicily's sea) in 1984 (Fig. 1). Comprised

Figure 1. The PV plants, comprised of different silicon-based solar modules, installed in the island of Vulcano, Sicily, in 1984. After more than 30 years, the plant has recorded a modest performance deterioration [Reproduced from Ref 11, with kind permission].

of 9% efficient PV modules in monocrystalline silicon manufactured in Italy in 1983, after 21 years the solar PV plant had lost only 6% of its original production capacity [11], namely about one half of the expected reduction (10%) based on the 0.5%/year degradation rate median value obtained from field testing throughout the last 40 years [12].

Another issue, heavily debated until recent times, has been the energy payback time (EPBT) of the PV technology, and the related energy return on energy invested (EROI) values of the various technologies based on different semiconductors employed to make solar cells.

Most recently, Apul et al. conducted a thorough study based on a systematic review of the EPBT and EROI metrics for the two main crystalline Si and the three main thin film PV technologies taking into account scientific articles published in 2000–2013 [13]. The team harmonized several parameters (performance ratio: 0.75; system lifetime: 30 years; insolation: 1700 kWh m^{-2} yr^{-1}; module efficiency: 13% mono-Si; 12.3% poly-Si; 6.3% a:Si; 10.9% CdTe; 11.5% CIGS. The mean harmonized EROI varied from 8.7 to 34.2 (Fig. 2), whereas the EPBT values vary from 1 to 4.1 years in the following order (from lowest to highest): cadmium telluride (CdTe), copper indium gallium diselenide (CIGS), amorphous silicon (a:Si), poly-crystalline silicon (poly-Si), and monocrystalline silicon (mono-Si).

The economic convenience of utility scale PV generation, briefly discussed above with electricity prices having reached negative values in Germany on April 2015, is evident even in the case of relatively small, ~ 1 MW plants. For example, in Israel-Palestine the recent opening of a 710 kW plant (Fig. 3) which transmits to Jerusalem the energy produced near Jericho, 278 m below the sea level, has resulted in an immediate, significant drop in electricity prices for households that were formerly dependent on high utility rates [14].

A Scalable Technology for Decentralized and Utility Scale Electricity Generation

In 2001, solar electricity provided less than 0.1% of the world's electricity consumption (13,000 TWh). In 2014, the figure had grown by one order of magnitude to reach 1% of the global demand, which in the meanwhile increased so much to exceed 20,000 TWh [15].

Moreover, the PV technology is easily scalable with minimal land consumption. For example, covering 0.6% of the EU territory with 10% efficient PV modules would

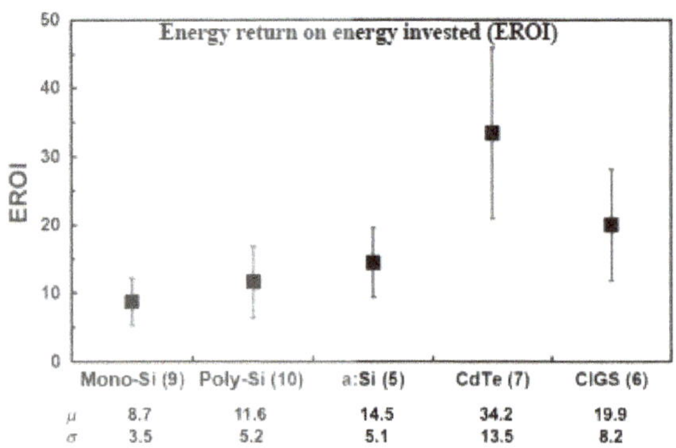

Figure 2. Mean harmonized EROI of the main five PV technologies. The number of values for each module type is included in parentheses. Mean (μ) and standard deviation (σ) are shown at the bottom of the graph [Reproduced from Ref 13, with kind permission].

Figure 3. The Dead Sea Photovoltaic Generating Plant, near Jericho, since late 2014 produces electricity dispatched to Jerusalem. Low-cost solar electricity has significantly reduced the cost of the kWh paid by households [Reproduced from Ref 14, with kind permission].

theoretically meet its electricity consumption [16]. Today's modules, however, approach and even exceed 20% efficiency (e.g.,, the silicon modules of Sun Power with no front contacts; or Panasonic–Sanyo's using the HIT technology) [17], so that the amount of required land will actually be much lower.

In this respect, Fthenakis et al. have shown that in most cases, ground-mounted PV systems in areas of high insolation (Fig. 4A) transform *less* land than the coal-fuel cycle coupled with surface mining (Fig. 4B) [18]. Obviously, the land-use requirements for wide-scale deployment of PV are further lowered by taking into account the huge built surfaces already existing, namely the roofs of all sort of buildings and of other constructions such as parking lots, highway and bridge walls, which once integrated with PV modules could supply a large fraction of industrial society's energy needs.

This aspect leads to the well-known unique advantage of solar photovoltaics, namely the decentralized power generation capable to locally produce power at low cost that is currently reshaping the electricity market in countries with significant PV penetration.

As emphasized by Varadi [19], after years of conflict and delayed interconnection to the grid, all four main Germany's utilities now offer new storage systems to the owners (over a million) of homes and farms equipped with PV systems. In this way, the utilities do not sell electricity any longer, becoming instead "PV + storage systems" installing and maintenance companies (for 20 or 30 years through a dedicated contract specifying the terms of the new business). It should be emphasized that such distributed storage approach enhances the value of PV generated electricity while dwarfing the negative impact of its increasing penetration.

Figure 4. Difference in land impact between (A) a solar photovoltaic plant (in Springerville, Arizona: land requirement for PV: 310 m^2/GWh) and (B) a mountaintop removal coal mining site (in Rawl West, Virginia: Land requirement for surface coal mining: 320 m^2/GWh) [Reproduced from Ref 18, with kind permission].

The Global Solar Boom

Today's low-cost PV technology is experiencing unprecedented growth on a truly global scale. At the end of 2014, the global installed capacity reached 185 gigawatts (GW) [20].

Driven by the ultralow cost of the technology, photovoltaic modules are now used in the vast majority of the world's countries, including remote islands in the Pacific Ocean, such as in the case of Tokelau island chain (comprising the three atolls of Atafu, Nukunonu and Fakaofo), between New Zealand and Hawaii, that in 2012 became the world's first territory able to meet all of its electricity needs with solar PV power (Fig. 5), coupled to a battery storage system to supply power at all times [21].

Followed by China, Germany is currently the world's largest producer of PV energy, while Italy is the country with the largest fraction (7.5% in 2014 and still growing) [22] of national electricity demand met by PV energy. Table 1 shows the rank of top ten countries for installed PV power capacity at the end of 2014.

In India, where some 80% of 2014 electricity demand was met by coal, the cumulated PV power hosted at the end of 2014 slightly exceeded 3 GW. The government, however, in early 2015 announced an ambitious 100 GW installation target for 2020 that, if met, will bring the country among those with the largest fraction of solar PV power in their energy mix [23].

In China, where almost two-thirds of the overall electricity demand is met by coal, in 2010 there was almost no PV power installed, out of an overall total installed capacity of 1360 GW. At the end of 2014, the country had an installed solar capacity exceeding 28 GW (ground-mounted capacity, 23.38 GW and distributed generation, 4.67 GW).

Figure 5. The PV plant in the Atafu atoll of Tokelau island chain in the Pacific Ocean. The world's first territory to become energy-independent with solar PV coupled to storage via lithium ion batteries [Reproduced from Ref 21, with kind permission].

Table 1. Top ten countries for installed PV power capacity at the end of 2014 [Sources: China's National Energy Administration, EurObserver, US Solar Energy Industries Association].

Rank	Country	Installed power (GW)
1	Germany	36.34
2	China	28.05
3	Japan	23.3
4	Italy	18.42
5	United States	18.3
6	Spain	4.74
7	France	4.61
8	Australia	3.3
9	Belgium	3.04
10	United Kingdom	2.78

In 2014, 10.6 GW of new capacity was connected to the country's grid, whereas total electricity generated by PV reached 25 billion kWh, an year-to-year increase exceeding 200% [24]. Only in the first 3 months of 2015, China added another 5 GW, with the government aiming to add further 17.8 GW only in the course of 2015.

In Japan, PV energy capacity has rapidly grown to more than 23 GW, from slightly more than 3 GW in 2011 when all 43 nuclear reactors were shut down following an earthquake and a tsunami that caused meltdown at the Fukushima power plant [25]. By March 2016, Japan plans to shut down nearly 2.4 GW of expensive and polluting oil-fired energy plants, replacing them with solar and wind power.

In Germany, after the 2011 Fukushima accident, the government decided to immediately close eight nuclear power reactors, and all the remaining reactors by 2022. The government reinforced its option for renewable energy (mostly wind power and PV) to replace nuclear and coal energy [26]. In February 2012, a winter month, during several days the country was exporting 4.5 GW of clean electricity to France [27].

The latter country relies on 58 nuclear power plants for most of its electricity needs. Yet, new energy companies in France understood the economic impact of low-cost PV generation, and France already hosts a number of utility scale PV plants including the 300 MW plant in Cestas (Fig. 6), the largest in Europe.

This plant, built in <12 months, will profitably sell electricity at €105/MWh, namely below the price of the electricity generated by the nuclear power plant that France's largest utility (EDF) is currently building in United Kingdom's Somerset under a minimum price of €117/MWh for a period of 35 years guaranteed by the government [28].

In Italy, the deployment of the FiT incentive scheme between 2006 and 2013 caused an impressive surge in the PV installed power that in 2 years (between 2011 and 2013) grew from few hundreds of MW to 18.42 GW at the end

Figure 6. The 300 MW plant in construction in Cestas, near Bordeaux. The modules have a East-West orientation to exploit the sun radiation from early morning to late evening [Reproduced from Ref 28, with kind permission].

of 2013. As a result of a concomitant vigorous increase in wind power, in 2014 Italy met 43.3% of its large electricity production (268 TWh) with renewable energy sources [21], without either significant or specific upgrades of the grid.

The United States closes the list of the >10 GW countries, where the installed PV power grew from slightly more than 100 MW in 2006 to over 18 GW as of late 2014; more than 6 GW were erected only in the course of 2014 [29].

Numerous other countries are emerging in the global PV rush, well beyond Europe, China, Japan and the United States. South Africa has already reached the milestone of 1 GW of installed capacity. Chile will cross the threshold in 2015.

As the price and availability of the PV technology keep falling, there are no reasons why most other large countries or regions such as Brazil, Jordan, the Philippines, Thailand, Malaysia, or Mexico should not adopt clean solar energy as a convenient option to lower the cost of the electricity, and meet rapidly increasing daily energy demand.

Indeed, large (>100 MW) solar PV plants are currently being built in few Brazilian states, while the whole Latin America is becoming a significant PV market with a forecasted 350% annual growth in 2015 (from 600 MW in 2014 to > 2.2 GW of installed power in 2015 along with 34 GW in the pipeline: 6.8 GW contracted, 1.9 under construction and 25 announced) [30].

As it happens in India and in all unsubsidized solar markets, all this energy will be produced and sold either to the state-owned grid or to large utilities, following auctions aimed to select the solar company capable to sell solar power at the lowest cost. In one of these auctions, in early 2015, the Dubai Electricity & Water Authority signed a contract with a company that will now build and undertake maintenance of a 200 MW solar PV plant (using thin film solar modules) selling electricity at $5.98 cents per kWh, which at that time was the lowest tariff quoted for any solar power project in the world [31].

The PV Industry

Originally small and concentrated in Japan and in the United States, the PV industry has undergone a full reshaping in the first decade of the 2000s [32], first by new German manufacturers and then by the massive market entrance of Chinese PV companies [33]. Today, seven out of the top ten largest manufacturers are based in China (Table 2) [34].

From a technical viewpoint, nine out of the top ten companies utilize silicon as photoactive materials, with only one (First Solar) relying on a different semiconductor (cadmium telluride). The largest companies are vertically integrated, namely they manufacture polysilicon, solar cells, and PV modules.

The industry is rapidly consolidating. For example, in 2014 the top 20 PV module suppliers reached out an overall market share of 68%, up from 60% in 2011 [35].

A quick review of the figures of the 2014 leading PV module manufacturer, which shipped 3.66 GW of solar modules recording total net revenues of $2.29 billion, allows to conclude that the industry is now profitable (a 16.9% margin, in this case), despite an average sale price of $0.62/W [36].

Such a low sale price, compared to the $7–6/W price of 2003, explains why the PV industry in Europe, largely based in Germany, was practically wiped out in less than a decade [37]. Almost all Europe's manufacturers of solar cells and PV modules indeed either closed, or were sold to foreign investors. However, hyper-competition affects also China's manufacturers. Indeed, the former two largest module makers (LDK Solar and Suntech) suffering from high debt and low profit margins, became insolvent in 2013–2014, before lately restarting production either under new ownership or following debt restructuring.

The industry, though, is entering financial and technology maturity as shown by its 13% average profitability estimated in 2015 ($5 billion profits out of $38 billion revenues from

Table 2. Top 10 PV manufacturers in 2014, technologies and location of production sites [Source: EurObserv'ER, 2015].

Company	Modules delivered in 2014 (in GW)	Country	Location of the production sites	Technologies
Trina solar	3.660	China	China	Silicon wafers, solar cells, modules
Yingli green energy	3.361	China	China	Silicon wafers, solar cells, modules
Canadian solar	3.105	China, Canada	China, Canada	Silicon wafers, cells, modules
Jinko solar	2.944	China	China	Silicon wafers, cells, modules
JA solar	2.407	China	China	Silicon wafers, cells, modules
Renesola	1.970	China	China, Japan, Malaysia, South Africa, Poland, South Korea	Silicon wafers, cells, modules
Sharp	1.900	Japan	Japan, USA	Silicon modules, thin film (a-Si, mc-Si)
Motech	1.632	Taiwan	Taiwan, China, Japan, USA	Silicon solar cells, modules
First solar	1.500	USA	Malaysia, USA	Thin film modules (CdTe)
Sun power	1.254	USA	USA, Philippines	Silicon solar cells, modules

module sales). In 2015, global shipments of modules will reach 61 GW, up 27% from 48 GW in 2014, with growth expected in at least 89 countries, and profits increasing 117% from $2.3 million in 2014 to $5 billion in 2015 [38].

Accordingly, companies are currently expanding their capacity at fast pace. For example, the world's largest manufacturer (Trina Solar) is currently building a plant in Thailand, whereas the third (Canadian Solar) and fifth global producer (Jinko Solar) are building new factories in China and in Malaysia, respectively. Almost all other manufacturers are similarly boosting their solar module production with new production lines, or by running existing production lines at full capacity.

Numerous manufacturers of excellent solar modules exist beyond the top ten ranked in Table 2. On a global scale, there are hundreds of PV module suppliers even though only 3% of them is vertically integrated and uses robotic lines to manufacture high quality solar modules (Tier 1 manufacturers, Fig. 7; even though clear quality differences exist among the modules of different Tier 1 suppliers) [39], with some using a business model integrating upwards the value chain so as to become supplier of energy, rather than of a product (the solar panel).

For example, Enel Green Power manufactures in Sicily (Italy) about 200 MW of thin film (mc-Si) solar modules, which are not sold to customers in the marketplace but rather used by the same company to realize utility scale PV plants across the globe, with which the company is able to win bids in Brazil for profitably supplying energy for 20 years at a tariff as low as $8.7 cent per kWh [41].

Similarly, First Solar not only manufactures and sells the most widely adopted thin film solar modules, but it also builds large PV plants scattered throughout the world and selling solar energy to widely different energy buyers.

Nor one should think that scientifically advanced, huge countries such as India or Russia will not develop their own PV industry while photovoltaic energy is becoming a strategic energy resource.

Indeed, Russia in early 2015 witnessed the debut of its first large PV manufacturer (Hevel, manufacturing thin film modules using the silicon-based micromorph technology at a new plant with 130 MW per year capacity) [42].

In India, where a relatively large PV industry was present (15 companies with cell manufacturing equipment and 48 with module equipment, even though capacity is far from being fully exploited: Table 3) [43], foreign PV companies currently installing large PV plants in the country are as well establishing advanced manufacturing facilities based upon Chinese technology in order to lower the production cost.

The overall picture arising from PV industry trend shows that business maturity has been largely achieved,

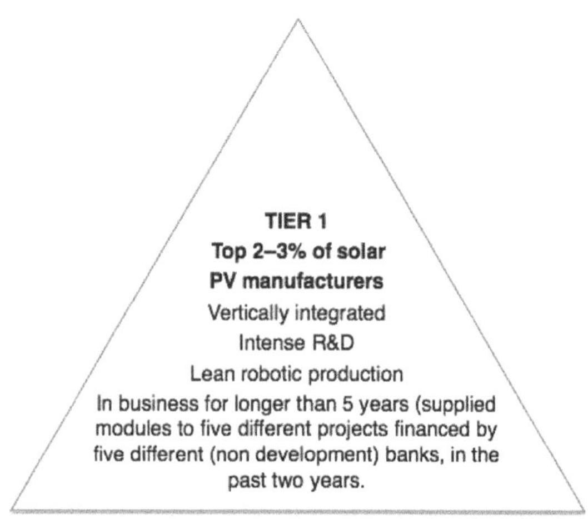

Figure 7. Common differentiation among manufacturers of PV modules uses the Tier 1 nomenclature [Adapted from Ref 40, with kind permission].

Table 3. India PV manufacturing capacity and actual production in 2014 [Source: India: PV Manufacturing Survey 2014].

Solar cell capacity (MW)	1386
Solar cell production (MW)	297
PV module capacity (MW)	2756
PV module production (MW)	1305

in terms both of size and location of the top manufacturing facilities.

Barriers to Overcome

The main barrier to further penetration of PV generation in electricity markets relates to the value of generated electricity, which is declining with increasing penetration, because PV power is intermittent, and electricity in most industrialized countries is traded on energy-only markets. This can be clearly observed, for example, in Germany where, as mentioned in Section 1, the spot market price turns negative on days with very high PV penetration.

While noting that such prices can be only nominally negative, because customers' power bills are still charged with PV subsidies, undoubtedly – if this trend continues and electricity markets in these countries is not reformed fundamentally – PV energy will barely become economically self-sufficient. For instance, in markets where PV electricity provides a significant part of the electricity supply, such as in Italy or in Germany, the need for increased self-consumption to maximize the value of the PV electricity during the sunny spring and summer days when market prices are very low or even negative becomes self-evident.

The technical solutions available that are technically and economically feasible on a large scale, include distributed storage (with batteries capable of absorbing and delivering power with sub-second response times) [44], centralized storage [45], geographical distribution [46], and demand-side management [47]. All are developing quickly with the cost of lithium-ion battery packs having fallen 8% annually between 2007 and 2014 to reach the unprecedented low value of $300 per kWh for battery packs used by leading electric vehicle manufacturers [48], a much lower cost than previously reported that will shortly translate into massive deployment of electric vehicles, a market in its turn growing at almost exponential rate since 2011 [49].

Turning to the regulatory solutions, Hildmann et al. have lately shown that energy-only power markets seem to work even for high renewable energy sources (RES) penetration scenarios (» 25%) if the day-ahead market's share of overall load demand is increased and the true marginal costs of RES units in the merit-order is used (negative prices disappear in most days, and the RES production volume settled) [50].

In order to achieve genuine parity with power generation via combustion of fossil fuels, an energy source should be functionally equivalent. Natural gas, coal, and uranium fossil fuels effectively store chemical or nuclear energy that is released exactly when needed.

This need will lead to the second innovation wave of the solar energy revolution, namely the need to overcome the difficulty of storing electricity in the huge amounts required to meet electricity demand 24 h a day, 7 days a week, 365 days a year. The development of cost-effective storage technologies, in other words, is the last significant technical hurdle to solve the heavy intra-daily intermittence and seasonal cycle of solar energy. For example, 80% of the energy output of a typical PV plant installed in Europe is concentrated in 4–5 months [16].

Both solar hydrogen [51] produced by water electrolysis, and newly developed batteries [52], will be massively used to store and release the electricity produced in large excess during the sunny months of the year.

The cost of hydrogen, indeed, is up to 97% defined by the cost of the PV component, while the material selection for the electrolysis components has vanishingly low effect [53]. Now that is available at low cost, thus, solar hydrogen can be massively used either to power fuel cell cars and trucks, or burned to generate power as it happens since years at Italy's Porto Marghera (Venice) hydrogen-fueled thermoelectric power unit [54].

A thorough net energy analysis [45] referring to the best and most widely installed battery (Li-ion) and hydrogen technology (alkaline water electrolyzer and a PEM fuel cell) for grid storage suggests that the round-trip efficiency of PEM fuel cells must improve dramatically before they can offer the same overall energy efficiency as batteries, which have round-trip efficiencies of 75–90% vs. 30% for PEM cells. From such a net energy analysis, batteries are preferable a PEM fuel cell to store photovoltaic over-generation.

On the other hand, the latter technology is perfectly suited to store over-generation from wind turbines, because its high electrical energy stored on invested (ESOI) ratio and the high energy return on invested (EROI) of wind generation offset the low round-trip efficiency.

In general, a similar dynamic net energy analysis [55] applied to the main PV technologies (mono-, poly-crystalline, amorphous and ribbon silicon Si, CdTe, and CIGS) reveals that energetically expensive solar PV can afford about 24 h of storage before the industry operates at an energy deficit. Said another way, PV systems could be deployed with enough storage to back up the natural day-night cycle and the PV industry could still operate at a surplus, supplying a net electricity yield to society even after taking into account the electricity to deploy new generation and storage capacity.

Outlook and Conclusions

The niche-size photovoltaic industry relying until the late 1990s on waste semiconductor-grade silicon used to manufacture integrated circuits (2100 tons of rejected material, or 10% of silicon used by the integrated circuit industry, in 1997) [56] has evolved into a 40 billion industry growing at >25% annual rate, supplying low-cost solar modules to customers across the world either for decentralized electricity generation, or for utility scale generation via several hundreds of MW (or even >1000 MW) large solar PV plants.

The consequences of this dramatic growth are that the price of solar modules as well as of the inverters [57] has decreased to a such low levels that in early 2015 the price of the solar kWh has reached and gone below the $6c threshold, namely around the cost of electricity produced by burning coal. The largest solar PV plants are no longer supported by generous subsidies, but sell power at low cost for 20 or 25 years in unsubsidized markets according to power purchase agreement (PPA) contracts.

Similarly, residential PV systems for decentralized electricity generation are now a global reality, both in economically developed and in developing countries [58].

This study clarifies in a single report several practically relevant aspects, and identifies the main barriers to the transition of the global economy from fossil fuels to solar electricity generated by low-cost photovoltaic technology. The role of low-cost PV solar energy in solving the world's energy and environmental concomitant crises appears likely to be significant.

For the entire economy running on solar PV (and wind) electricity within the next three decades, three main barriers – namely the rapidly decreasing value of PV electricity with increasing penetration due to intermittency and electricity trading in most industrialized countries on energy-only markets, along with accelerated electrification of transport and heating – must be overcome.

Eventually, along with rapidly increasing installation of PV power, the two central energy technologies of the 19th and 20th century, namely the internal combustion engine and the fuel-powered heater, will be replaced by electric motors and electricity-driven heat pumps. Such ongoing end-use energy electrification [59] will afford the prolonged economic and environmental benefits that will solve the current energy and environmental crises, opening the route to the solar-based economy: the helionomics.

Acknowledgments

This article is dedicated to Peter F. Varadi, co-founder of Solarex and solar pioneer, for all he did to advance the solar energy cause in the United States and worldwide. We thank two anonymous reviewers for their outstanding comments and suggestions.

Conflict of Interest

None declared.

References

1. Green, M. A. 2005. Silicon photovoltaic modules: a brief history of the first 50 years. Prog. Photovolt: Res. Appl. 13:447–455.
2. Varadi, P. F. 2014. Sun above the Horizon. Pan Stanford Publishing, Singapore.
3. Scheer, H. 2006. Energy autonomy. Earthscan/James & James, London.
4. Suri, M. Bankable performance assessment of PV projects using satellite-based solar data, 3rd Annual Conference "PV Power Plants 2012", Vienna, 29–30 March 2012.
5. Wesoff, E. "First solar reaches 16.3% efficiency in production PV modules", greentechmedia.com, 4 May 2015.
6. Morris, N. "The $1 per watt solar illusion", businessspectator.com.au, 31 March 2014.
7. Breyer, C., and A. Gerlach. 2013. Global overview on grid-parity. Prog. Photovolt: Res. Appl. 21:121–136.
8. Meneguzzo, F., F. Zabini, R. Ciriminna, and M. Pagliaro. 2014. Assessment of the minimum value of photovoltaic electricity in Italy. Ener. Sci. Engineer. 2:94–105.
9. Hall, M. "German renewables pushing wholesale electricity further into negative territory", pv-magazine, 17 April 2015.
10. IHS Technology. Top solar power industry trends for 2015, El Segundo, CA, January 8, 2015.
11. Previ, A., A. Iliceto, G. Belli, A. Buonarota, S. Guastella, and R. Patanè. 1995. Long term operational experience at Vulcano PV plant. Proceedings of the 13th European Photovoltaic Solar Energy Conference, Nice, France, 351–355.
12. Jordan, D. C., and S. R. Kurtz. 2013. Photovoltaic Degradation Rates—an Analytical Review. Progr. Photov. Res. Appl. 21:12–29.
13. Bhandari, K. P., J. M. Collier, R. J. Ellingson, and D. S. Apul. 2015. Energy payback time (EPBT) and energy return on energy invested (EROI) of solar photovoltaic systems: A systematic review and meta-analysis. Renew. Sust. Energ. Rev. 47:133–141.
14. Hall, M. "Dead Sea PV plant brings affordable energy to Palestinian households", pv magazine, 14 January 2015.
15. Global Energy Statistical Yearbook 2015. Enerdata, Grenoble, France, 2015. Available at https://yearbook.enerdata.net/.
16. Suri, M., T. A. Huld, E. D. Dunlop, and H. A. Ossenbrink. 2007. Potential of solar electricity generation

in the European Union member states and candidate countries. Sol. Energy 81:1295–1305.
17. SunPower makes the highest efficiency solar panels you can buy for your roof. Source: Highest of over 3200 silicon solar panels, Photon Module Survey, Feb 2014.
18. Fthenakis, V., and H. Chul Kim. 2009. Land use and electricity generation: A life-cycle analysis. Renew. Sust. Energ. Rev. 13: 1465–1474.
19. Varadi, P. F. The future of the large German utilities: it's already here, Energy Post, November 11, 2013.
20. REN21. 2015. Renewables 2015 global status report, Paris.
21. Pene, R. (General Manager Tokelau Energy). "Tokelau solar energy - where we are at...", Tokelau: 28 August 2014 http://tokelau.org.nz/site/tokelau/files/SIDS/SPREP-EnergyPresentation28Aug14.pdf
22. Terna. Rapporto Mensile sul Sistema Elettrico. Consuntivo di Dicembre, Rome: January 2015.
23. Mittal, S. "Coming soon: detailed plan on India's 100 GW solar power ambition", cleantechnica.com, 18 May 2015.
24. Shaw, V. "China: PV installed capacity grows to almost 30 GW in 2014", 16 February 2015.
25. Vivoda, V. 2012. Japan's energy security predicament post-Fukushima. Energy Pol. 46:135–143.
26. Huenteler, J., T. S. Schmidt, and N. Kanie. 2012. Japan's post-Fukushima challenge – implications from the German experience on renewable energy policy. Energy Pol. 45:6–11.
27. Shahan, Z. "Clean-energy-loving germany increasingly exporting electricity to nuclear-heavy France", cleantechnica.com, 9 February 2012.
28. Meza, E. "Yingli to supply 300 MW Neoen PV plant in France", pv magazine, 24 November 2014.
29. U.S. solar market insight 2014 in review, Solar Energy Industries Association, Washington, USA. 2015. Available at http://www.seia.org/research-resources/solar-market-insight-report-2014-q4
30. Latin America PV Playbook: Q1 2015, GTM Research, Boston, USA. 2015. Available at http://www.greentechmedia.com/research/subscription/latin-america-pv-playbook.
31. Chadha, M. "ACWA power secures $344 million loan for solar PV project with World's the lowest tariff", cleantechnica.com, 4 March 2015.
32. Grau, T., M. Huo, and K. Neuhoff. 2011. Survey of photovoltaic industry and policy in Germany and China. CPI Report, Berlin.
33. Sun, H., Q. Zhi, Y. Wang, Q. Yao, and J. Su. 2014. China's solar photovoltaic industry development: The status quo, problems and approaches. Appl. Energ. 118:221–230.
34. Photovoltaic barometer 2015, EurObserv'ER, Paris, France, 2015. Available at http://www.eurobserv-er.org/photovoltaic-barometer-2015/.
35. IHS. "Revenue for leading solar PV module suppliers reached three-year high in Q4 2014", Beijing: 10 February 2015.
36. Osborne, M. "Trina Solar increases module shipments 41.9% to 3.66 GW in 2014", pv-tech.org, 4 March 2015.
37. SPV Market Research. 2014. Photovoltaic manufacturers shipments: capacity, price & revenues 2013/2013, San Francisco.
38. IHS. "PV module gross profits will top $5 billion in 2015, which is more than double the profits garnered in 2014", London: 6 May 2015.
39. REC Solar. "Independent test results confirm all REC solar panels are PID-Free", Munich: October 2014.
40. Pike Research. Solar demand dynamics: cost structures, policy factors, and competitive differentiators for suppliers: market analysis and forecasts, boulder, CO: April 2011.
41. Bloomberg New Energy Finance. "Solar energy makes landfall in Brazil at record-setting low price", 4 November 2014.
42. Hevel. "Hevel to launch full-cycle PV module factory in Russia", 17 February 2015.
43. Osborne, M. "The sorry state of PV manufacturing in India", pv-tech.org, 25 June 2014.
44. Hill, C. A., M. C. Such, C. Dongmei, J. Gonzalez, W. M. Grady. 2012. Battery Energy Storage for Enabling Integration of Distributed Solar Power Generation, in Smart Grid, IEEE Transactions on, 3:850–857.
45. Pellow, M. A., C. J. M. Emmott, C. J. Barnhart, and S. M. Benson. 2015. Hydrogen or batteries for grid storage? A net energy analysis. Energy Environ. Sci. 8:1938–1952. doi:10.1039/c4ee04041d.
46. Marcos, J., L. Marroyo, E. Lorenzo, and M. García. 2012. Smoothing of PV power fluctuations by geographical dispersion. Prog. Photovolt. Res. Appl. 20:226–237.
47. Widén, J., J. Munkhammar. 2013. eceee Summer Studies Proceedings, 1223–1233.
48. Nykvist, B., and M. Nilsson. 2015. Rapidly falling costs of battery packs for electric vehicles. Nat. Clim. Change. 5:329–332.
49. Zentrum für Sonnenenergie-und Wasserstoff-Forschung Baden-Württemberg (ZSW), "More than 740,000 Cars Worldwide Powered by Electricity", press release, 23 March 2015.
50. Hildmann, M., A. Ulbig, and G. Andersson. 2014. Revisiting the merit-order effect of renewable energy sources, department of information technology and electrical engineering discussion paper, Swiss Federal Institute of Technology (ETH). arXiv:1307.0444.
51. Pagliaro, M., and A. G. Konstandopoulos. 2012. Solar hydrogen. RSC Publishing, Cambridge.
52. Cho, J., S. Jeong, and Y. Kim. 2015. Commercial and research battery technologies for electrical energy storage applications. Prog Energ. Combust. 48:84–101.

53. Rodriguez, C. A., M. A. Modestino, D. Psaltis, and C. Moser. 2014. Design and cost considerations for practical solar-hydrogen generators. Energy Environ. Sci. 7:3828–3835.
54. Pagliaro, M., A. G. Konstandopoulos, R. Ciriminna, and G. Palmisano. 2010. Solar hydrogen: fuel of the near future. Energy Environ. Sci. 3:279–287.
55. Carbajales-Dale, M., C. J. Barnhart, and S. M. Benson. 2014. Can we afford storage? A dynamic net energy analysis of renewable electricity generation supported by energy storage. Energy Environ. Sci. 7:1538–1544.
56. Tsuo, Y. S., G. M. Gee, P. Menna, D. S. Strebkov, A. Pinov, and V. Zadde. Environmentally benign solar cell manufacturing, 2nd World Conference and Exhibition on Photovoltaic Solar Energy Conversion, Vienna, 6–10 July 1998.
57. Gifford, J. IHS: Inverter price declines to boos BoS cost reductions, *pv-magazine*, 5 March 2015.
58. Alstone, P., D. Gershenson, and D. M. Kammen. 2015. Decentralized energy systems for clean electricity access. Nat. Clim. Change. 5:305–314.
59. McCollum, D., V. Krey, P. Kolp, Y. Nagai, and K. Riahi. 2014. Transport electrification: A key element for energy system transformation and climate stabilization. Clim. Change. 123:651–664.

10

Irradiance adjustment system developed for various types of solar cells and illumination conditions

Yoshihiro Nishikawa[1], Kiyoshi Imai[1], Keiji Miyao[1], Satoshi Uchida[2], Daisuke Aoki[3], Hidenori Saito[3], Shinichi Magaino[3] & Katsuhiko Takagi[3]

[1]KONICA MINOLTA, INC., 3-91 Daisen Nishi-machi, Sakai-ku, Sakai City, Osaka 590-8551, Japan
[2]The University of Tokyo, 4-6-1 Komaba, Meguro-ku, Tokyo 153-8904, Japan
[3]Kanagawa Academy of Science and Technology, 3-2-1 Sakado, Takatsu-ku, Kawasaki-shi, Kanagawa 213-0012, Japan

Keywords
DSR method, *I–V* characterizations, organic solar cell, reference cell, spectral responsivity, spectroradiometer

Correspondence
Yoshihiro Nishikawa, KONICA MINOLTA, INC. Osaka 580-8551, Japan.
E-mail:
yoshihiro.nishikawa@konicaminolta.com
Shinichi Magaino, Kanagawa Academy of Science & Technology (KAST), Kanagawa 213-0012, Japan.
E-mail: magaino@newkast.or.jp

Funding Information
The authors express their sincere appreciation to the Japan Science and Technology Agency (JST) Development of Advanced Measurement and Analysis Systems funding program for their financial support.

Abstract

A new irradiance adjustment system for the light source named "Programmable Reference cell system for Irradiance adjustment by Spectral Measurement (PRISM)" has been developed for photovoltaics evaluation. It can be applied for various kinds of solar cells and illumination conditions without the preparation of conventional reference cells. This system consists of a device for absolute spectral responsivity measurements, that is, the spectral responsivity unit, and a device for irradiance adjustment of the light source, that is, the spectral reference cell. First, the short circuit current of the target solar cell (I_{ref}) is calculated by convolution of the absolute spectral responsivity with the spectral irradiance under STC (AM1.5 G, 1 kW/m^2, 25°C) as defined in IEC 60904-3. The spectral irradiance of the light source is then adjusted for *I–V* measurements so that the short-circuit current of the spectral reference cell (I_{cal}), calculated by convolution of the absolute spectral responsivity with the spectral irradiance of the light source, agrees with I_{ref}. The difference between I_{ref} and the short circuit current, as determined at an internationally recognized test center under STC, was within ±2% for crystalline silicon solar cells and ±3% for organic solar cells. The difference between I_{cal} and the short-circuit current measured under a solar simulator or LED in which the irradiance was adjusted by this system was within ±4% when the irradiance was higher than 4 W/m^2.

Introduction

Crystalline silicon solar cells are used in the largest quantities for solar energy systems, however, their high manufacturing cost and heavy weight prevent the widespread use of solar energy. To address this problem, various types of solar cells, for example, compound semiconductors, organic thin films, dye-sensitized, and perovskite [1] solar cells are under investigation by various groups with some of them being already commercialized.

Further development of materials and cell structure as well as appropriate evaluation methods for cell functions will be required to gauge and improve the overall performance of solar cells. International standards for the evaluation of silicon solar cells have been established by the IEC as the 60904 Series, however, they define only test methods under standard test conditions (STC) although, when compared to crystalline silicon-based cells, higher performance can be expected for some types of solar cells under weak irradiation such as faint morning, evening, indoor or oblique light [2]. Moreover, new types of solar cells sometimes have characteristics different from that of silicon solar cells such as nonlinearity between the current and light intensity, a light soaking effect, and slow response time, which are not taken into consideration in the IEC 60904 standards [3, 4].

The reference cell method is widely used for irradiance adjustment of a solar simulator [5] and this method requires the preparation of a reference cell as defined in IEC 60904-2, 4 [6, 7]. However, preparation of a reference cell with a spectral response that matches the solar cell being tested is difficult and extremely time-consuming [8, 9], especially, for such unstable cells as dye-sensitized solar cells (DSCs). It is necessary to prepare a filtered reference cell with its relative spectral responsivity coincident with that of the organic PV to be tested, which should then be calibrated at a recognized institute. However, this may take several months to 1 year since the filtered reference cell must also be checked for long-term stability. In this work, we have developed a new irradiance adjustment system for the light source named, "Programmable Reference cell system for Irradiance adjustment by Spectral Measurement (PRISM)" which can be applied for any type of solar cell under various illumination conditions without the use of conventional reference cells [10–19]. In this work, the performance of this system has been investigated for various solar cells and illumination conditions.

Principle

Basic procedures

The PRISM system consists of a device for absolute spectral responsivity measurements (spectral responsivity unit: SK-1150) which meets IEC 60904-4, 8 standards [7, 8] and a device for light source irradiance adjustment (spectral reference cell: SRC-1100), as shown in Figure 1. The basic procedures for measurements at STC (AM1.5G, 1 kW/m^2, 25°C) are as follows:

1 Spectral responsivity of the target solar cell ($S(\lambda)$) is measured with the SK-1150 unit.
2 The short-circuit current of the target solar cell (I_{ref}) is calculated by convolution of the absolute spectral responsivity $S(\lambda)$ with the spectral irradiance at STC (AM1.5G, 1 kW/m^2, 25°C) as defined in IEC 60904-3 ($E_{AM1.5G}(\lambda)$) [9], using the SRC-1100 reference cell device, as follows:

$$I_{ref} = \int S(\lambda) \cdot E_{AM1.5G}(\lambda)\, d\lambda \qquad (1)$$

3 The spectral irradiance of the solar simulator ($E_{ss}(\lambda)$) is adjusted until the short-circuit current of SRC-1100 (I_{cal}), calculated from equation (2), agrees with I_{ref}.

$$I_{cal} = \int S(\lambda) \cdot E_{SS}(\lambda)\, d\lambda \qquad (2)$$

4 Current-voltage characteristics (*I–V* curves) of the target cell are measured by a solar simulator, the irradiance of which is adjusted through the above procedures.

Measurement of nonlinear cells

The short-circuit current of some types of organic solar cells is not always proportional to the irradiance. In this case, equation (1) should be modified to

$$I_{ref} = \int S(\lambda, E_b) \cdot E_{AM1.5G}(\lambda)\, d\lambda, \qquad (3)$$

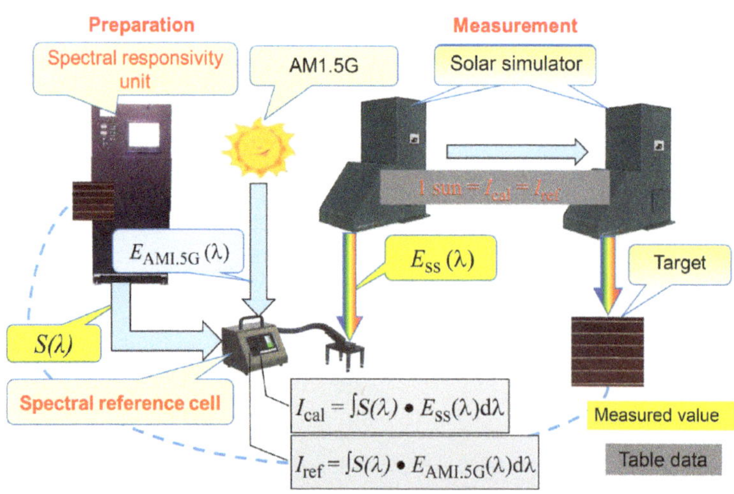

Figure 1. Schematic diagram of the PRISM system. PRISM, Programmable Reference cell system for Irradiance adjustment by Spectral Measurement.

where $S(\lambda, E_b)$ indicates the spectral responsivity which may vary with the irradiance ($E_b = \int E_{ss}(\lambda)\, d\lambda$). However, it is difficult to measure $S(\lambda, E_b)$ by commonly available devices since they are not able to regulate the irradiance level to desired values. To address this problem, we have developed the PRISM system which uses the differential spectral responsivity (DSR) method developed by Physikalisch-Technische Bundesanstalt (PTB) and defined in IEC 60904-4, 8 and ISO 15387 [7, 8, 20, 21].

Measurements under various illumination conditions

The PRISM system employing the DSR method can be applied for measurements under any illumination conditions. Thus, the short-circuit current (I_{sc}) for any solar cell under any illumination conditions can be calculated as follows:

$$I_{sc} = \int S(\lambda, E_b) \cdot E(\lambda)\, d\lambda, \qquad (4)$$

where $E(\lambda)$ is the spectral irradiation from any light source.

Key Devices for the PRISM System

Spectral responsivity unit (SK-1150)

The SK-1150 unit has been constructed to simultaneously illuminate the bias light irradiance (E_b) and monochromatic light irradiance onto a solar cell. Both lights are combined in the lens and irradiated on the solar cell. The irradiation unit consists of a halogen lamp, Xenon lamp, grating and chopping units, and light monitors to compensate for any instability of the lamps. The current measurement unit consists of a source meter and lock-in amplifier, both of which are controlled by a computer.

The absolute spectral responsivity of this device is measured using a standard detector calibrated at an internationally recognized test center. Differential spectral responsivity is measured at various irradiances of bias light and $S(\lambda, E_b)$ is then automatically calculated by a computer [7, 8, 20, 21].

When the current response to the applied light irradiation is slow, for example, in the case of dye-sensitized solar cells (DSCs), the chopping frequency of the monochromatic light should be low enough to obtain a steady-state current. Aoki et al. have reported that $S(\lambda, E_b)$ decreases as the chopping frequency is increased [3], indicating that it should be lower than 3.3 Hz to obtain accurate data under 1 sun bias light irradiation. However, spectral responsivity measurement using the lock-in amplifier is unsuitable for low-frequency measurements since long and repetitive measurements are required to obtain accurate data, for example, several hours at 0.3 Hz.

To address this problem, this system has two methods to determine $S(\lambda, E_b)$, that is, the selective amplification of the current caused by monochromatic light irradiation using the lock-in amplifier and the calculation of the difference between currents measured when the shutter of the monochromatic light is open and closed using the DC current amplifier. The latter method is similar to the AC-a method reported by Guo et al. [22], and requires less measuring time because the lock-in amplifier method takes a chopping frequency of 10 cycles.

Spectral reference cell device (SRC-1100)

The SRC-1100 device consists of a spectroradiometer for spectral irradiance measurement of the light source, an optical receiver and computer. With this device, irradiance adjustment of the solar simulator can be performed according to the procedures described in the Basic procedures section.

The spectroradiometer equipped in this device has a high sensitivity image sensor of back-thinned CCD to measure very low irradiance, such as indoor light, as compared to AM 1.5 G (1 kW/m^2). Additionally, it has a unique optical system to reduce stray light which causes inaccurate readings. Figure 2 shows the transmittance level of the stray light measured in accordance with JIS Z8724 (Japan Industrial Standards, Methods of measurement Light-source colour) [23].

The irradiance measured by the spectroradiometer should have a linear relationship with the short-circuit current of the linear cell under a wide irradiation range, for example, from indoor light to 1 sun. The short-circuit current (I_{sc}) and linearity error (ε) were determined for the linear cell (Konica Minolta, AK-200) by the spectroradiometer and were plotted against the *Irradiance* ($\int E_{ss}(\lambda) d\lambda$), as shown in Figure 3 (Konica Minolta is hereafter referred to as KM). The ε was calculated as follows:

$$\varepsilon = (I_{sc}/I_{mes} - 1) \times 100, \qquad (5)$$

where

$$I_{sc} = \int S(\lambda) \cdot E_{ss}(\lambda)\, d\lambda.$$

The linearity error (ε) was within ±0.2% at an irradiation range of 10–1000 W/m^2, as shown in Figure 3.

The sensitivity of a spectroradiometer is known to change with time [24] by vibration and degradation of its optical components, etc., while instrument-to-instrument variations also exist [25, 26]. No change in the relative spectral responsivity of the standard lamp was measured, but about a 5% decrease in irradiance was measured by the spectroradiometer in 100 days.

The SRC-1100 is first calibrated at the manufacturing factory using a standard lamp traceable to that of a recognized national institute; however, as mentioned above, the measured value changes with time. To reduce these effects, SRC-1100 has a user calibration function to resolve changes before measurements. This apparatus has a stable reference cell (KM, AK-200) with absolute spectral responsivity ($S_{rc}(\lambda)$) traceable to the calibrated value of a national institute. First, the reference cell is placed under the light source and the short-circuit current (I_{mes_rc}) is measured. Next, the irradiance ($E_{ss}(\lambda)$) is measured and the absolute irradiance calculated by multiplying ($E_{ss}(\lambda)$) by a correcting factor ($K = I_{mes\_rc}/E_{ss}(\lambda) \cdot S_{rc}(\lambda) d\lambda$). If the sensitivity of SRC-1100 shows no change with time, K equals 1.

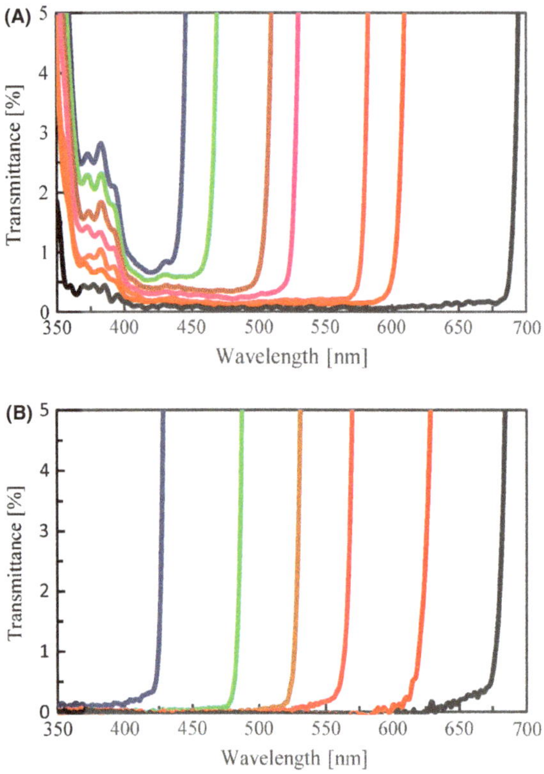

Figure 2. Transmittance level of stray light determined for: (A) a conventional spectroradiometer; and (B) the spectroradiometer equipped in this device.

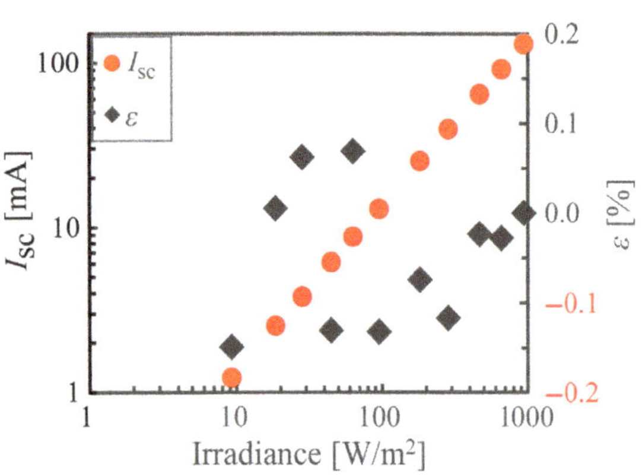

Figure 3. Short-circuit current (I_{sc}) and linearity error (ε) of SRC-1100 plotted against the irradiance. SRC, spectral reference cell

Intercomparison of Short Circuit Currents Among Calibration Centers

Intercomparisons of the short circuit currents were carried out at three internationally recognized calibration centers, that is, the National Institute of Advanced Industrial Science & Technology (AIST), National Renewable Energy Laboratory (NREL), and PTB [27] for multiple

Table 1. Types of reference cells.

Reference cells	Type	Centroid wavelength
AK-100	Amorphous Si-type	521 nm
AK-110	Microcrystalline Si-type	791 nm
AK-120	Top layer of triple junction	499 nm
AK-130	Middle layer of triple junction	664 nm
AK-140	Bottom layer of triple junction	831 nm
AK-200	Crystalline Si-type	772 nm
AK-300	DSC N719 type	553 nm

DSC, dye-sensitized solar cells.

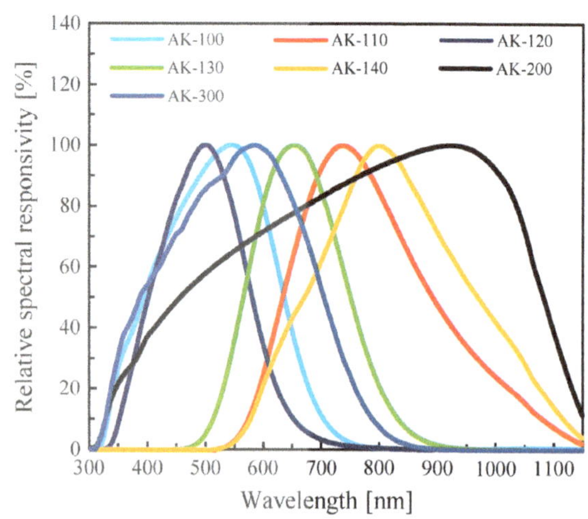

Figure 4. Relative spectral responsivities of the reference cells.

reference cells (KM, AK series) to investigate variations in their measured values. Table 1 and Figure 4 show the type of cells in the AK series and their relative spectral responsivities, respectively. Table 1 also summarizes the centroid wavelengths (W_c) of the reference cells for the AK series which were calculated by equation 6. Here, it should be noted that W_c is the representative wavelength of the spectral responsivity at STC.

$$W_c = \int S(\lambda) \cdot \lambda d\lambda \Big/ \int S(\lambda)\, d\lambda. \qquad (6)$$

The centroid wavelength can be employed as the index of each reference cell to indicate the representative spectral responsivity, thus, allowing comparisons of multiple reference cells. For example, the centroid wavelengths of AK-100 (Amorphous Si-type cell), AK-200 (Crystalline Si-type cell), and AK-110 (Microcrystalline Si-type cell) are 521 nm, 772 nm, and 791 nm, respectively. First, the short-circuit current of each cell was measured at KM, then the secondary reference solar cell was calibrated at AIST. The cells were then sent to PTB and NREL, and variations in the short circuit currents among the three recognized calibration centers were investigated. The difference in short-circuit current with the elapse of time was found to be within ±0.3% in a 2-year period from 2011 to 2013.

Figure 5 shows the differences in the short-circuit current among the test centers (ΔI_{sc_tc}) plotted against the centroid wavelength of the spectral responsivity (W_c) for each cell. The ΔI_{sc_tc} were calculated as follows:

$$\Delta I_{sc\_tc} = (I_{sc\_tc}/I_{sc\_km} - 1) \times 100, \qquad (7)$$

where I_{sc_tc} is the short-circuit current determined at PTB or NREL, and I_{sc_km} is that determined at KM with the secondary reference solar cell calibrated at AIST. Dirnberger et al. have reported intercomparison of the short-circuit current for crystal Si-type and amorphous Si-type modules performed by the Fraunhofer Institute for Solar Energy Systems (ISE), European Solar Test Installation (ESTI of the European Commission, Joint Research Centre, Institute for Energy and Transport, Renewable Energy Unit), and the NREL and AIST [28]. They reported that deviations in the short-circuit current of the crystal Si-type module was 2% (−1.1 to 0.9%), but that of the amorphous Si-type module was −2.9% for the data measured at AIST and 1.6% for that measured at NREL. The value from NREL was found to be 4.5% higher than that of AIST. In this paper, deviation of the short-circuit current was 0.9% for crystalline Si-type cells and 2.8% (0–2.8%) for amorphous Si-type cells. Thus, the value from NREL was 2.8% higher than that of AIST. The tendency for greater deviation of the short-circuit current for amorphous Si-type cells than crystalline Si-type cells was also recognized in this paper. As shown in Figure 5, the ΔI_{sc_tc} values of cells having W_c in the long wavelength range were negative and those of cells having W_c in the short wavelength range were positive, although the reason for this tendency is not yet clear at present. Further investigation is required to analyze this tendency. On the other hand, intercomparison results of this paper were found to be reasonable since the total estimated uncertainty to determine the short-circuit current was 3.8%. The estimated uncertainty of a recognized calibration center was within 1% and within 1.8% at KM. Taking these results into consideration, the maximum deviation under STC for the PRISM system was set within 4%, equivalent to the error range allowance in the above results.

Experimental

Performance verification of SK-1150

The performance of the SK-1150 unit was verified by comparing the short circuit currents of the solar cells at STC with those determined at a recognized test center. That is, the difference in the short-circuit current (ΔI_{sc_stc}) between that determined by SK-1150 (I_{ref}) and

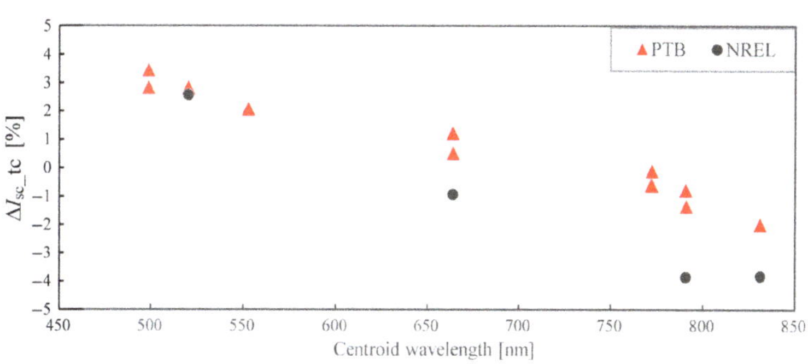

Figure 5. Difference in the short-circuit current among the calibration centers plotted against the centroid wavelength.

that determined at a test center (I_{ref_tc}) was calculated as follows:

$$\Delta I_{sc\_stc} = (I_{ref}/I_{ref\_tc} - 1) \times 100. \quad (8)$$

Performance verification of SRC-1100

Performance of the SRC-1100 device was verified by comparing the short circuit currents calculated from equation (2) (I_{cal}) with those measured under a solar simulator (I_{mes}) or LED. The difference in the short-circuit current between I_{cal} and I_{mes} (ΔI_{sc}) was calculated as follows:

$$\Delta I_{sc} = (I_{cal}/I_{mes} - 1) \times 100. \quad (9)$$

The three types of solar simulators and LED light source used in the evaluations are shown in Table 2. The solar simulators were all class AAA but with differing spectral irradiance distributions in order to examine the spectral irradiance dependence. The source meters were Agilent B2901A, ADCMT 6242, and Advantest R6246.

Spectral responsivity measurements of DSC

Two different types of dye-sensitized solar cells (DSCs) were tested: one cell containing 3-Methoxypropionitrile (MPN) as the solvent for the electrolyte (DSC-A) and the other containing an ionic liquid electrolyte (DSC-B). The viscosity of the ionic liquid electrolyte was about 30 times higher than that containing MPN at 20°C. A mesoscopic TiO_2 semiconductor electrode sensitized by cis-RuLL'-$(SCN)_2$ (L = 2,2′-bipyridyl-4,4′-dicarboxylic acid, L' = 4, 4′-dinonyl-2,2′-bipyridyl) (Z907) dye was used for both cells. The active area for each cell was 0.196 cm^2 [3]. Additionally, DSCs supplied by the National Institute for Materials Science (DSC-NIMS-1 and DSC-NIMS-2) were also used. For these DSCs, a carbazole dye with hexyl-substituted oligothiophenes, MK-2 (Soken Chemical & Engineering Co., Ltd., Tokyo, Japan), was used as a sensitizer. An ionic liquid based on imidazolium iodide was used as the electrolyte. Transparent nanocrystalline TiO_2 films were prepared as follows: The films were formed by a screen printing method, that is, TiO_2 paste (Nanoxide-20N, Nikki Syokubai Kasei) was screen-printed onto a transparent conducting oxide (TCO) glass substrate (F-doped SnO_2, sheet resistance 10 ohm/cm^2, Nippon Sheet Glass), followed by sintering at 500°C for 60 min in air.

Before measurements, the cells were stabilized by light soaking at the same irradiance as that for the bias light. In the case of DSC-NIMS-1, the short-circuit current reached a steady-state value in about 96 h during light soaking at an irradiance of 1 sun. The cell was kept in the dark for about 100 h after 96 h light soaking, and then irradiated again under the same conditions. The short-circuit current reached a steady-state value immediately after irradiation.

The absolute spectral responsivity ($S(\lambda)$) was measured by SK-1150 at 25°C with bias light irradiance levels between 0 to 1.2 sun. Wavelength resolution was 10 nm and the chopping frequency of the monochro-

Table 2. Solar simulators and LED light source.

Type	Manufacturer	Irradiance level	Target cells
XES-40S	SAN-EI ELECTRIC	10–1000 W/m^2	Reference cells/Nonlinear cells/DSC-NIMS/OPV
94023A	NewPort	1000 W/m^2	Reference cells
YSS-T150A	YamashitaDenso	10–1000 W/m^2	DSC-A/DSC-B
LED	Imac	10–300 W/m^2	Reference cells/Nonlinear cells/DSC-NIMS/OPV

DSC-NIMS-1, dye-sensitized solar cells-National Institute for Materials Science; OPV, organic thin film solar cell.

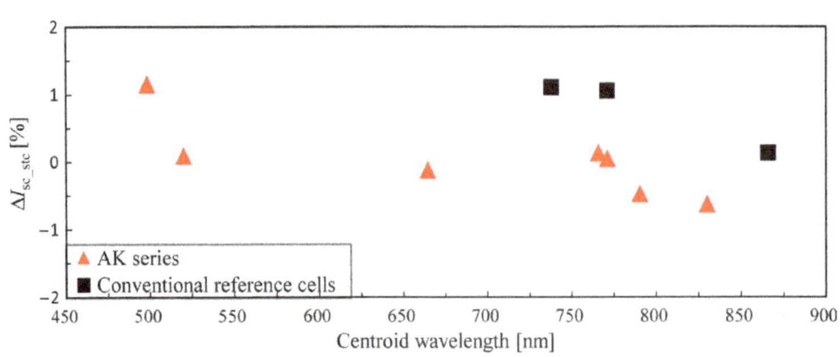

Figure 6. Difference in the short-circuit current (ΔI_{sc_stc}) between that determined by the PRISM SK-1150 unit and that determined at a calibration center for linear cells under STC plotted against the centroid wavelength of the spectral responsivity (W_c). PRISM, Programmable Reference cell system for Irradiance adjustment by Spectral Measurement; SK-1150, Spectral responsivity unit; STC, standard test conditions.

matic light was set low enough to obtain a steady-state current.

I–V measurements

Solar cell efficiencies are, at present, measured at STC, however, by adjusting the light irradiance using the PRISM method, *I–V* characteristics were measured under various irradiances with a solar simulator (San-ei Electric XES-40S) and LED light source (Imac). The voltage sweep rates were set so that an *I–V* curve determined by the forward scan (from short circuit to open circuit) agreed with that by reverse scan. The solar cell efficiency (η) and fill factor (*FF*) were calculated as follows:

$$\eta = \frac{V_{oc} \times J_{sc} \times FF}{P_{rad}} \times 100, \tag{10}$$

$$FF = \frac{P_{max}}{V_{oc} \times J_{sc}}, \tag{11}$$

where V_{oc}, J_{sc}, P_{rad}, and P_{max} represent the open circuit voltage, short circuit current density, incident light intensity and maximum power, respectively.

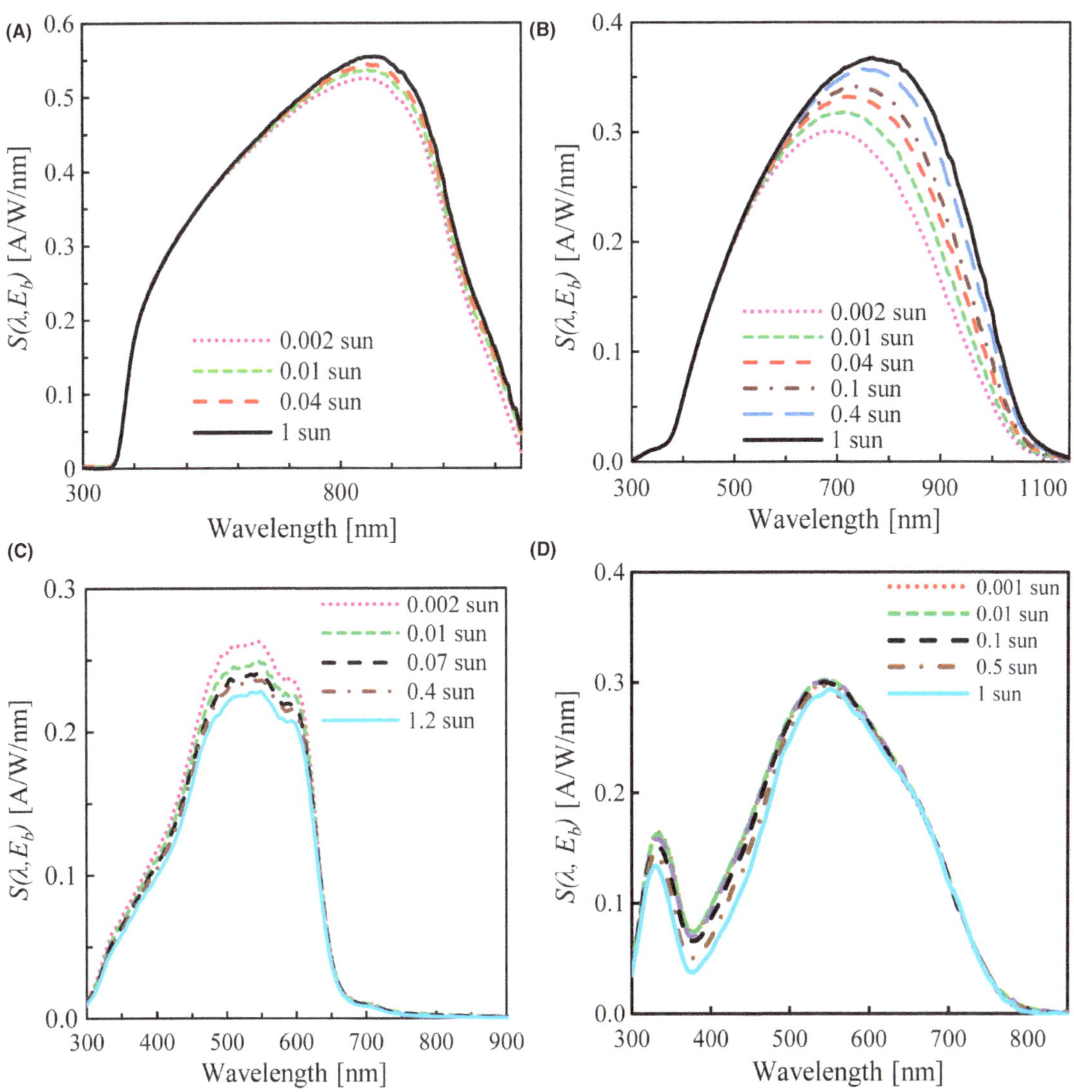

Figure 7. Spectral responsivity ($S(\lambda, E_b)$) determined for two nonlinear cells (cells A and B), OPV and DSC-B: (A) Cell A, (B) Cell B, (C) OPV, (D) DSC-B. DSC, dye-sensitized solar cells; OPV, organic thin film solar cell.

Results and Discussions

Performance of the SK-1150 unit

Figure 6 shows the differences in the short-circuit current (ΔI_{sc_stc}) between that determined by SK-1150 and that determined at a recognized calibration center for linear reference cells at STC plotted against the centroid wavelength (W_c). As shown in Figure 6, the ΔI_{sc_stc} was within ±1.5%.

The spectral responsivity ($S(\lambda, E_b)$) determined for the two nonlinear cells (cells A and B), an organic thin film solar cell (OPV), and a dye-sensitized solar cell containing an ionic liquid electrolyte (DSC-B) are shown in Figure 7. The ΔI_{sc_stc} values determined for these cells are shown in Tables 3 and 4. Nonlinear cells are those in which the spectral responsivities change with the bias light intensity [8]. The $S(\lambda, E_b)$ of both cells increased with an increase in the bias light intensity in a wavelength range over 500 nm, while the degree of increase was larger for cell B. The ΔI_{sc_stc} was 0.1% for cell A and −0.3% for cell B.

Table 3. ΔI_{sc_stc} determined for two nonlinear cells (Cell A and Cell B).

Cell	ΔI_{sc_stc} (%)
A	0.1
B	−0.3

Table 4. ΔI_{sc_stc} determined for OPV and DSC-B.

Cell	ΔI_{sc_stc} (%)
OPV	−1.1
DSC-B	−2.6

DSC, dye-sensitized solar cells; OPV, organic thin film solar cell.

Unlike the nonlinear cells, the $S(\lambda, E_b)$ of the OPV and DSC-B decreased with an increase in the bias light intensity at wavelengths under 650 nm. Similar results have been reported for DSCs by Hara et al. [29], Guo et al. [22] and Aoki et al. [3]. Hara et al. have suggested that this decrease in the spectral responsivity may be related to light absorption in the wavelength range of 400–500 nm caused by I_3^- accumulation in the vicinity of the electrode due to oxidation of I^- and accelerated by bias light irradiation. Table 4 shows that ΔI_{sc_stc} was −1.1% for the OPV and −2.6% for DSC-B.

The maximum deviation under STC for the PRISM system was set within 4%, as mentioned before. The ΔI_{sc_stc} values determined for the linear and nonlinear reference cells, OPV, and DSC were found to be within this maximum deviation.

Performance of the SRC-1100 device

First, the differences in the short-circuit current (ΔI_{sc}) between I_{cal} and I_{mes} calculated from equation (9) for the linear reference cells under STC were determined, and the results are shown in Figure 8. Two sets of the SRC-1100 device with different solar simulators (San-ei Electric XES-40S and Newport Corp 94023A) were employed. As shown in Figure 8, ΔI_{sc} was within ±3%, which is within the maximum deviation set for the PRISM system.

Next, the ΔI_{sc} of the nonlinear cells, OPV, and DSCs were determined by changing the irradiance. A San-ei Electric XES-40S solar simulator was used for the nonlinear cells and the OPV, and a Yamashita Denso YSS-T150A simulator was used for the DSCs. Figure 9 shows the ΔI_{sc} determined for these cells. The ΔI_{sc} of the nonlinear cells and the OPV was within ±4% and larger for cell B than cell A. The ΔI_{sc} of the DSCs was within ±3%

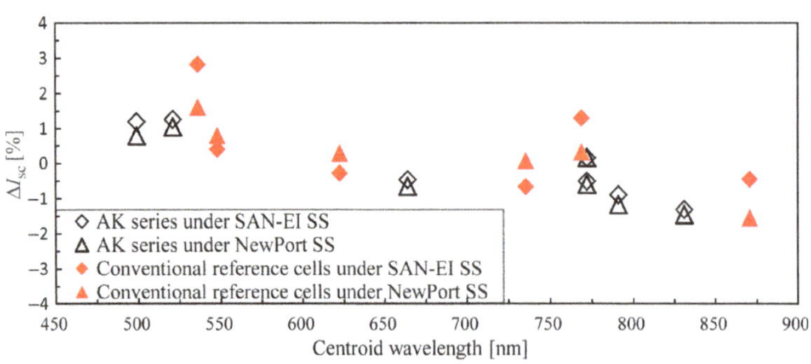

Figure 8. Difference in the short-circuit current (ΔI_{sc}) between I_{cal} and I_{mes} calculated from equation (9) for the reference cells (Konica Minolta AK series and conventional reference cells made by other companies) under the solar simulators.

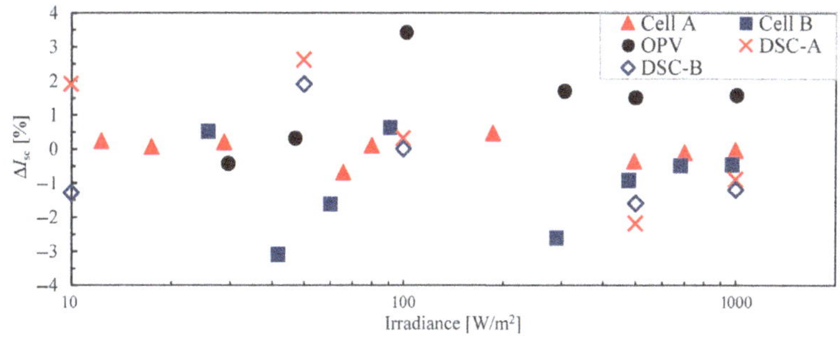

Figure 9. ΔI_{sc} determined for two nonlinear cells (cells A and B), OPV, DSC-A and DSC-B under the solar simulator. OPV, organic thin film solar cell; DSC, dye-sensitized solar cells.

and larger for DSC-A than DSC-B. The ΔI_{sc} determined for DSC-NIMS-1 and DSC-NIMS-2 were within ±4% under the solar simulator with an irradiance range of 10–1000 W/m^2. These results show that the ΔI_{sc} is within the maximum deviation set for the PRISM system even under such low irradiance as 10 W/m^2.

The ΔI_{sc} under LED was also investigated. The ΔI_{sc} for the reference cells under LED with an irradiance of 3.97 W/m^2 and that determined for nonlinear cells by changing the irradiance of LED are shown in Figure 10. The ΔI_{sc} of the reference cells was within ±4% which is within the maximum deviation set for the PRISM system. The ΔI_{sc} of the nonlinear cells was within ±3% when the irradiance was higher than 4 W/m^2. These results show that the PRISM system is applicable not only to sunlight but also LED light when the irradiance is higher than 4 W/m^2. Since an increase in ΔI_{sc} under low irradiance may result from the S/N ratio of the equipment, improvement in the equipment components will also need to be implemented for less uncertainty.

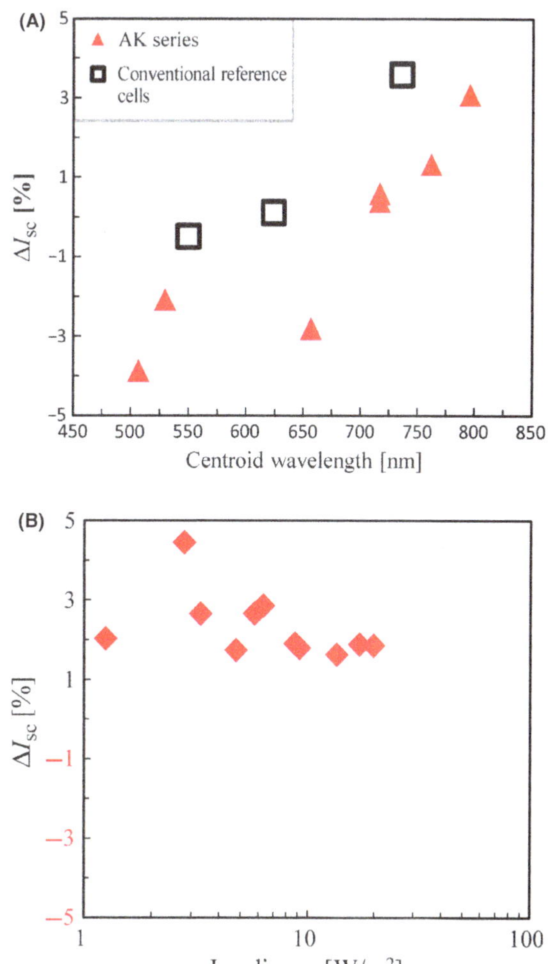

Figure 10. (A) ΔI_{sc} determined for the reference cells under LED with an irradiance of 3.97 W/m^2 and (B) ΔI_{sc} determined for nonlinear cells by changing the irradiance of LED.

I–V characteristic measurements

Changes in solar cell efficiency and open circuit voltage upon irradiance with a light source

The PRISM system was applied to determine changes in the solar cell efficiency (η) and open circuit voltage (V_{oc}) upon irradiance with a solar simulator. The target cells were DSC-NIMS-2 and AK-100, and the solar simulator was a San-ei Electric XES-40S. Figure 11 shows the changes in η and V_{oc} upon irradiance with the solar simulator. The η of the DSC increased with an increase in irradiance at a range from 0.01 to about 0.2 sun, then gradually decreased, whereas the V_{oc} monotonously increased with an increase in irradiance. Contrarily, in the case of AK-100, both η and V_{oc} decreased with a monotonous decrease in irradiance. The rates of decrease in η

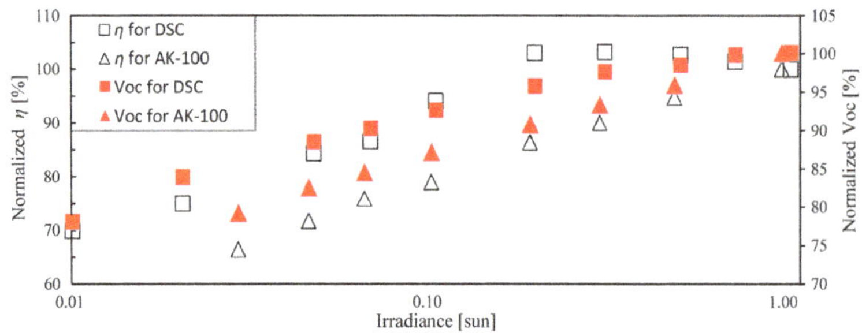

Figure 11. Changes in the solar cell efficiency (η) and open circuit voltage (V_{oc}) determined for DSC-NIMS-2 and AK-100 upon irradiance with a solar simulator. DSC-NIMS-2, dye-sensitized solar cells-National Institute for Materials Science.

Table 5. Performance of OPV under a solar simulator (SAN-EI ELECTRIC XES-40S) and LED (Imac).

Light source	J_{sc} mA/cm^2	V_{oc} V	P_{max} W/m^2	Irradiance W/m^2	η %
Solar simulator	7.05	0.46	18.0	1000 (1 sun)	1.8
LED	0.0815	0.34	0.176	4.57	3.9

OPV, organic thin film solar cell.

Table 6. Performance of DSC-NIMS-1 under a solar simulator (SAN-EI ELECTRIC XES-40S) and LED (Imac).

Light source	J_{sc} mA/cm^2	V_{oc} V	P_{max} W/m^2	Irradiance W/m^2	η %
Solar simulator	13.8	0.60	61.4	1000 (1 sun)	6.1
LED	0.120	0.47	0.345	4.10	8.4

DSC-NIMS-1, dye-sensitized solar cells-National Institute for Materials Science.

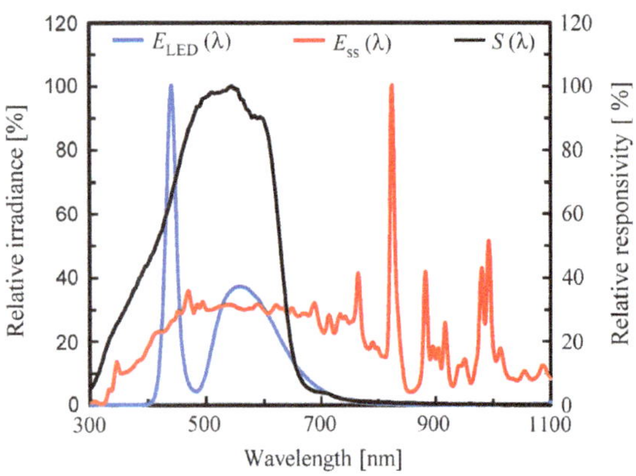

Figure 12. Relationship between the spectral responsivity of the OPV ($S(\lambda)$) and spectral irradiance of the light sources ((E_LED(λ) for LED, E_ss(λ) for the San-ei Electric XES-40S (1 sun)). OPV, organic thin film solar cell.

and V_{oc} for the DSC with a decrease in irradiance were less than of those for AK-100.

Determination of solar cell efficiency under weak LED irradiation

The results shown in Figure 11 suggest that, under weak irradiation, higher performance can be expected for some types of solar cells when compared to crystalline solar cells. For example, organic solar cells can be specifically considered for indoor use due to their flexibility, lightness and high designability, however, a standard method to verify their performance has yet to be established.

The PRISM system was also applied to determine the photoelectric conversion efficiencies of OPV and DSC under LED which exhibits irradiance similar to indoor light. Table 5 shows the performance of the OPV under the solar simulator and LED. The solar cell efficiency (η) under LED was more than twice that obtained under 1 sun. Figure 12 shows the relationship between the spectral responsivity of the OPV and spectral irradiances of the light sources. As shown in Figure 12, the light emission wavelength region of the LED was approximately within the absorption wavelength region of the OPV, however, the wavelength region of natural sunlight is much wider than the absorption wavelength region of the OPV. This may explain why the η obtained under LED was greater than that obtained under 1 sun. The performance of DSC-NIMS-1 under the solar simulator and LED are shown in Table 6. As with the OPV, the η under LED was larger than that under 1 sun.

Conclusions

The newly developed PRISM system, consisting of a SK-1150 unit which measures the absolute spectral responsivity of a target cell and the SRC-1100 device which stores this data along with the reference spectral irradiance and that of the light source in order to adjust its irradiance, was found to be applicable for any type of solar cell under various illumination conditions. Significantly, the SRC-1100 acts as the reference cell, allowing comparison studies without the preparation of conventional reference cells as defined in IEC 60904-2, 4, thus, dramatically reducing the preparation time for *I–V* characterization measurements of a newly developed cell. The PRISM system makes it unnecessary to prepare a filtered reference cell and then confirm its stability over months. It is thus possible to avoid these tedious and lengthy procedures before measurement of organic solar cell performance. This is because the data of the absolute spectral responsivity itself serves the same function as a reference cell.

Regarding accuracy, differences in the short-circuit current between that determined by the SK-1150 and at a recognized test center under STC were within ±2% for the filtered reference cells and nonlinear cells, and within ±3% for organic solar cells. The typical calibration uncertainty at a secondary or working calibration institute is around 2%, meaning that SK-1150′s accuracy is comparable to calibrations at these facilities. Upon irradiance adjustment with this PRISM system, the difference in the short-circuit current between that obtained by convolution of the absolute spectral responsivity with the spectral irradiance of a solar simulator or LED and that measured with the adjusted light source was within ±3% under STC and ±4% under various illumination conditions when the irradiance was higher than 4 W/m^2.

In the case of calibration of the light irradiance under STC conditions using a conventional method with a reference cell, the adjustment accuracy was also comparable the present PRISM method. However, in the case of the calibration of various kinds of light sources with even low irradiance, the PRISM system enables a traceable measurement method for irradiance adjustment which had so far been difficult due to the lack of a suitable reference cell with a certified value. This system will be particularly useful for the preparation and *I–V* measurement of nonlinear organic solar cells which have a variety of spectral responsivities.

Acknowledgments

The authors express their sincere appreciation to the Japan Science and Technology Agency (JST) Development of Advanced Measurement and Analysis Systems funding program for their financial support and encouragement. They thank Dr M. Yanagida of the National Institute for Materials Science for supplying the stable DSC cells used in this work. Sincere gratitude is also extended to the National Institute of Advanced Industrial Science & Technology (AIST, Japan), National Renewable Energy Laboratory (NREL, USA), and Physikalisch Technische Bundesanstalt (PTB, Germany) for their useful advice on the traceable calibration method.

Conflict of Interest

None declared.

References

1. Lee, M. M., J. Teuscher, T. Miyasaka, T. N. Murakami, and H. J. Snaith. 2012. Efficient hybrid solar cells based on meso-superstructured organometal halide perovskites. Science 338:643–647. doi: 10.1126/science.
2. Ito, S., H. Matsui, K. Okada, S. Kusano, T. Kitamura, Y. Wada, et al. 2004. Calibration of solar simulator for evaluation of dye-sensitized solar cells. Solar Energy Mater. Solar Cells 82:421–429. doi: 10.1016/j.solmat.
3. Aoki, D., T. Aoki, H. Saito, S. Magaino, and K. Takagi. 2012. Methods for spectral responsivity measurements of dye-sensitized solar cells. Electrochemistry 80:640–646. doi: 10.5796/electrochemistry.
4. Takagi, K., S. Magaino, H. Saito, T. Aoki, and D. Aoki. 2013. Measurements and evaluation of dye-sensitized solar cell performance. Photochem. Rev. 14:1–12. doi: 10.1016/j.photochemrev.
5. Emery, K. A., and C. R. Osterwald. 1989. Solar cell calibration methods. Solar Cells 27:445–453.
6. IEC 60904-2 2nd edition 2007–03: photovoltaic devices - part 2; Requirements for reference solar devices.
7. IEC 60904-4 1st edition 2009–06: photovoltaic devices - part 4; Reference solar devices - procedures for establishing calibration traceability.
8. IEC 60904-8 3rd edition 2014–05: photovoltaic devices - Part 8; Measurement of spectral responsivity of a photovoltaic (PV) device.
9. IEC 60904-3 2nd edition 2008–04: photovoltaic devices - Part 3; Measurement principles for terrestrial photovoltaic (PV) solar devices with reference spectral irradiance data.
10. Gevorgyan, S. A. 2011. An inter-laboratory stability study of roll to roll flexible polymer solar modules. Sol. Energy Mater. Sol. Cells 95:1398–1468.
11. Hinsch, A. 2013. Status of the dye solar cell technology (DSC) as a guideline for further research. Proceeding of the 28th European PV Solar Energy Conference and Exhibition 2013; 2239–2247.
12. Nishikawa, Y., K. Miyao, S. Tochimoto, K. Imai, and S. Winter. 2012a. The application of spectral reference cell to various photovoltaic cells. The Japan Society of Applied Physics Spring Meeting; March 2012: 16P-GP9-1.
13. Nishikawa, Y., K. Miyao, S. Tochimoto, K. Imai, and S. Winter. 2012b. The application of spectral reference cell to

various photovoltaic cells. The Japan Society of Applied Physics Spring Meeting; March 2012: 16P-GP9-2.
14. Nishikawa, Y., K. Imai, S. Tochimoto, K. Miyao, and S. Winter. 2013. The application of spectral reference cells to non-linear photovoltaic cell. Konica Minolta Technol. Rep. 10:81–87.
15. Uchida, S., and Y. Nishikawa. New evaluation method of photovoltaic by using spectral technology. International Conference on the Evaluation & Standardization of Organic Solar Cells 2013 (ICES2013); January 2013: IL3.
16. Nishikawa, Y. Organic photovoltaic evaluation method by utilizing spectral reference cell. Proceeding of the 93rd Annual Meeting of The Chemical Society of Japan; March 2013: 3H1-31.
17. Uchida, S., Y. Nishikawa, K. Miyao, K. Imai, S. Tochimoto, S. Magaino, et al. The new generation's PV evaluation method by utilizing spectral technology. Proceeding of the 28th European PV Solar Energy Conference and Exhibition 2014; 3552-3556.
18. Uchida, S., Y. Nishikawa, K. Miyao, K. Imai, S. Tochimoto, S. Magaino, et al. The differential spectral responsivity (DSR) technology for evaluation of new generation PV. Proceeding of the 23rd International Photovoltaic Science and Engineering Conference; October 2013: 5-O-5.
19. Nishikawa, Y. New traceable evaluation method for organic photovoltaics. International Conference on the Evaluation & Standardization of Organic Solar Cells 2014 (ICES2014); February 2014: IL5.
20. ISO 15387: 2005 space systems - single-junction solar cells - measurement and calibration procedures.
21. Metzdorf, J., W. Möller, T. Wittchen, and D. Hünerhoff. 1991. Principle and application of differential spectroradiometry. Metrologia 28:247–250.
22. Guo, X.-Z., Y.-H. Luo, Y.-D. Zhang, X.-C. Huang, D.-M. Li, and Q.-B. Meng. 2010. Study on the effect of measuring methods on incident photon-to-electron conversion efficiency of dye-sensitized solar cells by home-made setup. Rev. Sci. Instrum. 81:103106.
23. JIS Z 8724: 1997 methods of color measurement - Light-source colour.
24. Sperfeld, P., D. Dzafic, F. Plag, F. Haas, C. Kröber, H. Albert, et al. Usability of compact array spectroradiometers for the traceable classification of pulsed solar simulators. Proceeding of the 27th European Photovoltaic Solar Energy Conference and Exhibition 2012; 3060–3065.
25. Habte, A., A. Andreas, L. Ottoson, C. Gueymard, G. Fedor, S. Fowle, et al. Indoor and outdoor spectroradiometer intercomparison for spectral irradiance measurement. Technical Report NREL/TP-5D00-61476. 2014: i–21.
26. Galleano, R., W. Zaaiman, P. Morabito, A. Minuto, A. Spena, S. Bartocci, et al. 2012. Intercomparison of spectroradiometers for solar spectral irradiance measurements. Preliminary results. AIP Conf. Proc. 1477:139–142. doi: 10.1063/1.4753853.
27. Green, M. A., K. Emery, Y. Hishikawa, and W. Warta. 2009. Solar cell efficiency tables (Version 33). Prog. Photovoltaics Res. Appl. 17:85–94.
28. Dirnberger, D., U. Kräling, H. Müllejans, E. Salis, K. Emery, Y. Hishikawa, et al. 2014. Progress in photovoltaic module calibration: results of a worldwide intercomparison between four reference laboratories. Meas. Sci. Technol. 25:105005.
29. Hara, K., S. Igari, S. Takano, and G. Fujihashi. 2005. Characterization of photovoltaic performance of dye-sensitized solar cells. Electrochemistry 73:887–896. doi: 10.5796/Electrochemistry.

A three-dimensional multiphysics modeling of thin-film amorphous silicon solar cells

Ahmadreza Ghahremani & Aly E. Fathy

Electrical Engineering Department, University of Tennessee Knoxville, Knoxville, Tennessee

Keywords

3D multiphysics modeling, thin-film amorphous silicon, nanoparticles, plasmonic, solar cells

Correspondence

Ahmadreza Ghahremani, Electrical Engineering Department, University of Tennessee Knoxville, Min H. Kao Building, Suite 609 1520 Middle Drive, Knoxville, TN 37996-2250, USA.

E-mail: aghahrem@utk.edu

Funding Information

This work was supported by the grant from the National Science Foundation of USA (Grant No. NSF EPS-1004083).

Abstract

A 3D multiphysics simulation toolbox for thin-film amorphous silicon solar cells has been developed. The simulation is rigorous and is based on developing three modules: first to analyze light propagation using electromagnetic techniques, second to account for charge generation and transportation based on the physics of the semiconductor device, and third including electrode modeling by applying electrostatic techniques. Published results of a P-I-N thin-film amorphous silicon solar cell fabricated on a thick glass and experimentally evaluated was used as a vehicle to validate our 3D multiphysics toolbox and demonstrate its capabilities. The toolbox utilizes COMSOL for solving the partial differential equations describing the three modules, and MATLAB to input data, control the solver, and provides the coupling between the three modules. The developed toolbox was used to investigate both the effect of embedding Metallic Nanoparticles (MNPs) and the impact of defects on the external quantum efficiency. The simulator, besides being rigorous, is suitable to model various types of solar cells (organic, inorganic, thick film, thin film, heterojunction, or plasmonic) as well.

Introduction

Currently thin-film amorphous silicon solar cell has relatively poor efficiency due to low photocurrent generation. Although efficiency can be significantly increased upon enhancing light management and absorption mechanisms for absorber layers [1]. Fortunately, there are different methods to achieve such enhancement. For instance, applying a randomly textured layer inside the device is a standard approach to gain more effective scattered rays inside the device [2–5]. Additionally, introducing a periodic structure as a reflector to enable the increase of light path inside the absorber [6–8]; thus enhancing light absorption. Alternatively, scientists have tried to utilize metallic nanoparticles (MNPs) to improve the cell efficiency upon creating a high-intensity localized fields around these MNPs [9–11], but unfortunately not much success has been reported yet. To investigate such effects, several groups carried out many studies on solar cells performance enhancement using 1D and 2D analysis [12–16]. However, for accurate modeling, rigorous 3D analysis [17–19] is indeed required to resolve many associated real practical fabrication problems. For example, only 3D multiphysics tools can be used to analyze the effects of placing spherical MNPs, also accounting for the random polarization of rays (incident waves), and considering their different multipaths behaviors after being scattered from these randomly shaped metallic nanoparticles. Table 1 shows different commercial CAD tools, their features and compares their capabilities and analysis methods to address solar cells multiphysics problems. However, solving such coupled 3D multiphysics problem (solar cell) is very challenging as it includes: light propagation—an optics problem, energy absorption—a semiconductor problem, and light conversion to photocurrent—an electrostatic problem. These three distinct physical fields need to be understood very well and accurately modeled by utilizing a fast and accurate simulator and account for their coupling. In other words, we need to first accurately predict the propagation of light within the different regions like dielectrics (passive region), semiconductors (active region) plasmonic nanoparticles (parasitic region), and effects of metals as reflectors/scatterers. Second, we need to solve the physics of the semiconductor devices so that we estimate the number of free electrons and holes. Finally, the third step is applying electrostatic formulas to calculate the photocurrent that is collected by the two electrodes placed at both sides of the device.

In this study, a 3D multiphysics toolbox has been developed to accurately model various types of solar cells. The toolbox main features are given in Table 1, and its novelty is the 3D Model of plasmon nanoparticles embedded inside semiconductors and the analysis of the effect of defects created by the plasmon layer. The main analysis concepts are explained in section Background, equations for the 3D problem are explained in detail in Appendix A1, supplemented by a description of the simulator initialization given in Appendix A2 and a list of some major parameters required for electrical characteristic of solar cells is derived in Appendix A3. Finally, the results of this study are presented for solar cells-no plasmon, simulation of solar cell-no plasmon with 3D gratings, electromagnetic field analysis in the presence of MNPs, and finally the effect of utilizing MNPs in sections (Simulated and Measured Results-no Plasmon, Analysis of Simulation Results of Solar Cell-No Plasmon With 3-D Gratings, Electromagnetic Field Analysis of MNPs, and Simulated and Measured Results-With Plasmon), respectively. Our conclusion is given in section Conclusion.

Background

To numerically analyze a solar cell structure using finite element of finite difference methods, first the 3D structure is drawn and meshed using a nonuniform grid. Definitely, the accuracy and speed of simulation are completely dependent on the selected mesh density. For instance, the critical regions like electrodes, semiconductor junctions, plasmon, and sharp or tiny structures should be generated using very dense mesh ($\sim\lambda/50$). Next, initial conditions for all variables and boundary conditions are set, respectively, to numerically solve the partial differential equations (i.e., Maxwell's equations, and the physics of the device transport equations). Given that the solution of the nonlinear PDEs system is sensitive to these initial and boundary conditions. For example, initializing the intensity of the impinging light (based on the latest published data for the Sun spectrum), setting up correctly all electric parameters including the electro-optical material properties (like complex refractive indices for semiconductors), and using a fast PDE solver are essential ingredients to get an accurate realistic results.

In general, electromagnetic modeling is carried out by numerically solving Maxwell's equations using one of the available techniques [20–24], like a finite element method that has been applied here. Meanwhile, Drude–Lorentz dispersion model [25] has been utilized here to model metallic nanoparticles. Subsequently, the result of this EM simulation is used to calculate the light power intensity distribution inside the semiconductor region (active layer). Then, the

Table 1. Introduce some commercial CAD Tools were used to solve solar cells' model.

Reference	Tools	Analysis Method	Type of modeling	Validation	Problem solved
[17]	HFSS	FEM	Optical model	Validated by another CAD tool (ASA)	3D modeling of Single junction thin-film silicon Solar cells with flat surface and 1D, and 2D modeling of trapezoid-shaped surface
[15]	SCAPS	Numerical methods (Newton)	Semiconductor and Electrostatic		1D modeling of Polycrystalline thin-film solar cells, The problem of transients of current to time dependent bias conditions is solved
[16]	D-AMPS-1D (1D simulation program Analysis of Microelectronic and Photonic Devices)	Finite differences (Newton Raphson iteration method)	Semiconductor and Electrostatic	Validated by measurement	1D modeling of structures such as homojunctions, heterojunctions, multijunctions, and Ge photovoltaic devices
[18]	COMSOL	FEM	Multiphysics	Validated by measurement	3D modeling of plasmonic solar cells; existing metallic nanoparticles at front of P-I-N
[19]	FDTD Solution TCAD Sentaurus	FDTD FEM	Multiphysics		Thin-film amorphous silicon; existing metallic nanoparticles at back of P-I-N
Our approach	COMSOL+MATLAB	FEM	Multiphysics	Validated by measurement	Thin-film amorphous silicon; existing metallic nanoparticles inside P-I-N and 3D modeling of trapezoid-shaped surface

number of electron-hole pairs is estimated as a function of the previously calculated light intensity using chemical absorption data [26]. In our analysis, the free charge carriers' recombination due to different factors (free carrier life time, energy trap level, and density of carrier concentrations) are considered, as they could cause significant impact in predicting the final results. Keep in mind that the recombination rate is a strong function of both the doping concentration and density of defects inside the semiconductor. As a final step, the number of electron-hole pairs captured by the two electrodes at both sides of the active region is calculated using Poisson's Electrostatics Equation. Meanwhile, other major parameters like external quantum efficiency, solar cell efficiency, J–V curve, and fill factor can be extracted as well as by-products [27–29]. A step-by-step illustration of the above calculations are given in the Appendix A1.

Simulated and Measured Results-no Plasmon

For model validation, the simulation results for a thin-film P-I-N solar cell model were compared to previously published measured results [10]. However, some material properties were not cited on this publication and their values were assumed here based on well-known references like [13] and [30–34].

The structure of the modeled solar cell shown in Figure 1A consists of five stacked layers: a Glass (for protection) on top, a SnO:F TCO (transparent conductive oxide), followed by absorber layers (a-Si:H) as a P-I-N structure, then a TCO on the back, and all on top of a back reflector. The Amorphous silicon has 260-nm-thick intrinsic layer, the front electrode has a 75 nm-thick (SnO:F) layer, and the back electrode (AZO) is 75-nm-thick layer on top of the reflector. The N+ and P+ layers are modeled here too as 40-nm-thick layer each. Figure 1B shows a comparison between the simulated and measured EQE results, where only a slight discrepancy is seen.

Given that most of the material parameters were obtained from well-known references (listed in Table 2) but their

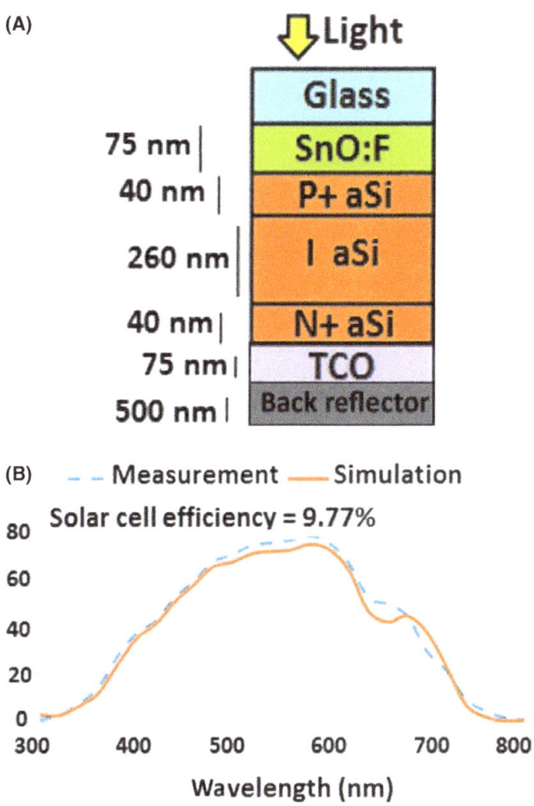

Figure 1. (A) Schematic of the P-I-N device; (B) EQE curve (comparison between measurement [10], and simulation result).

Table 2. The utilized value of each parameter for the utilized validation example.

Parameter – Name	Value
T (temperature)	300 [K]
Ni -Ref [34]	0.949×10^6 [cm^{-3}]
Doping (N+)	1×10^{20} [cm^{-3}]
Doping (P+)	10^{21} [cm^{-3}]
Thickness (N+, a-Si)-Ref [10]	40 [nm]
Thickness (intrinsic, a-Si)- Ref [10]	260 [nm]
Thickness (P+, a-Si)- Ref [10]	40 [nm]
Thickness (AZO) -(TCO in front)- Ref [10]	75 [nm]
Thickness (glass)- Ref [10]	200 [μm]
Thickness (Air)- Ref [10]	20 [μm]
Thickness -(TCO BACK)- Ref [10]	75 [nm](AZO)
Thickness (Silver) Reflector back- Ref [10]	500 [nm]
Electron mobility, a-Si -intrinsic-Ref [13]	20 [$cm^2/(V\ s)$]
Hole mobility, a-Si -intrinsic-Ref [13]	2 [$cm^2/(V\ s)$]
Electron mobility, a-Si N+ -Ref [13]	20 [$cm^2/(V\ s)$]
Hole mobility, a-Si N+ -Ref [13]	2 [$cm^2/(V\ s)$]
Electron mobility, a-Si P+ -Ref [13]	20 [$cm^2/(V\ s)$]
Hole mobility, a-Si P+ -Ref [13]	2 [$cm^2/(V\ s)$]
Electron life time, a-Si -intrinsic-Ref [33,34]	20 [ns]
Hole life time, a-Si -intrinsic-Ref [33,34]	20 [ns]
Electron life time, a-Si N+ -Ref [33,34]	0.0001 [ns]
Hole life time, a-Si N+ -Ref [33,34]	10 [ns]
Electron life time, a-Si P+ -Ref [21, 22]	10 [ns]
Hole life time, a-Si P+ -Ref [21, 22]	0.0001 [ns]
Density of state valence band, a-Si -Ref [31]	2.5×10^{20} [cm^{-3}]
Density of state conduction band, a-Si -Ref [31]	2.5×10^{20} [cm^{-3}]
Difference between defect level and intrinsic level N+, P+ -Ref [27]	0.7
Difference between defect level and intrinsic level intrinsic-Ref [27]	0.3
Energy band gap a-Si-Ref [34]	1.74
Diameter silver NPs-Ref [10]	20 [nm]
Affinity, a-Si (electro affinity)-Ref [13]	4.00 eV
Incident light angle	0 [deg]

real values could be within a specified uncertainty range; also the physical layers dimensions were guessed based on the intended design but could deviate slightly from the fabricated ones. In addition, the dopant distribution and the density of the traps (recombination) were not listed in the experimental description and were assumed as well. Considering the approximations, the simulated results are still indicating a similar behavior to the experimentally demonstrated ones. The solar efficiency and fill factor FF were used as a base line for comparison and their values were 9.77%, and 74%, respectively, and the maximum EQE occurs at 580 nm. The calculated current density as a function of voltage is shown in Figure 2. Additionally, the measured and simulated short circuit current J_{sc} and the open circuit voltage V_{oc} were listed too.

Additional parameters including the reflection and absorbance coefficient were calculated for the whole solar cell as a function of wavelength and are shown in Figure 3. The transmission coefficient, however, is almost zero.

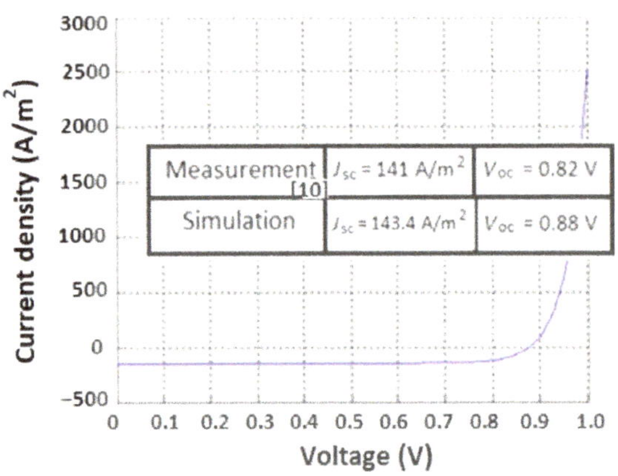

Figure 2. J–V curve (simulation), and a comparison between simulation and measured data [10] is shown in the embedded Table.

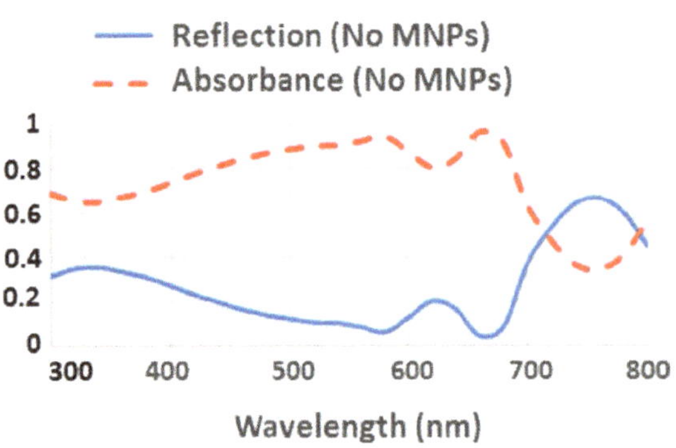

Figure 3. Simulation results for solar cells without MNPs; the blue solid line represents reflection coefficient; and red dash line shows the absorbance coefficient versus wavelength.

When comparing Figures 3 and 1B, the level of light absorbance coefficient in some regions of the spectrum is higher than what is indicated by the EQE curve. This difference demonstrates that the energy was not totally converted to current and this drop is due to internal device losses. Figure 4 represents the extinction coefficient of amorphous silicon calculated based on [35], and it can be observed that the absorbance of light inside the amorphous silicon increases at higher frequencies. For this reason in the UV region most of the impinging solar energy on the cell is absorbed at the top layer of the semiconductor (P+ region) before approaching the depletion region. It turns out that the probability of creating free electron-hole pairs outside the depletion region increases; thus unfortunately increasing the probability of free carrier recombination immediately and this would be an energy loss. If the light could propagate for a longer path inside the semiconductor, it can generate free electron-hole pairs inside the depletion region, subsequently the electrostatic electric field (created based on majority carriers in the depletion region) could cause charge separation (electrons and holes) creating electricity. In other words, the presence of material defects inside the semiconductors can elevate the recombination probability of free carriers as well and reduce the overall conversion efficiency.

Analysis of Simulation Results of Solar Cell-no Plasmon with 3D Gratings

Surface roughness of solar cell layers can affect the solar cell performance. Where most of the time, the surface roughness is created by the morphology of the transparent conductive oxide (TCO) located at the top and bottom of the semiconductor layers. Use of large grains in the TCO film too can increase surface roughness [36–38]. At the same time, the size of the grains is correlated to the thickness of the thin-film TCO layer, that is, a thinner film would have less surface roughness. For instance, the thickness of the thin-film TCO considered for this model here is 75 nm, and its surface roughness is estimated to be <10 nm. To model the surface roughness effect, in general, a 3D device model can be used where a trapezoidal grating can be assumed. A periodic structure of a trapezoidal shape (like that of [17]) with a 30° degree slope in each side, and a 10 nm height (with periodicity of 200 nm) is assumed and implemented to model the 3D gratings of the solar cells (see Fig. 5A). Figure 5B shows that the results of the simulation for the EQE for both cases (flat and with gratings) are very similar for the thin-layered case. After analyzing the simulation's raw data, it was clear that the surface roughness at low frequencies (IR region) is very small compared

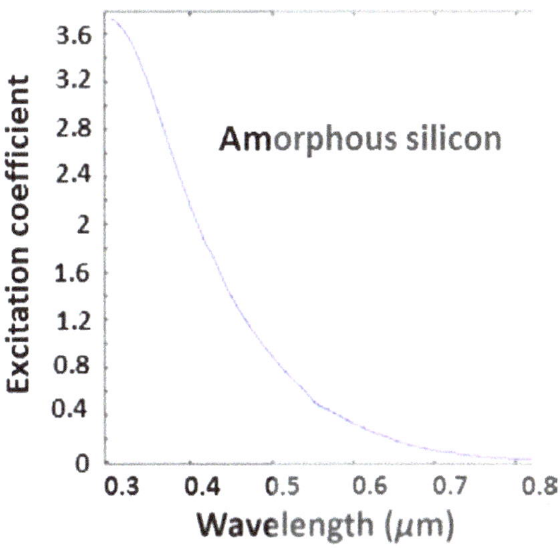

Figure 4. Extinction coefficient of amorphous silicon.

to the wavelength of solar rays. So, at these frequencies, it is a good approximation to assume that it behaves like a flat surface (from optics point of view). On the other hand, at higher frequencies (UV region) the gratings cannot be neglected, because its size is close to a wavelength. The gratings lead to ray scattering and it could cause substantial increase in light path length inside the absorber (amorphous silicon), and it is anticipated that longer path would improve solar cell efficiency. Although, no efficiency improvement has been observed in the UV region, because the absorption of the amorphous silicon in this region is very high (see Fig. 4), and the penetration depth (see Appendix A1, equations 3–4) is rather very small compared to the path length of the ray inside the absorber. Hence, the effect of the gratings could only be observed weakly over a narrow bandwidth (510–620 nm). Apart from slight ripples seen in Figure 5B, the assumption of a flat surface for very thin layers has been validated as the simulation results are very similar to the measurements of [10], and for simplification it will be implemented in our analysis of the effect of MNPs.

Electromagnetic Field Analysis of MNPs

One way to prove that our electromagnetic 3D model for a MNP works properly, was to validate our simulation results with an analytical method. Here, we use the well-known conic problem of scattered fields from a metallic sphere in a vacuum space (shown in Ref [39]) to validate our models for MNPs. If an electric field of a uniform plane wave is polarized in the x direction, and is traveling along the z-axis, then the monostatic radar cross section can be expressed by

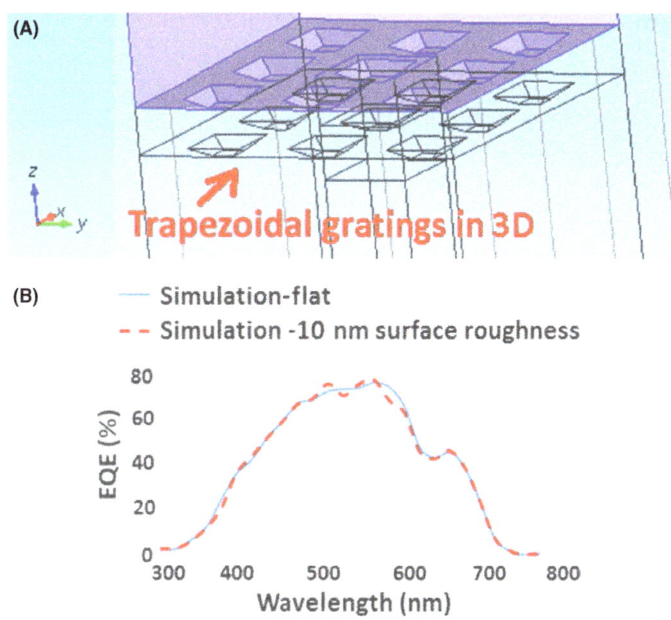

Figure 5. (A) schematic of the gratings in 3D inside the TCO; (B) EQE curve (comparison between simulation flat, and simulation with gratings (surface roughness = 10 nm).

$$\sigma_{3-D}(\text{monostatic}) = \lim_{r\to\infty}\left[4\pi r^2 \frac{|E^s|^2}{|E^i|^2}\right] = \frac{\lambda^2}{4\pi}\left|\sum_{n=1}^{\infty}\frac{(-1)^n(2n+1)}{H'^{(2)}(\beta a)\,H^{(2)}(\beta a)}\right|^2 \quad (1)$$

where r, E^s, E^i, λ, β, a, and $H^{(2)}$ are represent the range, scattered electric field, incident electric field, wavelength, wave propagation constant, radius of the metallic sphere, and spherical Hankel function, respectively. A plot of equation 1 as a function of the radius of the sphere is shown in Figure 6.

The results can be divided into three regions; the *Rayleigh*, the *Mie* (or *resonance)*, and the *optical* region. The Rayleigh region represents the part of the curve for small radii values ($a < 0.1\lambda$). Hence, in the Rayleigh region, equation 1 can be reduced to

$$\sigma_{3-D}(\text{monostatic}) \cong \frac{9\lambda^2}{4\pi}(\beta a)^6 \quad (2)$$

Hence, we compared our 3D model for a sphere in vacuum and the analytical method and results are shown in Figure 7. A good accuracy is seen for the long wavelengths region. Some slight deviation is, however, seen for longer wavelength but still adequate for our modeling efforts here. To improve our model at long wavelength region, finer mesh can still be used but will significantly reduce computation speed.

Simulated and Measured Results-with Plasmon

Our next step was to analyze the effect of adding silver nanoparticles in each layer; one layer at a time to determine the optimum location and position that would enhance the solar cell efficiency. In our model, silver nanoparticles are modeled as spheres with 10 nm radius [10] and are arranged in a random 2D array (in xy-plane) with a maximum center to center spacing of 50 nm. Initially, these silver NPs were placed 2 nm above the absorber layer along the SnO:F P-type A-Si interface as shown in Figure 8A as suggested by [10]. Although this resulted in a pronounced drop in solar cell efficiency

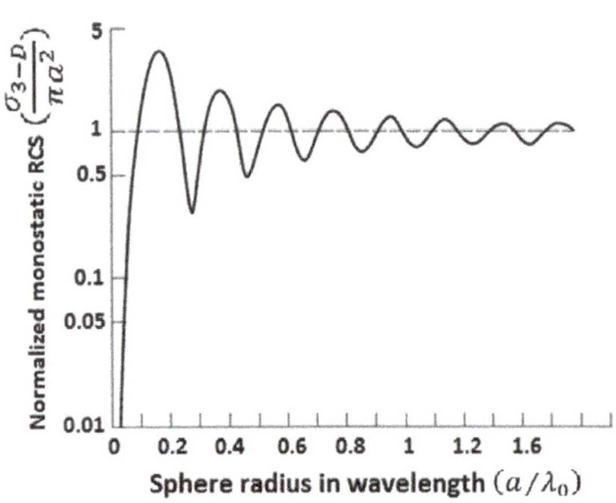

Figure 6. Normalized monostatic radar cross section for a conducting sphere as a function of sphere radius [40].

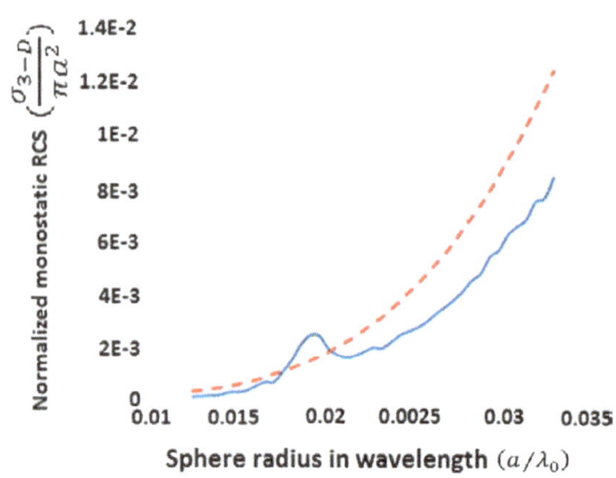

Figure 7. Normalized monostatic RCS for a single silver nanoparticle (radius = 10 nm) inside vacuum; solid blue line represents simulation result and red dash line shows the analytical method.

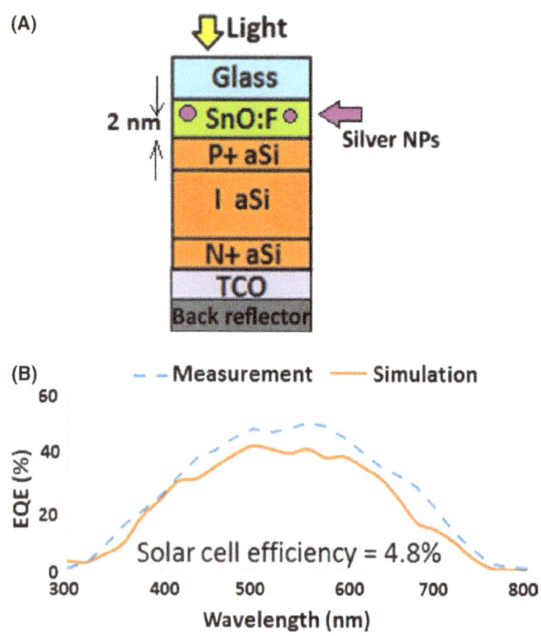

Figure 8. (A) Schematic of a PIN device; (B) EQE curve using MNPs embedded inside SnO:F.; blue dash line represents measurement [10]; orange solid line shows simulation.

(A)

Particles	J_{sc} (mA/cm^2)	V_{oc} (V)	FF	Efficiency (%)
Measurement				
Ag	9.06	0.806	72.7	5.32
No particles	14.10	0.823	73.2	8.51
Simulation				
Ag	7.2	0.87	73	4.8
No particles	14.34	0.878	74	9.77

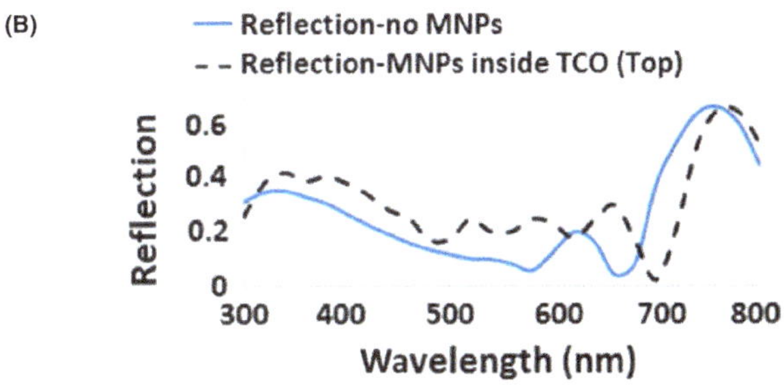

Figure 9. (A) A comparison between the measurements of [10] and simulation data for the electrical outputs; (B) Simulation results for the Reflection coefficient; solid blue line represents reflection from the solar cell without MNPs; dash black line shows reflection from the solar cell with MNPs embedded inside the TCO layer on top of the PIN device.

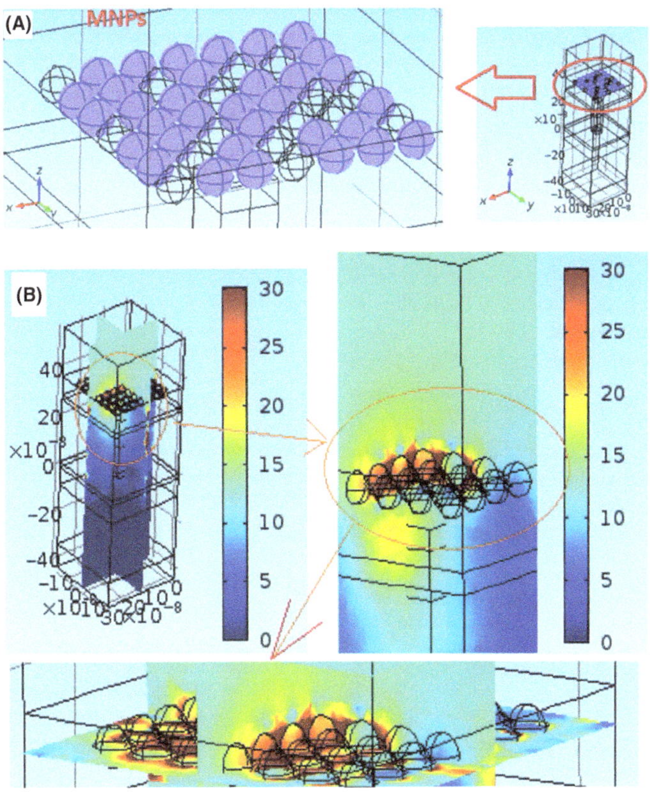

Figure 10. (A) schematic of MNPs are shown as blue spheres embedded inside TCO (B) the intensity of power flow of the light (wavelength@720 nm).

as noticed in Figure 8B, which is also consistent with the observations of Ref. [10]; where the efficiency has dropped to only 4.8%. Figure 9A shows a comparison between the simulated and measured electrical outputs of a solar cell with and without MNPs. The simulated reflection coefficient for these cases are also shown in Figure 9B.

Next step, the optical behavior of the solar cell excited by a polarized plane wave is analyzed in 3D. Figure 10A shows a schematic of a random 2D array of MNPs in xy-plane. Silver nanoparticles are embedded inside TCO (on top of the PIN device). An x-polarized electromagnetic plane wave (wavelength@720 nm) is propagating in the z direction toward the structure (assuming normal incidence). The power intensity of the propagating plane wave is dropped just after passing through the MNPs (see Figure 10B). Two MNPs (next to each other) create strong localized fields (see Figure 10B) as expected.

The above test was repeated, but this time with these silver nanoparticles embedded in the absorber layer (t = 50 nm under P+ layer) as seen in Figure 11A. However, it is believed that when MNPs are integrated into the amorphous silicon, they will cause an increase in the amount of defects inside the amorphous silicon structure. Unfortunately, it turns out that if this degradation phenomenon is not accounted for, then the simulation result is totally very optimistic and unrealistic. In summary, this misleading simulation result is shown in Figure 11B and indicates that significant efficiency improvement can be achieved, and as a matter of fact initially has led to a considerable confusion and a myth that adding MNPs should automatically lead to efficiency enhancement; but this has not been experimentally materialized yet due to significant fabrication defects. Practically, we needed to revise our model too to account for the impact of defects on performance, hence we modified the device physics equations as suggested by [33] to include these defects effects as well.

The modified model results after including the defects, gives results that are now in a fairly good agreement with measured data (Fig. 12). The results indicate a rather significant drop in efficiency which is consistent again with the observation of [10]. For t = 50 nm, the efficiency, for example, has dropped to only 3.5% in contradiction to the common belief that efficiency should have been enhanced. The pronounced efficiency drop was even worsening upon moving the MNPs from the top of the intrinsic layer to its bottom, where the solar cell efficiency dropped from 3.5% to even 1.89%. A plausible reason to explain this phenomenon is that; by varying the position of the MNPs inside the absorber and moving further down toward the bottom of the structure, the intensity of the localized field created around the MNPs decreases exponentially, because most part of the solar energy in the UV region has been absorbed significantly after passing the top layers.

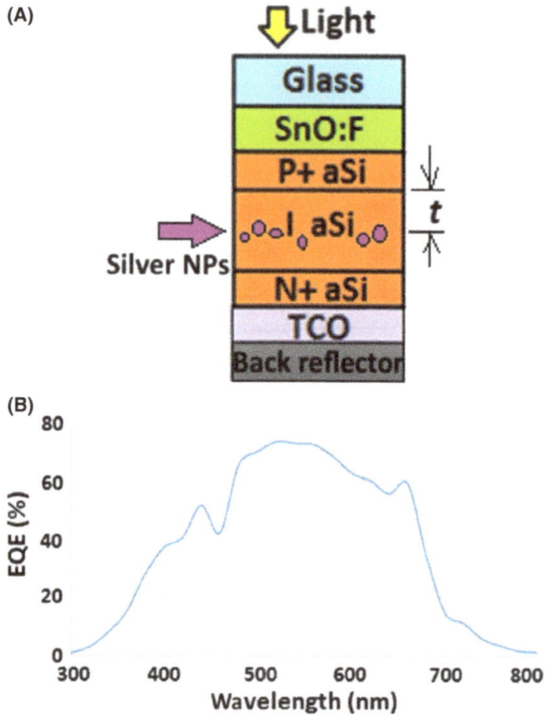

Figure 11. (A) Schematic of the PIN device; (B) EQE curve (simulation), using MNPs embedded inside intrinsic layer (*t* = 50 nm) without considering of defects.

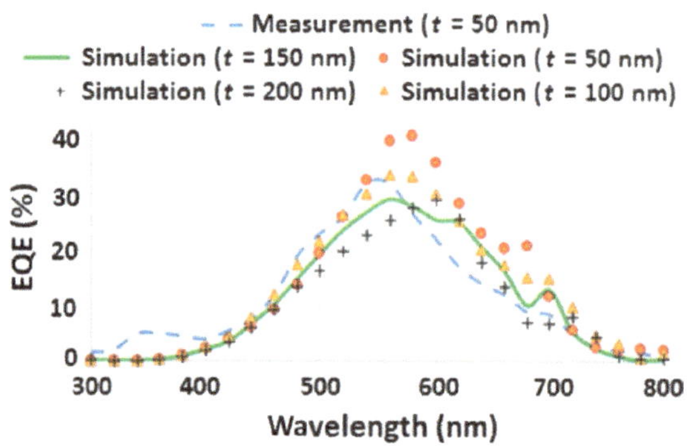

Figure 12. EQE curve in case of existing MNPs embedded inside intrinsic layer (a-Si) blue dash line measurement (*t* = 50 nm) [10]; other curves represents simulation results (after considering of defects).

Hence, fields would become very weak to excite the MNPs. For this reason the generation rate compared to the recombination rate in the intrinsic layer goes down

(A)

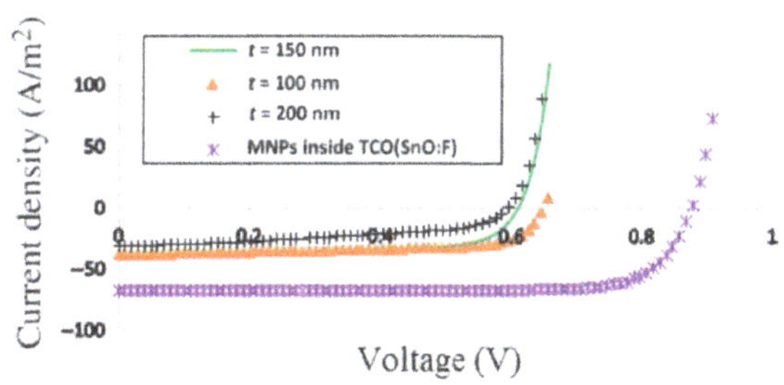

(B)

t_{top} (nm)	J_{sc} (mA/cm^2)	V_{oc} (V)	FF	Efficiency (%)
Measurement				
50	6.38	0.882	55.3	3.11
100	4.18	0.899	57.0	2.14
150	3.96	0.881	56.5	1.97
200	3.14	0.876	53.7	1.48
Simulation				
50	6.4	0.67	78	3.5
100	4.3	0.64	81	2.34
150	4.1	0.61	77	2.02
200	3.8	0.59	80	1.89

Figure 13. (A) J–V curve of the solar cell; for existing MNPs inside the structure; (B) comparison electrical outputs from measurement [9], and simulation.

upon moving the MNPs toward the bottom. The simulation result for the JV curves, with existing MNPs in different locations inside the solar cells are shown in Figure 13A. Figure 13B shows the electrical outputs for the measured [10] and simulated data. Finally, the silver NPs were relocated to the back (embedded inside the TCO), and that represented the most suitable location as seen in Figure 14A. The maximum solar efficiency now is 9.8% which is very close to the solar cell case with no NPs (see Fig. 14B). Again this observation is consistent with [10], but definitely much lower than our expectation of enhancing the efficiency upon adding MNPs.

Conclusion

In this study, an effective rigorous 3D multiphysics modeling of solar cells was presented. Our developed simulator is a real multiphysics modeling toolbox that is comprised of three coupled modules: Optics, carrier transport in semiconductors, and Electrostatic. To solve their associated nonlinear partial differential equations (PDEs) in 3D, we used two commercial tools (COMSOL, and MATLAB). One of the main reasons to carry out this 3D simulation is to accurately predict the electric field distribution due to the light scattering of the 3D plasmonic particles excited by the randomly polarized sun light. Our 3D tool has

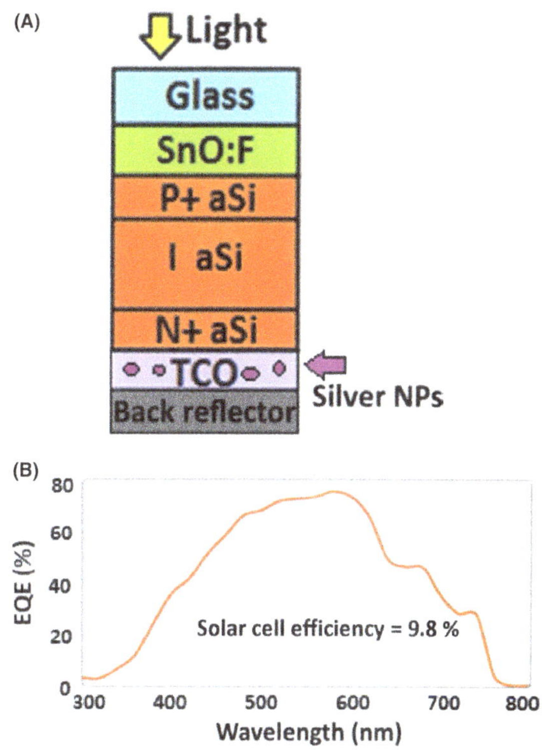

Figure 14. (A) Schematic of the PIN device; (B) EQE curve (simulation), using MNPs embedded inside TCO (back).

been validated by comparing its results with published measured ones.

The comparison between both measured and simulated results indicates a very good agreement even though some material parameters were assumed like the dopant distribution, and the density of traps (recombination), in addition to some assumed layers' thicknesses. The developed toolbox has been used to observe some major phenomena like strong localized fields around the MNPs and relate intense recombination to defects. The presence of MNPs might be helpful to enhance light absorption, although it would not guarantee improving solar cells' efficiency. Our simulation results demonstrated too that the trap density caused by defects around MNPs is the main reason for efficiency degradation that was seen experimentally. Here, we have verified that embedding MNPs randomly can lead to a significant external quantum efficiency degradation which is consistent with various experimental observations. Meanwhile, relocating the MNPs to the TCO layer close to the bottom electrode could provide some efficiency enhancement even though not pronounced. So, work is still under way to simulate other cases using our 3D toolbox to get significant efficiency improvement and to address the defect degradation issue, preliminary results are encouraging but would require more validation at this point.

Acknowledgments

This work was supported by the grant from the National Science Foundation of USA (Grant No. NSF EPS-1004083).

Conflict of Interest

None declared.

References

1. Zeman, M. 2006. Advanced amorphous silicon solar cell technology. Pp. 173–236 *in* J. Poortmans, V. Archipov, eds. Thin film solar cells: fabrication, characterization and applications. Wiley, Chichester. ISBN: 978-0-470-09126-5.
2. Deckman, H. W., C. B. Roxlo, and E. Yablonovitch. 1983. Maximum statisticalincrease of optical absorption in textured semiconductor films. Opt. Lett. 8:491–493.
3. Isabella, O., F. Moll, J. Krč, and M. Zeman. 2010. Modulated surface textures using zinc-oxide films for solar cells application. Phys. Status Solidi A 207:642–646.
4. Kambe, M., A. Takahashi, N. Taneda, K. Masumo, T. Oyama, and K. Sato. 2008. Fabrication of a-Si:H solar cells on high haze SnO2:F thin films. p. 1, *in* Photovoltaic Specialist Conference, San Diego, USA. 33rd IEEE (doi: 10.1109/PVSC.2008.4922507)
5. Borri, C., and M. Paggi. 2015. Topological characterization of antireflective and hydrophobic rough surfaces: are random process theory and fractal modeling applicable? J. Phys. D: Appl. Phys. 48:045301.
6. Haug, F.-J., T. Söderström, O. Cubero, V. Terrazzoni-Daudrix, X. Niquille, S. Perregeaux, et al. 2008. Periodic textures for enhanced current in thin film silicon solar cells. Mater. Res. Soc. Symp. Proc. 1101:KK13–KK. doi: 10.1557/PROC-1101-KK13-01.
7. Zeman, M., O. Isabella, K. Jäger, R. Santbergen, S. Solntsev, M. Topic, and J. Krc. 2012. Advanced Light management approaches for thin-film silicon solar cells. Energy Procedia 15:189–199.
8. Pahud, C., O. Isabella, A. Naqavi, F.-J. Haug, M. Zeman, H. P. Herzig, et al. 2013. Plasmonic silicon solar cells: impact of material quality and geometry. Opt. Express 21:A786–A797. doi: 10.1364/OE.21.00A786.
9. Toroghi, S., L. Chatdanai, and P. G. Kik. 2013. Cascaded plasmon resonances multi-material nanoparticle trimers for extreme field enhancement. Proc. SPIE 8809, Plasmonics: Metallic Nanostructures and Their Optical Properties XI, 88091M. doi: 10.1117/12.2024709.
10. Santbergen, R., A. H. M. Smets, and M. Zeman. 2011. Silver nanoparticles for plasmonic light trapping in A-Si:H solar cells. Pp 000673–000678 *in* Photovoltaic Specialists Conference (PVSC), 37th IEEE, Seattle, WA.
11. Stuart, H. R., and D. G. Hall. 1998. Island size effects in nanoparticle-enhanced photodetectors. Appl. Phys. Lett. 73:3815.
12. Malm, U., and M. Edoff. 2009. 2D device modelling and finite element simulations for thin-film solar cells. Sol. Energy Mater. Sol. Cells 93:1066–1069.
13. Ihalane, E., M. Meddah, A. Elfanaoui, L. Boulkaddat, E. El Hamri, X. Portier, et al. 2011. Numerical simulation of photocurrent in a solar cell based amorphous silicon. Moroccan J. Condens. Matter. 13:83–87.
14. Basore, P. A. 1990. Numerical modeling of textured silicon solar cells using PC-1D. IEEE Trans. Electron Devices 37:337–343.
15. Degrave, S., M. Burgelman, and P. Nolle. Modelling of polycrystalline thin film solar cells: new features in scaps version 2.3. 3rd World Conference on Pholovolroic Lnirergv Conversion Mov 11–18, 2003 Osaka, Japan.
16. Barrera, M., F. Rubinelli, I. Rey-Stolle, and J. Pla. 2012. Numerical simulation of Gesolar cells using D-AMPS-1Dcode. Phys. B 407:3282–3284.
17. Isabella, O., S. Solntsev, and M. Zeman. 2013. 3-D optical modeling of thin-film silicon solar cells on diffraction gratings. Prog. Photovoltaics Res. Appl. 21:94108.

18. Li, X., N. P. Hylton, V. Giannini, K.-H. Lee, N. J. Ekins Daukes, and S. A. Maier. 2011. Bridging electromagnetic and carrier transport calculations for three-dimensional modelling of plasmonic solar cells. Opt. Express 19:A888–A896.
19. Deceglie, M. G., V. E. Ferry, A. P. Alivisatos, and H. A. Atwater. 2012. Design of nanostructured solar cells using coupled optical and electrical modeling. Nano Lett. 12:2894–2900. doi: 10.1021/nl300483y.
20. Yee, K. S. 1966. Numerical solution of initial boundary value problems involving Maxwell's equations. IEEE Trans. Antennas Propag. 14:302–307.
21. Weiland, T. 1977. A discretization method for the solution of Maxwell's equations for six-component fields. Int. J. Electron. Commun. AEU 31:116–120.
22. Moharam, M. G., and T. K. Gaylord. 1981. Rigorous coupled-wave analysis of planar-grating diffraction. J. Opt. Soc. Am. 71:811–818.
23. Gibson, W. C. 2008. The method of moments in electromagnetics. Chapman & Hall/CRC, Boca Raton. ISBN 978-1-4200-6145-1.
24. Jin, J.-M. 2002. The finite element method in electromagnetics, 2nd ed. John Wiley & Sons, New York. ISBN 978-0-471-43818-2.
25. Quinn, J. J., and K.-S. Yi. 2009. Solid state physics: principles and modern applications. Springer-Verlag, Berlin Heidelberg.
26.Dewan, R., S. Fischer, V. B. Meyer-Rochow, Y. Özdemir, S. Hamraz, and D. Knipp. 2012. Studying nanostructured nipple arrays of moth eye facets helps to design better thin film solar cells. Bioinspir. Biomim. 7:016003.
27. SZE. 1981. Physics of semiconductor devices, 2nd ed. Wiley-Interscience, New York.
28. Neamen, D. A. 2003. Semiconductor physics and devices: basic principles, 3rd ed. McGraw Hill, New York.
29. Madelung, O. 2004. Semiconductors: data handbook. Springer-Verlag Berlin Heidelberg.
30. Deng, X., and E. A. Schiff. 2005. Amorphous silicon–based solar cells, chapter 12. John Wiley & Sons, Ltd, Chichester, UK.
31. Bakr, N. A., A. M. Funde, V. S. Waman, M. M. Kamble, R. R. Hawaldar, D. P. Amalnerkar, et al. 2010. Determination of the optical parameters of a-Si:H thin films deposited by hot wire–chemical vapor deposition technique using transmission spectrum only. Pramana J Physics. 76:519–531.
32. Alamo, J. A., and R. M. Swanson. 1987. Modeling of minority-carrier transport in heavily doped silicon emitters. Solid-State Electron. 30:1127–1136.
33. Piprek, J. 2003. Semiconductor optoelectronic devices introduction to physics and simulation. Academic Press; 1 edition (January 21, 2003). ISBN-13: 978-0125571906.
34. Sakata, I., and Y. Hayashi. 1985. Theoretical analysis of trapping and recombination of photo generated carriers in amorphous silicon solar cells. Appl. Phys. A 37:153–164.
35. Pierce, D. T., and W. E. Spicer. 1972. Electronic structure of amorphous SI from photoemission and optical studies. Pierce Spicer Phys. Rev. B. 5:3017.
36. Gracia, M., F. Rojas, and G. Gordillo. Morphological and optical characterization of SnO_2:F thin films deposited by spray pyrolysis. 20th European Photovoltaic Solar Energy Conference, 6–10 June 2005, Barcelona, Spain.
37. Chaaya, A. A., R. Viter, M. Bechelany, Z. Alute, D. Erts, A. Zalesskaya, et al. 2013. Evolution of microstructure and related optical properties of ZnO grown by atomic layer deposition. Beilstein J. Nanotechnol. 4:690–698. doi: 10.3762/bjnano.4.78.
38. Kumar, V., N. Singh, R. M. Mehra, A. Kapoor, L. P. Purohit, and H. C. Swart. 2013. Role of film thickness on the properties of ZnO thin films grown by sol-gel method. ELSEVIER Thin Solid Films 539:161–165.
39. Balanis, C. A. 1989. Advanced engineering electromagnetics. Wiley, Hoboken, New Jersey, USA.
40. Aden, A. L. 1951. Scattering from spheres with sizes comparable to the wavelength. J. Appl. Phys. 22:601.
41. http://www.astm.org/Standards/G159.htm.
42. Hall, R. N. 1952. Electron-hole recombination in silicon. Phys. Rev. 87:387.
43. Shockley, W., and W. T. Read. 1952. Statistics of the recombinations of electrons and holes. Phys. Rev. 87:835–842.
44. Johnson, P. B., and R. W. Christy. 1963. Optical constants of the noble metals. Phys. Rev. Lett. 11:541.
45. OptiFDTD: Technical Background and Tutorials.
46. Terrestrial photovoltaics measurement procedures, NASA Tech. Memo. TM 73702, National Aeronautics and Space Administration, Cleveland, OH, 1977.

Appendix A1

Solving Solar Cell Problem in 3D

If α and $I_f(x, y, z)$ are the absorption coefficient and the intensity of the photons at position (x, y, z), respectively (Fig. 15), then the relationship for the optical absorption for the differential length (Dx, Dy, Dz) is given bellow.

$$\vec{I}_f(x,y,z) = I_{f_x}(x,y,z)\hat{x} + I_{f_y}(x,y,z)\hat{y} + I_{f_z}(x,y,z)\hat{z} \quad (3)$$

$$\frac{dI_{f_x}(x,y,z)}{\partial x}\hat{x} + \frac{\partial I_{f_y}(x,y,z)}{\partial y}\hat{y} + \frac{\partial I_{f_z}(x,y,z)}{\partial z}\hat{z} = -\alpha \vec{I}_f(x,y,z) \quad (4)$$

This equation clearly shows that the intensity of a photon decreases exponentially with the propagation distance through the semiconductor material due to the absorption [25,27–29]. Meanwhile, the number of electron-hole pairs generated by the light is:

Generation rate [28]:

$$G(x,y,z) = \frac{\alpha\left(\sqrt{I_{f_x}(x,y,z)^2 + I_{f_y}(x,y,z)^2 + I_{f_z}(x,y,z)^2}\right)}{hf} \quad (5)$$

where h is plank constant, and f is frequency of the wave.

Subsequently, to calculate the free electron density and free hole density, Poisson's Equation is used:

$$\vec{\nabla}\cdot\vec{\nabla}\varphi = -\frac{q}{\varepsilon}(p - n + c) \quad \text{PDE: } \frac{\partial^2\varphi}{\partial x^2} + \frac{\partial^2\varphi}{\partial y^2} + \frac{\partial^2\varphi}{\partial z^2} = -\frac{q}{\varepsilon}(p - n + c) \quad (6)$$

Shockley Read Hall recombination rate (7–11) [42,43]:

$$\Delta E_t = E_t - E_i \quad (7)$$

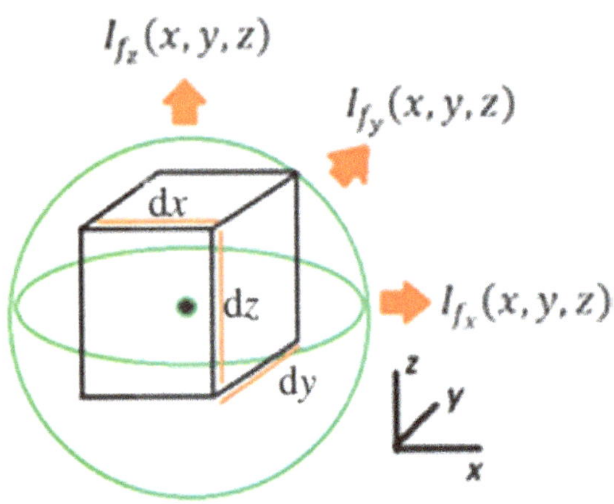

Figure 15. Photon intensity in a 3D cell.

$$n_1 = \gamma_n\sqrt{N_c N_v}\exp\left(-\frac{E_g - \Delta E_g}{2V_{th}}\right)\exp\left(\frac{\Delta E_t}{V_{th}}\right) \quad (8)$$

$$p_1 = \gamma_p\sqrt{N_c N_v}\exp\left(-\frac{E_g - \Delta E_g}{2V_{th}}\right)\exp\left(\frac{\Delta E_t}{V_{th}}\right) \quad (9)$$

$$n_i = \gamma_n\gamma_p\sqrt{N_c N_v}\exp\left(-\frac{E_g - \Delta E_g}{2V_{th}}\right) \quad (10)$$

$$R(x,y,z) = R_n(x,y,z) = R_p(x,y,z) = \frac{np - n_i^2}{\tau_p(n_1 + n)\tau_n(p_1 + p)} \quad (11)$$

$$\text{Total rate: } R_t(x,y,z) = G(x,y,z) - R(x,y,z) \quad (12)$$

Transport of Diluted species (Electron)

$$\text{PDE: } \vec{\nabla}\cdot\vec{J}_n - q\frac{\partial n}{\partial t} - q\frac{\partial \rho_n}{\partial t} = qR_t(x,y,z)$$

$$\frac{\partial J_{nx}}{\partial x} + \frac{\partial J_{ny}}{\partial y} + \frac{\partial J_{nz}}{\partial z} - q\frac{\partial n}{\partial t} - q\frac{\partial \rho_n}{\partial t} = qR_t(x,y,z) \quad (13)$$

$$\text{PDE: } \vec{J}_n = \vec{J}_{nx} + \vec{J}_{ny} + \vec{J}_{nz}$$

$$\vec{J}_n = \left(-qn\mu_n\frac{\partial\varphi}{\partial x} + qD_{nx}\frac{\partial n}{\partial x}\right)\vec{x} + \left(-qn\mu_n\frac{\partial\varphi}{\partial y} + qD_{ny}\frac{\partial n}{\partial y}\right)\vec{y} + \left(-qn\mu_n\frac{\partial\varphi}{\partial z} + qD_{nz}\frac{\partial n}{\partial z}\right)\vec{z} \quad (14)$$

$$\text{Space charge density (Electron): } \rho_n = -\frac{q}{\varepsilon}n \quad (15)$$

where φ (Electric Potential), p (Hole concentration), and n (Electron concentration) are variables, q (Electron charge), ε (Optical property of the material), c (Initial value for carrier concentration), n_i (intrinsic concentration), τ_n (Electron life time), τ_p (Hole life time), ϕ_0 (incident photon flux), α, (absorption coefficient of material), μ_n (Electron mobility), μ_p (Hole mobility), and D_n (Electron Diffusivity) are constants.

γ_n and γ_p are the electron and hole degeneracy factors, N_c and N_v are the effective densities of states for the conduction and valence band, E_g is the band gap, ΔE_g the band gap narrowing and E_t is the trap energy level. Energy difference between the defect level and the intrinsic level is ΔE_t.

The same equations can be derived for holes.

After coupling equations (13), (14), and (15), there will be only two nonlinear PDEs (16), (17).

$$-n\mu_n\left(-\frac{q}{\varepsilon}(p - n + c)\right) + D_{nx}\frac{\partial^2 n}{\partial x^2} + D_{ny}\frac{\partial^2 n}{\partial y^2} + D_{nz}\frac{\partial^2 n}{\partial z^2} + \left(-1 + \frac{q}{\varepsilon}\right)\frac{\partial n}{\partial t} = R_t(x,y,z) \quad (16)$$

$$p\mu_p\left(-\frac{q}{\varepsilon}(p-n+c)\right)-D_{px}\frac{\partial^2 p}{\partial x^2}-D_{py}\frac{\partial^2 p}{\partial y^2}-D_{pz}\frac{\partial^2 p}{\partial z^2}+\left(1+\frac{q}{\varepsilon}\right)\frac{\partial p}{\partial t}=-R_t(x,y,z) \quad (17)$$

Gaussian distribution (for both acceptor and donor) is considered as a dopant distribution. The junction depths in x, y, and z direction are set at 2 μm, 2 μm, and 300 [nm], respectively.

Figure A1 is shown in below that could be a good representation of our 3D model.

The flowchart clearly shows that the proposed solution is a staggered one, and not a fully coupled one. First, the optical problem is solved. Then the two other physics

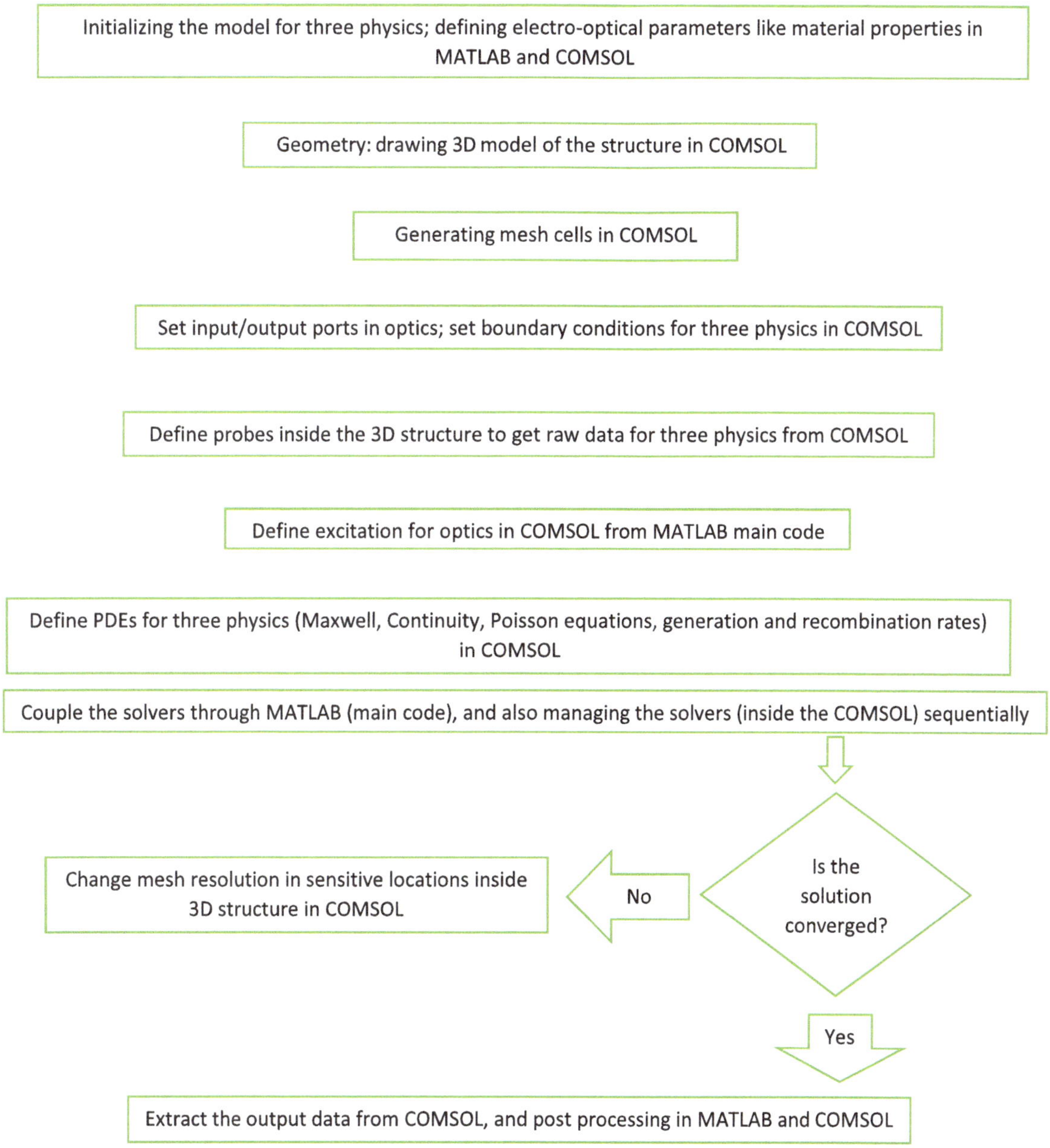

Figure A1. The flow diagram of the 3D model.

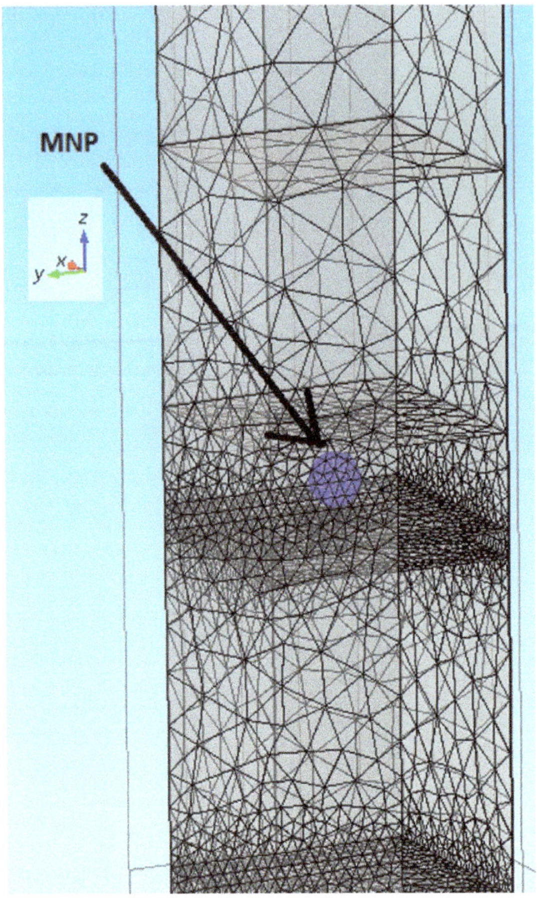

Figure 16. Tetrahedral Mesh Generation of the 3D device for one unit cell.

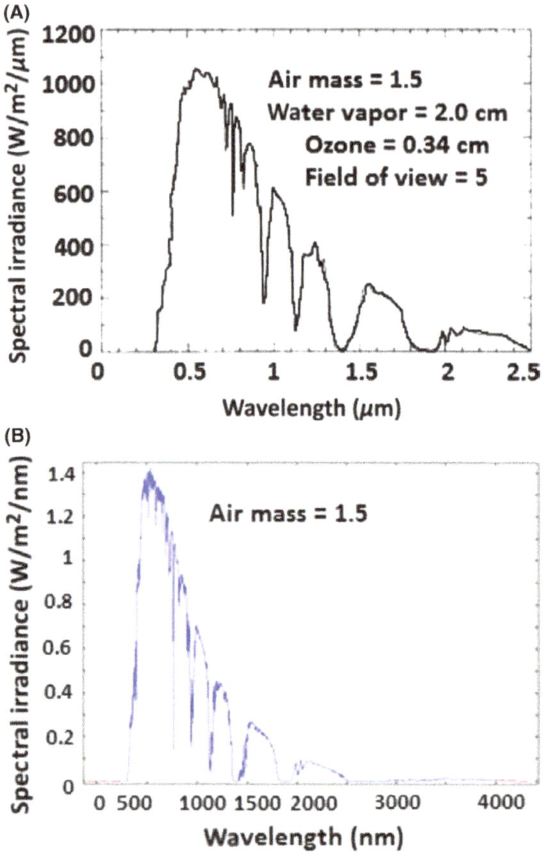

Figure 17. (A) Standard solar spectrum generated with the BRITE Monte Carlo Solar Insolation Model using the atmospheric conditions specified in [41]; (B) the utilized update of the spectral irradiance—downloaded from [26]..

(the semiconductor device and the electrostatic) are solved sequentially.

In this analysis, the PDEs are solved in steady state. So the transient solutions are not considered. It means that $((\partial n/\partial t)=0$ or $(\partial p/\partial t)=0)$. The excess minority carrier hole concentration increases with time with the time constant (which is the excess minority carrier lifetime). The excess carrier concentration reaches a steady-state value as time goes to infinity, even though a steady-state generation of excess electrons and hole exists [28]. The generation rate G a function of wavelength and time. However, here we are solving only for steady state and ignore any time dependence. Wavelength dependence is often included by summing the result obtained at one wavelength over all wavelengths of interest.

The RF interfaces formulate and solve the differential form of Maxwell's equations together with the initial and boundary conditions. The equations are solved using the finite element method. The steps are defined from MATLAB, and it can control COMSOL through a live link.

1-Define the geometry 2-Select materials 3-Select a suitable RF interface 4-Define boundary and initial conditions 5-Define finite element mesh 6-Define PDEs 7-Visualize the results.

To minimize the amount of physical memory (RAM) for computation and speed up computation we can use periodic boundary conditions along x, and y axis, which means that we solve equations for only one unit cell. Tetrahedral has been chosen as a type of mesh. Sharp edges or small particles like MNPs should be generated with high mesh density. Also the mesh size physically placed close to the critical edges (p and n type region) should be considered fine to make sure the PDEs solver is converged properly (Fig. 16).

Appendix A2

Initialization of the Simulator

In order to solve these partial differential equations, all parameters are needed to be initialized. The electro-optical properties like dielectric constant, refractive indices (real part and imaginary part), electron/hole mobility and the other electronic properties have been extracted from [13], [30–34]. In case of using MNPs or a plasmon layer the dielectric constant value has to be calculated for each optical wavelength of interest. For gold or silver NPs, optical constants were measured and given by Drude [44]. Drude-Lorentz dispersion model is a well-known model and is given by

$$\varepsilon_r = \varepsilon_\infty + \sum_{j=1}^{M} \frac{f_j \omega_p^2}{\omega_{oj}^2 - \omega^2 + i\Gamma_j \omega}, \sigma = 0 \tag{18}$$

Additional references can be used like [45,46] to double check materials properties.

Generally, the spectrum of incident light and intensity of impinging on the device is the most important initializing step to get accurate results from the simulation. Because some of the simulation outputs (like solar cell efficiency, and short circuit current) are directly related to the impinging light intensity as a function of wavelength, we needed then to utilize very accurate data. Hence, we utilized the spectral irradiance chart that has been measured several times and published by NASA [41] (Fig. 17A). Although, we need to keep in mind that the spectral irradiance depends on different factors like: the height from the sea level, water vapor, and air pollution. It is time dependent as well. For simplicity, we assume as a constant for the time being. Figure 17B shows the latest update that was found from [26], and utilized in our simulator.

Appendix A3

Electrical Outputs

Our model can be used to calculate many parameters like: External Quantum Efficiency (EQE), Solar Cell Efficiency, open circuit voltage, and short circuit [27,28].

The external quantum efficiency (EQE) is defined as:

$$\text{EQE (wavelength)} = \frac{\text{Number of electrons is collected by electrode}}{\text{Number of photons light source}} \tag{19}$$

The short-circuit current density can be computed as:

$$J_{sc} = \text{Charge of electron} \times \text{number of electrons collected by electrode} \tag{20}$$

$$\text{Solar Cell Efficiency is: Efficiency} = \frac{J_{sc} V_{oc} \text{FF}}{P_{in}} \tag{21}$$

FF is the filling factor, which can be derived from J–V curve of the device.

$$\text{FF} = \frac{I_m V_m}{I_{sc} V_{oc}} \tag{22}$$

12

Thermoeconomic and thermoenvironomic modeling and analysis of selected gas turbine power plants in Nigeria

Sunday O. Oyedepo[1], Richard O. Fagbenle[2], Samuel S. Adefila[3] & Md. Mahbub Alam[4]

[1]Mechanical Engineering Department, Covenant University, Ota, Nigeria
[2]Mechanical Engineering Department, Obafemi Awolowo University, Ile Ife, Nigeria
[3]Chemical Engineering Department, Covenant University, Ota, Nigeria
[4]Institute for Turbulence-Noise-Vibration Interaction and Control, Shenzhen Graduate School, Harbin Institute of Technology, Shenzhen, China

Keywords
Exergy, gas turbine, sustainability index, thermodynamic modeling, thermoeconomic

Correspondence
Sunday O. Oyedepo, Mechanical Engineering Department, Covenant University, Ota, Nigeria.
E-mail: sunday.oyedepo@covenantuniversity.edu.ng

Funding Information
Md. Mahbub Alam wishes to acknowledge the support given to him by the Research Grant Council of Shenzhen Government through grant KQCX20140521144423867.

Abstract

This study presents comprehensive thermoeconomic and thermoenvironomic modeling and analysis of selected gas turbine power plants in Nigeria using the first and second laws of thermodynamics (exergy) concept. Exergetic analysis was conducted using operating data obtained from the power plants to determine the exergy destruction and exergy efficiency of each major component of the gas turbine in each power plant. The results of the study showed that the combustion chamber (CC) is the most exergy destructive component compared to other cycle components. The percentage of exergy destruction in CC varies between 86.05% and 94.6%. By increasing the gas turbine inlet temperature (GTIT), the exergy destruction of this component can be reduced. The total exergy improvement potential of the selected plants varies from 54.04 to 159.88 MW. The component with the highest exergy improvement potential is the CC, which has the value that varies from 30.21 to 88.86 MW. Thermoeconomic analysis showed that the cost of exergy destruction is high in the CC, and an increase in the GTIT effectively decreases this cost. The exergy costing analysis revealed that the unit cost of electricity produced in the plants ranged from cents 1.99/kWh (₦3.16/kWh) to cents 5.65 /kWh (₦8.98/kWh). Thermoenvironomic analysis showed that the CO_2 emissions varied between 100.18 and 408.78 kg CO_2/MWh, while cost rate of environmental impact varied from 40.18 \$/h (6, 388.62 ₦/h) to 276.97 \$/h (44, 038. 23 ₦/h). The results further showed that CO_2 emissions and cost of environmental impact decrease with increasing GTIT. The sustainability index increase with increasing GTIT. Finally, this study will assist efforts to understand the thermodynamic losses in the gas turbine cycle, and to improve efficiency as well as provide future recommendations for better performance, sustainability, and lessening environmental impact of power plant.

Nomenclature

c	average unit cost (fuel or product)
$\dot{C}$ and $\dot{Z}$	cost rates
c_p	specific heat at constant pressure [kJ/kg]
D_p	depletion number
$\dot{E}$	exergy rate [kW]
$\dot{E}_L$	exergy loss rate
$\dot{E}_D$	exergy destruction rate
ExIP	exergetic improvement potential
eCO_2	emissions of CO_2 per unit mass of fuel
f_k	exergoeconomic factor
$\dot{m}$	mass flow [kg/sec]
P	power output [kW]
pe	potential energy [kJ]
PEC	purchase equipment cost
r_p	pressure compression ratio
R	gas constant [kJ/mol – K]
$\dot{S}$	entropy rate
T_{pz}	primary zone combustion temperature
TDI	thermal discharge index
y_D	exergy destruction rate ratio

₦	Nigerian Currency (NGN)
γ	adiabatic index
Δp_{cc}	pressure drop in combustion chamber (bar)
η_c	isentropic efficiency of compressor
η_T	isentropic efficiency of turbine
η_{th}	thermal efficiency
ε	exergetic efficiency
$\varnothing$	rational efficiency
δ	component efficiency defect
ψ	overall exergetic efficiency
ξ	exergetic performance coefficient
π	dimensionless pressure
Ψ	hydrogen–carbon ratio
Φ	fuel–air equivalent ratio
τ	residence time
ε	CO_2 emissions per unit net electricity (kg CO_2/MWh)
i	inlet
e	exit or outlet
p	pressure
a	air
pg	combustion product
f	fuel
T	turbine
cc	combustion chamber
th	thermal
sys	system
0	ambient
cv	control volume
D	destruction
gen	generation
ac	air compressor
gt	gas turbine
k	component
tot	total
PH	physical
KN	kinetic
PT	potential
CHE	chemical
T	thermal
P	mechanical
OM	operating and maintenance
CI	capital investment

Introduction

The present world energy scenario exhibits that most of the energy requirements are met by fossil fuels which cannot be newly formed at any significant rate; the present stocks are therefore finite. Also these fossil fuels are not environmental friendly and emits significant amount of pollutants causing serious environmental issues, such as global warming, ozone layer depletion, and climate change [1]. Awareness of limited hydrocarbon resources, environmental and economic concerns, and ever-increasing demand for electricity necessitates the design of optimal gas turbine (GT) power plants in terms of technical and cost aspects [2].

Thermodynamic analysis of thermal systems for design and optimization purposes is essential for effective utilization of limited available fossil fuel. In this regard, there are two essential tools available, such as energy analysis and exergy analysis [3]. The most commonly used method for analysis of the energy conversion process is the first law of thermodynamics (energy analysis). However, there is increasing interest in combined utilization of the first and second laws, using such concepts as exergy and exergy destructions in order to evaluate the efficiency with which the available energy is consumed [4]. So, by analyzing the exergy of different energy forms, we can reduce the wastage of energy and utilize it in a more efficient way thus making it available for future use [5].

Energy and exergy analyses provide insight into losses in various components of a power generating system. Unlike energy, exergy is generally not conserved but is destroyed in the system. The major causes of irreversibilities like heat transfer through a finite temperature difference, chemical reactions, friction, and mixing are accounted by exergy analysis [6]. Exergy analysis, therefore, has potential of identifying the source, magnitude, and location of exergy destruction (thermodynamic inefficiencies) in a thermal system. Based on this fact, exergy analysis usually predicts the thermodynamic performance of an energy system and the efficiency of the system components by accurately quantifying the entropy generation of the components [7].

Thus, exergy analysis plays an important role in developing strategies and in providing guidelines for more effective use of energy in the existing power plants. Another important issue for improving the existing power system is the origin of the exergy loss. Hence, a clear picture, instead of only the magnitude of exergy loss in each section, is required. Therefore, the exergy analysis has been widely used for the evaluation of the existing thermal power plants (TPPs) [3]. Increasing application and recognition of the usefulness of exergy methods in power plant design and optimization have been observed and reported in a number of articles in recent years [8–11]. Ofodu and Abam [12] applied exergy concept in analyzing the performance of Afam IV TPP. The analysis was presented in the form of Sankey and Grassman diagrams, which give energy and exergy values at prescribed points in the plant. The results showed that the greatest exergy loss occurred in the turbine and the exhaust diffuser followed by combustion chamber (CC) losses. These are traced to high temperature, incomplete combustion, and mechanical losses in the various parts of the plant. Abam and Moses [13] carried out computer simulation of the performance of a 33 MW GT power plant, using exergy analysis. The result obtained showed that the largest amount of exergy destruction occurred in the CC and the least in the GT. Ganapathy et al. [14] performed an exergy analysis on an operating 50 MWe unit of lignite fired steam power plant, aiming to find the energy efficiency and exergy efficiency of the power plant. The results revealed that maximum energy loss of 39% occurs in the condenser, whereas the maximum exergy loss of 42.73% occurs in the combustor.

Rajkumar and Ashok [15] carried out an energy and exergy study on 55 MW TPP at 40 and 55 MW output, respectively, to pinpoint the losses taking place in the plant. Over 57.822% exergy destruction took place in the boiler. The overall plant efficiency (first law) varied between 22.015% and 24.208% and the second law efficiency varied between 19.938% and 22.208%. It was concluded that exergy analysis is the realistic criteria for performance measurement of any thermal systems. Balkrishna [16] carried out study on energy and exergy analysis of an actual captive steam power plant using blast furnace gas, a by-product of pig iron production. The result showed that the energy loss as a percentage of input energy was found to be maximum in the steam condenser (64%), followed by boiler (19%). The percentage ratio of the exergy destruction to the total exergy destruction was found to be at maximum in the boiler system (75%), followed by steam turbine (11%), and then steam condenser (8%). The study concluded that the boiler was the major source of irreversibility in the power plant and the chemical reaction was the most significant source of exergy destruction in the boiler. Fagbenle et al. [17] carried out a thermodynamic first and second law analysis on a 53 MWth (net) biogas fired integrated gasification steam injected gas turbine (BIG/STIG) power plant with thermal efficiency of 41.5% (power based) and 45.0% (power and recovered heat based), respectively. The second law analysis and the energy utilization diagrams revealed that the CC has the largest irreversibility, with about 79% of the total system exergy loss. Suggestions for reduction of this large exergy loss were made. These include addition of spray water and preheating of the reactants in the CC, and the use of exergy left in the stacks from heat recovery steam generator (HRSG). Aljundi [4] presented the energy and exergy analysis of Al-Hussein power plant in Jordan. The performance of the plant was estimated by a component-wise modeling and a detailed breakup of energy and exergy losses for the plant was considered. The result of the study showed that energy losses mainly occurred in the condenser where 134 MW was lost to the environment, while only 13 MW was lost from boiler system. The percentage ratio of the exergy destruction to the total exergy destruction was found to be maximum in the boiler system (77%) followed by the turbine (13%), and then the forced draft fan condenser (9%). It was concluded that the boiler was the major source of irreversibilities in the power plant. Chemical reaction was the most significant source of exergy destruction in the boiler system which can be reduced by preheating the combustion air and reducing the air–fuel ratio. Ehsana and Yilmazoglub [18] carried out an exergy analysis on a 240 MW_{el} TPP to be operated with 10 different types of Turkish lignite. Exergy destruction of each component was investigated by using conservation of mass, conservation of energy, and exergy destruction in an open system at steady state. Net energy and exergy efficiencies of the plant were calculated as 37.16% and 34.84%, respectively. The analysis of the case study revealed that the boiler is the major source of exergy destruction with 299.10 MW and 83.29% of the total exergy destruction of the overall plant. Ighodaro and Aburime [19] carried out an exergetic appraisal of Delta IV Power Station, Ughelli in Nigeria. The study employed the first and second laws of thermodynamics. In the exergetic analysis, mass and energy conservation laws were applied to each component. The turbine was found to have the largest exergetic efficiency of 95.4% while that of the total plant was 45.7%. The CC had the largest exergetic destruction (56%) while that of the total plant was 58.5%. Exergy analysis of a steam cycle with double reheat and turbine extraction was presented by Rashidi et al. [20]. In this study, six heaters were used, three of them at high pressure and the other three at low pressure with deaerator. The results of the study showed that the biggest exergy loss occurred in the boiler followed by the turbine. The results also showed that the overall thermal efficiency and the second law efficiency decreased as the condenser pressure increased for any fixed outlet boiler temperature, however, these parameters increased as the boiler temperature increased for any condenser pressure.

In the assessment of existing TPPs, the exergy destruction within a component represents not only a thermodynamic inefficiency but, in general, also an opportunity to reduce the investment cost and sometimes also the environmental impact associated with the component being considered and, thus, with the overall system. Thus, assessment of economic and environmental impacts of existing thermal plants from exergy analysis point of view is worthy of consideration.

The relationship between exergy and economics, particularly the trade-offs that normally occur between efficiency and costs, has been an important concern for decades and continues to be so. Economic methods based on exergy, also known as thermoeconomics or exergoeconomics, have evolved and are applied in assessing power plants performance [21]. Also, the environmental impacts and nonsustainability of energy use have become of significant concerns recently, and exergy methods have been used as improvement efforts to (1) reduce environmental emissions and lengthening the lives of resources through increasing efficiencies and (2) assess the potential impacts of emissions [22, 23]. Thermoenvironomic or exergoenvironomic combines exergy and environment analyses to determine thermodynamic efficiency and formation of environmental impacts on plant components. The analysis reveals the interdependencies between thermodynamic behavior and environmental impacts between plant components [24].

In the literature, many studies related to the linkages between exergy, economics, and environmental impact to assess performance of TPPs are reported [25–28]. Thermoenvironmental and economic analysis of simple and regenerative GT cycles with regression modeling and optimization was performed by Memon et al. [29]. In this study, a comprehensive modeling of simple and regenerative cycle GT power plants were conducted. Result of the study showed that the power output and efficiencies increased while CO_2 emissions decreased, with decreasing compressor inlet temperature (CIT) and increasing turbine inlet temperature (TIT). With increasing pressure ratio (PR) to some initial values, power output and efficiencies increased and CO_2 emissions decreased, which vary oppositely with increasing PR further. Furthermore, five objective functions (OBFs) were optimized for maximization of net power output, energy and exergy efficiencies, and minimization of CO_2 emissions and costs of the cycles. The optimal operating parameters thus obtained exhibit a trade-off between thermoenvironmental benefits and additional costs. The study concluded that, from the thermoenvironmental point of view, TIT should be selected as high as possible, however, this leads to an increase in the capital cost and levelized cost of electricity. Seyyed et al. [30] carried out thermoenvironomic optimization of GT cycles with air preheat. In this study, the effect of air preheater (APH) on thermoeconomic and thermoenvironomic optimizations in the thermodynamic cycles was investigated. The results showed that the existence of APH increased exergetic efficiency of the cycles and environmental impacts cost flow rate, whereas it decreased the total cost flow rate. Also, in the cycles without APH, the optimum values of decision variables corresponding to thermoeconomic and thermoenvironomic OBFs do not change considerably, while in the cycles with APH, they change noticeably. Mert et al. [31] carried out exergoeconomic analysis of a cogeneration plant in an iron and steel factory, in Turkey. The result of the study showed that cost of compressed air exiting the compressor was 3696.85 \$/h, while the cost of power required for the compression was 3461.38 \$/h. The cost of product exiting the CC, namely combustion gas, was 8048.49 \$/h. The product of the CC was also the fuel of the GT, so the cost of the fuel of the GT was also 8048.49 \$/h. Power generated at GT was 6002.15 \$/h, while the unit exergy cost was 17.86 \$/GJ. The cost of combustion gas exiting from GT and entering HRSG as fuel was 2356.19 \$/h and the cost of steam produced at HRSG was 453.21 \$/h for the 14 bar steam and 1458.11 \$/h for the 45 bar steam. In other words, the costs of steam were 22.66 and 26.51 \$/ton, respectively. Gorji-Bandpy and Goodarzian [32] carried out exergoeconomic optimization of GT power plant operating parameters using genetic algorithms. In the study, exergoeconomic optimization and analysis were performed on a 140 MW GT power plant. The two objectives considered in the study are thermodynamic (e.g., maximum efficiency and minimum fuel consumption) and economic (e.g., minimum cost per unit of time and maximum profit per unit of production). The results of the study showed that the unit cost of product decreased from 5.25 to 4.73 \$/GJ, however, with 8.77% increase in capital investment, the exergy destruction cost decreased from 2947 to 2407 \$/h. Sahoo [33] studied exergoeconomic analysis and optimization of a cogeneration system using evolutionary programming. In this study, a cogeneration system that produced 50 MW of electricity and 15 kg/sec of saturated steam at 2.5 bar was optimized using exergoeconomic principles and evolutionary programming. The result showed that the product cost, cost of electricity and steam, was 9.9% lower with respect to the base case. Exergoeconomic analysis for Unit Gt14 of South Tripoli GT power plant was investigated by Fellah et al. [34]. The methodology used in the study was based on the Specific Exergy Costing approach. Results of the study showed that the average cost per unit exergy net power equal to 7.1 \$/GJ at 40% design load, and equal to 5.5 \$/GJ at 60% design load, and equal to 4 \$/GJ at full operating load. It was found that the cost of exergy destruction in the CC presented the main contribution to the total cost of exergy loss; its value varied in the CC from 1474 \$/h at 40% design load to 1123 \$/h at the full operating load. The contribution and the variation of cost of exergy destruction with load are lower for the other two main components.

Most of the past studies on combined exergy, exergoeconomic, and exergoenvironomic analyses of TPPs were based on a single power unit. In the present work, economic and environmental impacts of power plants from exergy point of view are performed on 11 GT units at three different stations in Nigeria.

The prime objectives of this study are:

1. To model a GT power plant and compare the simulation code with the actual existing GT power plants in Nigeria to ensure the correctness of a simulation code.
2. To perform exergy, thermoeconomic, and thermoenvironomic analyses of selected GT power plants in Nigeria.
3. To identify the most significant source of exergy destruction in the power plants and the location(s) of occurrence and to demonstrate exergy analysis as a powerful tool to assist in the quest for sustainable development.
4. To estimate unit cost of electricity (using exergy costing) in the selected power plants.
5. To quantify the carbon dioxide (CO_2) emissions and cost of environmental impact from selected GT stations.
6. To determine the sustainability index (SI) for the system and quantify its variation with exergy destruction.

System Description

GT power plants in Nigeria operate on simple GT engine. The simple GT power plant mainly consists of a GT coupled to a rotary type air compressor and a CC which is placed between the compressor and turbine in the fuel circuit. Auxiliaries, such as cooling fan, water pumps, etc., and the generator itself are also driven by the turbine. Other auxiliaries are starting device, lubrication system, duct system, etc. For ease of analysis, the steady-state model of simple GT is presented in Figure 1.

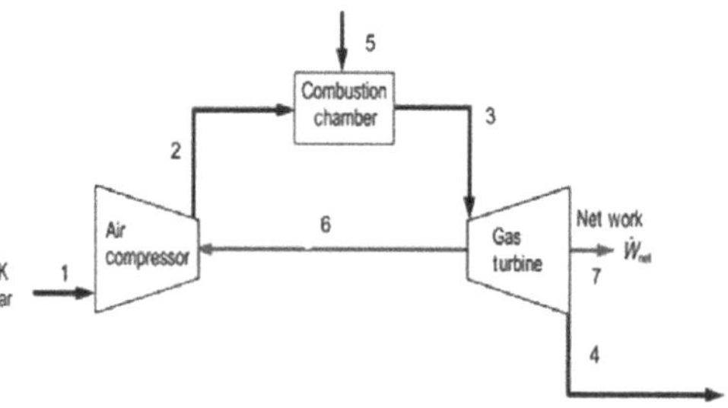

Figure 1. A schematic diagram for a simple gas turbine cycle.

Methodology

Based on the idea that exergy represents the only rational basis for assigning both costs and environmental impacts to the energy carriers and to the inefficiencies within the system, a methodological approach called thermoeconomic (exergoeconomic) analysis is applied together with thermoenvironomic (exergoenvironomic) analysis to evaluate the performance of selected GT power plants in Nigeria.

Modeling and simulation of GT power plant

The thermodynamic performance of power plants of 11 different capacities is estimated by a component-wise modeling followed by a system simulation. A flow-sheet computer program, "MATLAB R2010a Software," (The MathWorks, Inc., Natick, MA) is used for the study. It is designed for the steady-state thermodynamic (exergy, thermoeconomic, and thermoenvironomic) modeling and analysis of systems for the production of electricity.

The model employed in this study to simulate the performance of the simple GT system presented in Figure 1 comprises three different submodels: compressor, CC, and turbine. Under steady-state condition, application of the mass and exergy balance equations are used to determine the flow rate, exergy destruction rate, and the exergy efficiency for each component in the base case and for the whole system in the power plant.

Basic assumptions

The following basic assumptions are made to carry out thermodynamic modeling and simulation of the power plants [27, 35, 36]:

1. All the thermodynamic processes in GT cycle are considered based on the steady-state model.
2. The principle of ideal gas mixture was applied for the air and combustion products with variable specific heat.
3. The fuel injected to the CC is assumed to be natural gas.
4. The dead/reference state condition is P_0 = 1.013 bar and T_0 = 298.15 K.
5. Complete combustion of fuel occurs in CC.
6. In the CC, 2% pressure drop was assumed.
7. The exergy value of air entering the compressor was assumed zero.

The thermoeconomic (exergoeconomic) model

The thermoeconomic analysis developed in this work combines the second law of thermodynamics with the exergy concept in combination with economic engineering and the following four steps proposed by Tsatsaronis [37] were followed in this study. The first step is exergy analysis. The second step is economic analysis of each of the plant component. This step provides the monetary costs associated with investment, operation, and maintenance. The third step is the estimation of exergetic costs associated with each flow and finally, the fourth step is the exergoeconomic evaluation of each system component.

Exergy model

Exergy can be divided into four distinct components. The two important ones are the physical exergy and chemical exergy. In this study, the two other components which are kinetic exergy and potential exergy are assumed to be negligible, as the changes in them are negligible. The physical exergy is defined as the maximum theoretical useful work obtained as a system interacts with an equilibrium state. The chemical exergy is associated with the departure of the chemical composition of a system from its chemical equilibrium. The chemical exergy is an important part of exergy in combustion processes. Applying the first and the second laws of thermodynamics, the following exergy balance is obtained [27, 28, 36]:

$$\dot{E}_x = \sum_j \left(1 - \frac{T_0}{T_j}\right)\dot{Q}_j + \dot{W}_{CV} + \sum_i m_i e_i - \sum_e m_e e_e \quad (1)$$

The subscripts i, e, j, and 0 refer to conditions at inlet and exits of control volume (CV) boundaries and reference state. Equation (1) can be written as:

$$E_i^{tot} - E_e^{tot} - E_D = 0 \quad (2)$$

Equation (2) implies that the exergy change of a system during a process is equal to the difference between the net exergy transfer through the system boundary and the exergy destroyed within the system boundaries as a result of irreversibilities.

The exergy balance equations and the exergy destroyed during each process and for the whole GT plant are written as follows [38]:

Air compressor

$$\dot{E}^{WAC} = \left(\dot{E}_1^T - \dot{E}_2^T\right) + \left(\dot{E}_1^P - \dot{E}_2^P\right) + T_0\left(\dot{S}_1 - \dot{S}_2\right) \quad (3a)$$

$$\dot{E}_{DAC} = T_0\left(\dot{S}_2 - \dot{S}_1\right) = \dot{m}T_0\left[c_{p1-2}\ln\left(T_2/T_1\right) - R\ln\left(P_2/P_1\right)\right] \quad (3b)$$

Combustion chamber

$$\dot{E}^{CHE} + \left(\dot{E}_2^T + \dot{E}_5^T - \dot{E}_3^T\right) + \left(\dot{E}_2^P + \dot{E}_5^P - \dot{E}_3^P\right) + T_0\left(\dot{S}_3 - \dot{S}_2 + \dot{S}_5 + \frac{\dot{Q}_{CC}}{T_0}\right) = 0 \quad (4a)$$

$$\dot{E}_{DC} = T_0\left[\dot{S}_3 - \dot{S}_2 + \dot{S}_5 + \frac{\dot{Q}_{CC}}{T_0}\right] = \dot{m}_{T_0}\left\{\left(c_{p2-3}\ln(T_3/T_2) - R\ln(P_3/P_2)\right) + \left(c_{p5}\ln(T_5/T_0) - R\ln(P_5/P_0)\right) + \frac{c_{p2-3}(T_3 - T_2)}{T_{inCC}}\right\} \quad (4b)$$

Gas turbine

$$\dot{E}^{WGT} = \left(\dot{E}_3^T - \dot{E}_4^T\right) + \left(\dot{E}_3^P - \dot{E}_4^P\right) + T_0\left(\dot{S}_3 - \dot{S}_4\right) \quad (5a)$$

$$\dot{E}_{DGT} = \dot{m}T_0\left[c_{p3-4}\ln\left(T_4/T_3\right) - R\ln\left(P_4/P_3\right)\right] \quad (5b)$$

Exergy improvement potential of GT plant

The exergy improvement potential of an energy conversion system is a measure of how much and how easily the system could be improved for optimization purposes. It is a thermodynamic approach combining exergy losses and effectiveness to have a more complete parameter of the performance of the system [39]. The exergy improvement potential makes it possible to determine the critical points of the system stating a hierarchy on its components in such a way that the measure be applied in the places where they will be most effective.

The exergetic improvement potential is obtained from the exergy losses and the efficiency of the system. It is calculated by the equation (6) [40, 41]:

$$\text{ExIP} = (1 - \epsilon)I, \quad (6)$$

where ExIP is the exergetic improvement potential, ε is the exergetic efficiency (%), and I is the exergy loss or irreversibility rate.

$$I = \triangle E_{lost} = E_{in} - E_{out} < 0 \quad (7)$$

For a CV at steady state, the exergetic efficiency is

$$\epsilon = \frac{\dot{E}_P}{\dot{E}_F} = 1 - \frac{\dot{E}_D + \dot{E}_L}{\dot{E}_F} \quad (8)$$

where the rates at which the fuel is supplied and the product is generated are denoted by $\dot{E}_F$ and $\dot{E}_P$, respectively. $\dot{E}_D$ and $\dot{E}_L$ denote the rates of exergy destruction and exergy loss, respectively.

In any real engineering system (which is irreversible) exergy is degraded and the exergy efficiency is consequently less than unity. According to Van Gool [42], the maximum improvement in the exergy efficiency of a process or system is obviously achieved when ΔE_{lost} is minimized.

The ith component efficiency defect denoted by δ_i is given by equation (9) [43]:

$$\delta_i = \frac{\sum \Delta \dot{E}_{Di}}{\sum \Delta \dot{E}_{xin}} \quad (9)$$

The overall exergetic efficiency of the entire plant is given as [43]:

$$\varphi_i = \frac{\dot{W}_{net}}{\dot{E}_{xfuel}} \quad (10)$$

The amount of exergy loss rate per unit power output as important performance criteria is given as:

$$\xi = \frac{\dot{E}_{D\,Total}}{\dot{W}_{net}} \quad (11)$$

where ξ is the exergetic performance coefficient.

Exergy destruction rate and efficiency equations for the GT power plant components and for the whole cycle are summarized in Table 1.

Economic model

The economic analysis, conducted as part of the exergo-economic analysis, provides the appropriate monetary values associated with the investment, operation, maintenance, and fuel costs of the system being analyzed [44, 45]. These values are used in the cost balances [46].

The annualized (levelized) cost method of Moran [47] is used to estimate the capital cost of system component in this work.

The amortization cost for a particular component may be written as [48]:

$$\mathrm{PW} = \mathrm{PEC} - (\mathrm{SV})\,\mathrm{PWF}\,(i,n) \tag{12}$$

where the salvage value (SV) at the end of the nth year is taken as 10% of the initial investment for component (or purchase equipment cost, PEC). The present worth (PW) of the component may be converted to the annualized cost by using the capital recovery factor, CRF (i, n) [32, 48], that is,

$$\dot{C}\,(\$/\mathrm{year}) = \mathrm{PW} \times \mathrm{CRF}\,(i,n) \tag{13a}$$

$$\mathrm{CRF}(i,n) = i/1-(1+i)^{-n} \tag{13b}$$

$$\mathrm{PWF} = (1+i)^{-n} \tag{13c}$$

where i is the interest rate and it is taken to be 17% [32], n is the total operating period of the plant in years and was obtained from the selected plants. PEC is the purchased equipment cost.

Equations for calculating the PEC for the components of the GT power plant are as follows [27, 47]:

Air compressor

$$\mathrm{PEC}_{AC} = \left[\frac{71.1\dot{m}_a}{0.9-\eta_{sc}}\right]\left[\frac{P_2}{P_1}\right]\ln\left[\frac{P_2}{P_1}\right] \tag{14}$$

Combustion chamber

$$\mathrm{PEC}_{cc} = \left[\frac{46.08\dot{m}_a}{0.995-P_3/P_2}\right]\left[1+\exp\left(0.018T_3-26.4\right)\right] \tag{15}$$

Table 1. The exergy destruction rate and exergy efficiency equations for gas turbine.

Component	Exergy destruction rate	Exergy efficiency
Compressor	$\dot{E}_{DC} = \dot{E}_{in} - \dot{E}_{out} + \dot{W}_C$	$\epsilon = \frac{\dot{E}_{out}-\dot{E}_{in}}{\dot{W}}$
Combustion chamber	$\dot{E}_{DCC} = \dot{E}_{in} - \dot{E}_{out} + \dot{E}_{fuel}$	$\epsilon = \frac{\dot{E}_{out}}{\dot{E}_{in}-\dot{E}_{fuel}}$
Gas turbine	$\dot{E}_{DT} = \dot{E}_{in} - \dot{E}_{out} - (\dot{W}_{net} + \dot{W}_C)$	$\epsilon = \frac{\dot{W}_{net}-\dot{W}_C}{\dot{E}_{in}-\dot{E}_{out}}$
Total exergy destruction rate	$\dot{E}_{DTotal} = \Sigma\dot{E}_D = \dot{E}_{DC} + \dot{E}_{DCC} + \dot{E}_{DT}$	

Gas turbine

$$\mathrm{PEC}_{gt} = \left[\frac{479.34\dot{m}_g}{0.92-\eta_{st}}\right]\ln\left[\frac{P_3}{P_4}\right]\left[1+\exp\left(0.036T_3-54.4\right)\right] \tag{16}$$

Dividing the levelized cost by annual operating hours, N, we obtain capital cost rate for the kth component of the plant [49]:

$$\dot{Z}_k = \frac{\phi_k \dot{C}_k}{3600 \times N} \tag{17}$$

The maintenance cost is taken into consideration through the factor $\phi_k = 1.06$ for each plant component [27, 49].

Estimation of GT exergy costing

Exergoeconomics is based on exergy costing and is usually applied at the plant component [50].

In order to perform exergy costing calculations, GT components (Fig. 1) must be combined into suitable control volumes (CVs), on which exergetic cost balance equation was then applied, on an individual basis. The component in each CV with their input and output streams are given as follows:

CV 1: Air compressor (AC) – Input streams: 1, 6
Output stream: 2
CV 2: Combustion chamber (CC) – Input streams: 2, 5
Output stream: 3
CV 3: Gas turbine (GT) – Input stream: 3
Output streams: 4, 6, 7

For a component that receives heat transfer and generates power, cost balance equation may be written as follow [27, 32, 51]:

$$\sum\left(c_e\dot{E}_e\right)_k + c_{w,k}\dot{W}_k = c_{q,k}\dot{E}_{q,k} + \sum\left(c_i\dot{E}_i\right)_k + \dot{Z}_k \tag{18}$$

$$\dot{C}_j = c_j\dot{E}_j \tag{19}$$

The cost–balance equations for all the components of the system construct a set of nonlinear algebraic equations, which was solved for $\dot{C}_j$ and c_j.

The formulations of cost balance for each component and the required auxiliary equations are as follows [32]:

Air compressor

$$\dot{C}_2 = \dot{C}_1 + \dot{C}_6 + \dot{Z}_{ac} \tag{20}$$

where subscript 6 denotes the power input to the compressor.

Combustion chamber

$$\dot{C}_3 = \dot{C}_2 + \dot{C}_5 + \dot{Z}_{cc} \tag{21}$$

Gas turbine

$$\dot{C}_4 + \dot{C}_6 + \dot{C}_7 = \dot{C}_3 + \dot{Z}_{gt} \tag{22}$$

The auxiliary equation for GT is given as:

$$\frac{\dot{C}_3}{\dot{E}_3}=\frac{\dot{C}_4}{\dot{E}_4} \tag{23}$$

An additional auxiliary equation is formulated assuming the same unit cost of exergy for the net power exported from the system and power input to the compressor:

$$\frac{\dot{C}_6}{\dot{W}_{AC}}=\frac{\dot{C}_7}{\dot{W}_n} \tag{24}$$

The cost rate associated with fuel (methane) is obtained from [52]:

$$\dot{C}_f=c_f\dot{m}_f\times \text{LHV}, \tag{25}$$

where the fuel cost per energy unit (on an LHV basis) is c_f = 0.004\$/MJ [51], $\dot{m}_f$ is the mass flow rate of fuel and LHV is the lower heating value of fuel.

A zero unit cost is assumed for air entering the air compressor, that is,

$$\dot{C}_1=0 \tag{26}$$

In order to estimate the cost of exergy destruction in each component of the plant, the cost–balance equations were solved for each component. In application of the cost–balance equation (eq. 18), there is usually more than one inlet and outlet streams for some components. In this case, the numbers of unknown cost parameters are higher than the number of cost–balance equations for that component. Auxiliary exergoeconomic equations (eqs. 23 and 24) are developed to solve this problem. Implementing equation (18) for each component together with the auxiliary equations forms a system of linear equations as follows [36, 53]:

$$[\dot{E}_k]\times[c_k]=[\dot{Z}_k] \tag{27}$$

where $[\dot{E}_k]$, $[c_k]$, and $[\dot{Z}_k]$ are the matrix of exergy rate which were obtained in exergy analysis, exergetic cost vector (to be evaluated), and the vector of $\dot{Z}_k$ factors (obtained in economic analysis), respectively.

The above set of equations was solved using MATLAB R2010a to obtain the cost rate of each line in Figure 1.

Exergoeconomic variables for GT components evaluation

In exergoeconomic evaluation of thermal systems, certain quantities play an important role. These are the average cost of fuel ($c_{F,k}$), average unit cost of product ($c_{P,k}$), the cost rate of exergy destruction ($\dot{C}_{D,k}$), relative cost difference r_k, and exergoeconomic factor f_k.

Then the average costs per unit of fuel exergy ($c_{F,k}$) and product exergy ($c_{P,k}$) are calculated from [34]:

$$c_{F,k}=\frac{\dot{C}_{F,k}}{\dot{E}_{F,k}} \tag{28}$$

$$c_{P,k}=\frac{\dot{C}_{P,k}}{\dot{E}_{P,k}} \tag{29}$$

The cost rate associated with exergy destruction is estimated as:

$$\dot{C}_{D,k}=c_{F,k}\dot{E}_{D,k} \tag{30}$$

Relative cost difference r_k is given as [54]:

$$r_k=\frac{c_{p,k}-c_{F,k}}{c_{F,k}}=\frac{1-\varepsilon_k}{\varepsilon_k}+\frac{\dot{Z}_k}{c_{F,k}\dot{E}_{p,k}} \tag{31}$$

One indicator of exergoeconomic performance is the exergoeconomic factor, f_k. The exergoeconomic factor is defined as [32, 34]:

$$f_k=\frac{\dot{Z}_k}{\dot{Z}_k+\dot{C}_{D,k}} \tag{32}$$

The thermoenvironomic (exergoenvironomic) model

To minimize the environmental impacts, a primary target is to increase the efficiency of energy conversion processes and, thus, decrease the amount of fuel and the related overall environmental impacts, especially the release of carbon dioxide, which is one of the main components of greenhouse gas [27].

In this study, three steps were applied to carry out the exergoenvironomic analysis of GT system. The first step is the determination of pollutant emission (CO and NOx) in grams per kilogram of fuel, the estimation of the total cost rate of product, and environmental impact and CO_2 emission calculation.

Determination of pollutant emission

In order to determine the pollutant emission in grams per kilogram of the fuel, the adiabatic flame temperature in the CC has to be computed first.

The adiabatic flame temperature in the primary zone of the CC is derived from the expression given by [55]:

$$T_{PZ}=A\sigma^{\alpha}\exp\left(\beta(\sigma+\lambda)^2\right)\pi^x\theta^y\psi^z \tag{33}$$

where π is a dimensionless pressure P_2/P_{ref} (P_2 being the combustion pressure and P_{ref} = 1.013 bar); θ is a dimensionless temperature T_2/T_{ref} (T_2 being the inlet temperature and T_{ref} = 298.15 K); ψ is the H/C atomic ratio (ψ = 4); $\sigma = \phi$ for $\varphi \le 1$ (φ is the fuel to air equivalent ratio), and $\sigma = \varphi - 0.7$ for $\varphi \ge 1$. Moreover, x, y, and z are quadratic functions of σ based on the following equations [25, 56]:

$$x = a_1 + b_1\sigma + c_1\sigma^2 \quad (34)$$

$$y = a_2 + b_2\sigma + c_2\sigma^2 \quad (35)$$

$$z = a_3 + b_3\sigma + c_3\sigma^2 \quad (36)$$

where parameters A, α, β, λ, a_i, b_i, and c_i are constant parameters. All the parameters in equations (34–36) are listed in Table 2.

The amount of CO and NOx produced in the CC and combustion reaction depends on the adiabatic flame temperature [25, 27, 56]. Accordingly, to determine the pollutant emission in grams per kilogram of the fuel, the equations proposed by [55] are used in this study.

$$\dot{m}_{NOX} = \frac{0.15E16\tau^{0.5}\exp\left(-71100/T_{PZ}\right)}{P_2^{0.05}\left(\Delta P_2/P_2\right)} \quad (37)$$

$$\dot{m}_{CO} = \frac{0.179E9\exp\left(7800/T_{PZ}\right)}{P_2^2\tau\left(\Delta P_2\right)}, \quad (38)$$

where τ is the residence time in the combustion zone (τ is assumed constant and is equal to 0.002 sec); T_{pz} is the primary zone combustion temperature; P_2 is the combustor inlet pressure; $\Delta P_2/P_2$ is the nondimensional pressure drop in the CC.

Cost of environmental impact

The cost of environmental impact expresses the environmental impact as the total pollution damage ($/s) due to CO and NOx emission by multiplying their respective flow rates by their corresponding unit damage cost (C_{CO}, and C_{NOx} are equal to 0.02086 $/$kg_{CO}$ and 6.853 $/$kg_{NOx}$)

Table 2. Constants for equations (34–36).

	0.3 ≤ ϕ ≤ 1.0		0.3 ≤ ϕ ≤ 1.0	
Constants	0.92 ≤ θ ≤ 2	2 ≤ θ ≤ 3.2	0.92 ≤ θ ≤ 2	2 ≤ θ ≤ 3.2
A	2361.7644	2315.752	916.8261	1246.1778
α	0.1157	−0.0493	0.2885	0.3819
β	−0.9489	−1.1141	0.1456	0.3479
λ	−1.0976	−1.1807	−3.2771	−2.0365
a_1	0.0143	0.0106	0.0311	0.0361
b_1	−0.0553	−0.045	−0.078	−0.085
c_1	0.0526	0.0482	0.0497	0.0517
a_2	0.3955	0.5688	0.0254	0.0097
b_2	−0.4417	−0.55	0.2602	0.502
c_2	0.141	0.1319	−0.1318	−0.2471
a_3	0.0052	0.0108	0.0042	0.017
b_3	−0.1289	−0.1291	−0.1781	−0.1894
c_3	0.0827	0.0848	0.096	0.1037

[27, 45]. In the present work, the cost of pollution damage is considered to be added directly to the expenditures that must be paid.

Total cost rate:

$$\dot{C}_{Tot} = \dot{C}_f + \sum_k \dot{Z}_k + \dot{C}_D + \dot{C}_{env} \quad (39)$$

where

$$\dot{C}_{env} = C_{CO}\dot{m}_{CO} + C_{NOX}\dot{m}_{NOX} \quad (40)$$

$$\dot{C}_f = c_f\dot{m}_f \times LHV \quad (41)$$

where $\dot{Z}_k$, $\dot{C}_f$, $\dot{C}_D$, and $\dot{C}_{env}$ are the purchase cost of each component, fuel cost, cost of exergy destruction, and cost of environmental impact, respectively.

CO_2 emissions calculation

Using the combustion equations, the normalized CO_2 emission of the plant is expressed as [27, 57]:

$$\varepsilon = \frac{\dot{m}_{CO_2}}{\dot{W}_{net}} \quad (42)$$

where ε is the CO_2 emission per unit net electricity output (kg CO_2/MWh).

The effect of CO_2 emissions is of considerable significance, such that reduction of its harmful release is twofold. The first is obviously related to communal and environmental health. The second, as suggested in many references, is improvement in reduction of harmful emissions in the CC can lead to improvements of GT cycle efficiency.

Reduction of the harmful emissions in the CC to the environment has proven its benefits in increasing system efficiency, which in turn increases sustainability by lengthening the lives of the fuel resources. The relationships between sustainability, exergy efficiency, and environmental impact have been previously documented [22, 58]. They proposed that efficient fuel consumption could be characterized by a depletion number defined as:

$$D_p = \frac{\dot{E}_{dest}}{\dot{E}_{xin}} \quad (43)$$

where $\dot{E}_{dest}$ is the exergy destruction and $\dot{E}_{xin}$ is the exergy input by fuel consumption. The relationship between the depletion number and the exergy efficiency is described by:

$$\eta_{ex} = 1 - D_p \quad (44)$$

Furthermore, the sustainability of the fuel resource can be expressed by a sustainability index (SI) as the inverse of the depletion number [57]:

$$SI = \frac{1}{D_p} \quad (45)$$

Table 3. Average operating data for selected gas turbine power plants.

	AES station			Afam station				Delta station			
Plant/average operating data	PB204 (AES1)	PB209 (AES2)	PB210 (AES3)	GT17 (AF1)	GT18 (AF2)	GT19 (AF3)	GT20 (AF4)	GT9 (DEL1)	GT10 (DEL2	GT18 (DEL3)	GT20 (DEL4)
Ambient temperature, T_1 (K)	303.63	302.31	305.28	300.34	301.48	300.38	300.9	300.55	301.41	301.15	301.79
Compressor outlet temperature, T_2 (K)	622.31	627.48	636.28	593.73	595.82	610.90	618.32	613.73	619.07	634.32	630.32
Turbine inlet temperature, T_3 (K)	1218.62	1256.86	1222.45	1133.4	1192.82	1200.15	1215.65	1226.15	1224.73	1233.57	1234.73
Turbine outlet temperature, T_4 (K)	750.00	755.00	827.05	712.73	723.75	770.07	807.32	710.56	707.48	730.15	705.07
Temperature of exhaust gas, T_{exh} (K)	715.40	750.52	746.48	731.45	664.65	707.23	741.48	622.65	649.90	635.65	636.73
Compressor inlet pressure, P_1 (bar)	1.013	1.013	1.013	1.013	1.013	1.013	1.013	1.013	1.013	1.013	1.013
Compressor outlet pressure, P_2 (bar)	9.8	9.86	9.60	9.50	9.80	9.60	9.60	11.05	10.98	10.82	10.84
Pressure ratio	9.00	9.14	9.48	9.38	9.67	9.48	9.48	10.91	10.84	10.68	10.70
Mass flow rate of fuel (kg/sec)	2.58	2.54	2.81	6.50	6.40	8.10	8.40	3.08	3.10	8.15	8.13
Inlet mass flow rate of air (kg/sec)	122.16	122.20	121.93	359.00	359.00	470	470	140	140	375	375
Power output (MW)	29.89	29.37	31.52	49.90	58.00	132	135.4	19.42	20.8	92.8	93.42
Lower heating value of fuel (kJ/kg)	47,541.57	47,541.57	47,541.57	48,948.3	48,948.3	48,948.3	48,948.3	46,778	46,778	46,778	46,778

Results and Discussion

The average operating data for the selected GT power plants for the period of 6 years (2005–2010) are presented in Table 3.

Results of exergy analysis

The exergy flow rates at the inlet and outlet of each component of the plants were evaluated based on the values of measured properties such as pressure, temperature, and mass flow rates at various states. These quantities were used as input data to the computer program (MATLAB) written to perform the simulation of the performance of the components of the GT power plant and the overall plant.

Table 4 presents results of the net exergy flow rates crossing the boundary of each component of the plants, exergy destruction, exergy defect, exergetic performance coefficient, and exergy efficiency of each component of the plants. The two most important performance criteria, exergy efficiency and exergetic performance coefficient (ξ) vary from 15.66% to 30.72% and 1.45 to 2.44, respectively, for the considered plants. Since the condition of good performance is derived from a higher overall exergetic efficiency but lower exergetic performance coefficient for any thermal system, hence, it can be inferred that AF2, AF1, and DEL4 GT plants have good performance.

To illustrate the effect of operating parameters on the second law efficiency of the components of the GT, the AES1 (PB204) plant is considered as a typical case. The simulation of the performance of plant and components was done by varying the air inlet temperature from 290 to 320 K; and the TIT from 1000 to 1400 K, respectively. Figure 2 compares the second law efficiencies of the air compressor, CC, GT, and the overall plant when the ambient temperature increases. The exergy efficiency of the turbine component and the overall exergetic efficiency of plant decreased with increased ambient temperature, whereas the exergy efficiencies of the compressor and turbine increased with increased ambient temperature. The overall exergetic efficiency decreased from 18.53% to 17.26% for ambient temperature range of 290–320 K. It was found that a 5 K rise in ambient temperature resulted in a 1.03% decrease in the overall exergetic efficiency of the plant. The reason for the low overall exergetic efficiency is due to large exergy destruction in the CC [59].

The exergetic efficiency (or second law efficiency) of the plant was also found to depend significantly on a change in TIT. Figure 3 shows that the second law efficiency of the plant increases steadily as the TIT increases. The increase in exergetic efficiency with increase in TIT is limited by turbine material temperature limit. This can

Table 4. Result of exergy performance analysis.

Exergy performance indicator	PB204 (AES1)	PB209 (AES2)	PB210 (AES3)	GT17 (AF1)	GT18 (AF2)	GT19 (AF3)	GT20 (AF4)	GT9 (DEL1)	GT10 (DEL2)	GT18 (DEL3)	GT20 (DEL4)
Installed rated power (MW)	33.5	33.5	33.5	75.0	75.0	138.0	138.0	25.0	25.0	100.0	100.0
Fuel exergy flow rate (MW)	220.53	235.23	237.68	327.96	363.28	459.15	449.06	274.85	276.78	441.20	440.24
Exergy destruction rate of AC (MW)	4.69	4.98	5.64	8.62	8.09	13.14	14.80	3.14	3.63	13.36	12.48
Exergy destruction of CC (MW)	56.55	56.58	55.35	139.42	159.84	176.78	180.83	62.52	61.76	171.84	173.33
Exergy destruction rate of turbine (MW)	0.29	0.52	0.23	5.99	1.47	9.39	14.50	0.39	0.14	0.70	1.80
Total exergy destruction rate (MW)	61.54	62.09	61.23	154.02	169.40	199.31	210.13	66.04	65.53	185.91	187.61
Exergy destruction of AC (%)	7.62	8.03	9.21	5.59	4.78	6.60	7.04	4.75	5.54	7.19	6.65
Exergy destruction of CC (%)	91.90	91.13	90.39	90.51	94.36	88.70	86.05	94.67	94.25	92.43	92.39
Exergy destruction rate of turbine (%)	0.48	0.84	0.41	3.89	0.87	4.71	6.90	0.58	0.21	0.38	0.96
Efficiency defect of AC (%)	14.01	14.83	16.83	9.15	8.43	12.46	14.03	7.79	9.05	12.52	11.69
Efficiency defect of CC (%)	66.11	66.26	64.63	58.31	58.05	58.35	56.20	73.45	72.43	56.11	56.97
Efficiency defect of turbine (%)	0.38	0.68	0.32	3.05	0.68	4.08	6.50	0.42	0.15	0.29	0.74
Total efficiency defect (%)	80.50	81.77	81.78	70.51	67.16	74.89	76.73	81.66	81.63	68.92	69.40
Exergy efficiency of AC (%)	85.99	85.17	83.17	90.85	91.57	87.54	85.97	95.21	90.95	87.48	88.31
Exergy efficiency of CC (%)	74.36	75.95	76.71	57.49	56.00	61.50	59.73	77.25	77.69	61.05	60.63
Exergy efficiency of turbine (%)	99.62	99.32	99.67	96.86	99.32	95.75	93.09	99.57	99.85	99.71	99.25
Overall exergetic efficiency (%)	19.50	18.23	18.22	29.49	32.84	25.11	23.27	18.34	18.37	31.08	30.60
Exergetic performance coefficient (ξ)	1.43	1.45	1.46	1.59	1.32	1.73	2.01	1.47	1.59	1.36	1.39
Exergy improvement potential											
Air compressor (MW)	3.94	4.14	4.71	7.83	6.88	8.59	14.95	2.89	3.30	11.69	11.02
Combustion chamber (MW)	38.48	40.33	43.82	69.49	68.52	85.81	88.86	30.47	30.21	78.48	78.67
Turbine (MW)	0.015	0.023	0.13	0.19	0.22	1.05	2.50	0.15	0.17	1.00	0.38
Entire plant (MW)	119.44	147.15	159.28	98.30	89.99	124.18	59.99	54.04	56.46	159.88	143.28

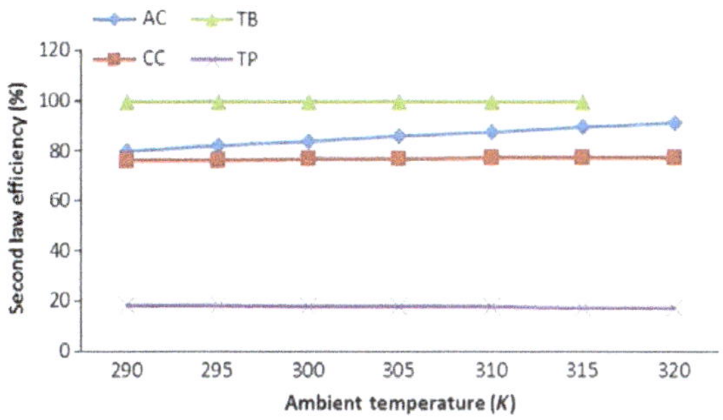

Figure 2. Variation in second law efficiency with ambient temperature. AC, second law efficiency of compressor; CC, second law efficiency of combustion chamber; TB, second law efficiency of turbine; TP, second law efficiency of entire plant.

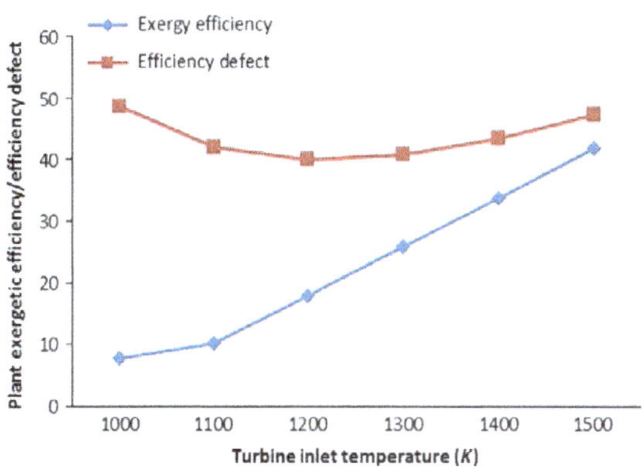

Figure 3. Variation in plant exergetic efficiency and efficiency defect with turbine inlet temperature.

be seen from the plant efficiency defect curve. As the TIT increases, the plant efficiency defect decreases to minimum value at certain TIT (1200 K), after which it increases with TIT. This shows degradation in performance of GT plant at a high TIT.

Exergy improvement potential

The results of the exergy improvement potential of the selected plants are presented in Table 5. The total exergy improvement potential of the selected plants varies from 54.04 to 159.88 MW. The component with the highest exergy improvement potential is the CC, which has value varies from 30.21 to 88.86 MW. This is followed by the air compressor which has value varies from 3.30 to 14.95 MW. The high improvement potential in the CC is due to the irreversibility associated with combustion and the large temperature difference between the air entering the CC and the flame temperature. These immense losses basically mean that a large amount of energy present in the fuel, with a greater capacity to generate useful work, is being wasted. Exergy improvement potential can be afforded in the CC by preheating the reactants and by reducing the heat loss and the excess air entering the CC. The lower improvement potential in the air compressor when compared with the CC is due to an additional heat loss from the air compressor through friction as compared to large temperature difference between the air entering the CC and the flame temperature. These results have made it possible to determine the critical points of the GT system stating hierarchy on its components in such a way that the measure be applied in the places where they will be most effective.

Results of thermoeconomic (exergoeconomic) analysis

The thermoeconomic performance of the selected GT plants in Nigeria was investigated considering the values of measured properties such as temperature, pressure, and mass flow rate at various points in the GT power plants. The period of operation for each plant investigated in this study are: AES station (12 years for PB204, PB209, and PB210), Afam station (30 years for GT17 and GT18; 11 years for GT19 and GT20), and Delta station (12 years for GT9, GT10, GT18, and GT20).

Solving the linear system of equations (14–20), the cost rates of the unknown streams of the system are obtained. For the selected GT power plants, the exergy costing method gave the unit cost of electricity produced in each plant and this varies as: from cents 3.90/kWh

Table 5. Exergy improvement potential.

Components	AES1	AES2	AES3	AF1	AF2	AF3	AF4	DEL1	DEL2	DEL3	DEL4
Air compressor (MW)	3.94	4.14	4.71	7.83	6.88	8.59	14.95	2.89	3.30	11.69	11.02
Combustion chamber (MW)	38.48	40.33	43.82	69.49	68.52	85.81	88.86	30.47	30.21	78.48	78.67
Turbine (MW)	0.015	0.023	0.13	0.19	0.22	1.05	2.48	0.15	0.17	1.00	0.38
Entire plant (MW)	119.44	147.15	159.28	98.17	89.99	124.18	159.59	54.04	56.46	159.88	143.28

(N6.20/kWh) to cents 5.07/kWh (8.07/kWh) for AES station, from cents 3.30/kWh (₦5.25/kWh) to cents 4.94/kWh (₦7.85/kWh) for Afam station and from cents 1.88/kWh (₦2.99/kWh) to cents 5.65/kWh (₦8.98/kWh) for Delta station.

The exergoeconomic parameters considered in this study include average costs per unit of fuel exergy C_F and product exergy C_P, rate of exergy destruction $\dot{E}_D$, cost rate of exergy destruction $\dot{C}_D$, investment and O&M costs rate $\dot{Z}$, and exergoeconomic factor f. In analytical terms, the components with the highest value of $\dot{Z}_k + \dot{C}_{Dk}$ are considered the most significant components from an exergoeconomic perspective. This provides a means of determining the level of priority a component should be given with respect to the improving of the system.

For all the plants considered, the CC and air compressor have the highest value of the sum $\dot{Z}_k + \dot{C}_{Dk}$ and are, therefore, the most important components from the exergoeconomic viewpoint. The low value of exergoeconomic factor, f, associated with the CC suggests that the cost rate of exergy destruction is the dominate factor influencing the component. Hence, it is implied that the component efficiency is improved by increasing the capital investment. This can be achieved by increasing gas turbine inlet temperature (GTIT). The maximum GTIT is limited by the metallurgical considerations [32, 57].

The air compressor has the highest f value in most of the plants investigated except plants AES1 (PB204), AES2 (PB209), and AES3 (PB210) with f value 50.44%, 45.94%, and 36.53%, respectively. The cost effectiveness of the entire system of the plants investigated can be improved if the $\dot{Z}$ value of GT is reduced. According to equation (17) of the cost model, the capital investment and O&M costs of the GT depend on temperature T_3, PR P_3/P_4, and turbine isentropic efficiency η_{st}. To reduce the high $\dot{Z}$ value associated with the GT, we need to consider a reduction in the value of at least one of the variables.

The results of the exergoeconomic analysis of the plants investigated show that the CC exhibits the greatest exergy destruction cost. The next highest source of exergy destruction cost is the air compressor. In comparing the results of exergy and exergoeconomic analyses, similar trends are revealed. Increasing GTIT effectively decreases the cost associated with exergy destruction. Further comparisons between related results are consistent with those reported previously [28], and confirm that the most significant parameter in the plant is GTIT. The finding establishes the concept that the exergy loss in the CC is associated with the large temperature difference between the flame and the working fluid. Reducing this temperature difference reduces the exergy loss. Furthermore, cooling compressor inlet air allows the compression of more air per cycle, effectively increasing the GT capacity.

To illustrate the effect of GTIT on the exergy destruction cost of CC of the selected plants, AES1 (PB204) plant is considered as sample. The simulation was done by varying the GTIT from 950 to 1500 K. Figure 4 shows the effect of variation in GTIT on CC exergy destruction cost. This figure shows that, like the exergy analysis results, the cost of exergy destruction for the CC decreases with an increase in the GTIT. This is due to the fact that the cost of exergy destruction is proportional to the exergy destruction. Hence, an increase in the GTIT can decrease the cost of exergy destruction. Furthermore, from Figure 3, an increase in the TIT of about 200 K can lead to a reduction of about 29% in the cost of exergy destruction. Therefore, TIT is the best option to improve cycle losses.

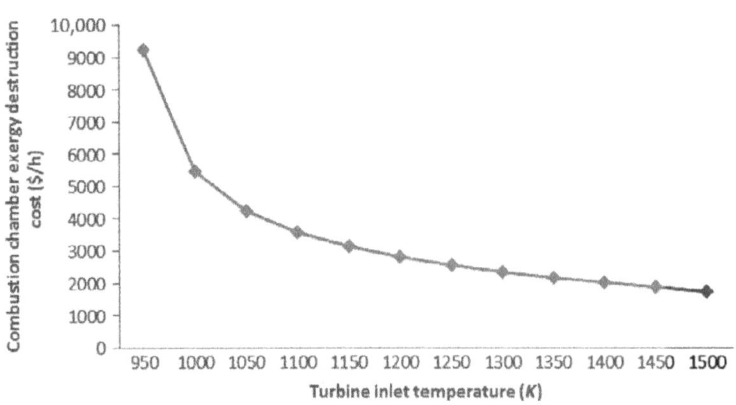

Figure 4. Combustion chamber exergy destruction cost and turbine inlet temperature.

Result of thermoenvironomic (exergoenvironomic) analysis

Table 6 shows the results of the exergoenvironomic analysis of this work. The exergoenvironomic parameters computed are CO_2 emission in kg per MWh of electricity generated, depletion number, SI, cost flow rate of environmental impacts ($\dot{C}_{env}$) in \$/h, and total cost rates of products ($\dot{C}_{Tot}$) in \$/h. For the period considered, the CO_2 emissions for the selected plants vary from 100.18 to 408.78 kg CO_2/MWh. AF2 (GT18) has the highest value and AES3 (PB210) has the least value. This result is justified by the fact that AES GT plants are barges mounted on lagoon, whereas other GT units are mounted on land. Moreover, units AF1 and AF2 were commissioned in the year 1982. The units have been in operation for over 30 years. Hence, these units have the highest CO_2 emissions.

For comparison purpose, CO_2 emissions of selected countries are presented in Table 7. The CO_2 emission in the present study (Nigeria) is comparable with other selected countries and the Intergovernmental Panel on Climate Change CO_2 emissions factor. Thus it can be

Table 6. Result of exergoenvironmental performance analysis.

Exergoenvironmental parameters	AES1	AES2	AES3	AF1	AF2	AF3	AF4	DEL1	DEL2	DEL3	DEL4
CO_2 emissions (kg CO_2/MWh)	112.73	109.92	100.18	406.18	408.78	393.68	248.38	181.36	226.33	327.02	287.95
Depletion number, D_p	0.69	0.69	0.72	0.81	0.81	0.84	0.82	0.77	0.77	0.82	0.80
Sustainability index, SI	1.44	1.44	1.39	1.24	1.23	1.20	1.22	1.31	1.29	1.21	1.25
Cost flow rate of environmental impact, $\dot{C}_{env}$ (\$/h)	51.91	51.98	46.29	257.90	243.42	276.97	209.37	61.12	76.32	252.39	247.40
Total cost rates of products, $\dot{C}_{Tot}$ (\$/h)	2991.59	2974.18	2935.69	9473.10	9225.40	12,232.84	11,885.62	3580.25	3700.39	11,772.48	11,414.39

Table 7. CO_2 emissions per MWh (kg CO_2/MWh) from electricity generation using natural gas for selected countries kg CO_2/MWh.

Country	Average 2008–2010[1]	This study (2013)
Canada	483	
Japan	437	
Spain	353	
Turkey	371	
United Kingdom	387	
USA	405	
Ukraine	379	
Egypt	490	
China	504	
India	436	
Cuba	502	
IPCC emission		
Factor[2]	225	
Nigeria		
AES1		112.73
AES2		109.92
AES3		100.18
AFI		406.18
AF2		408.78
AF3		393.68
AF4		248.38
DEL1		181.36
DEL2		226.33
DEL3		327.02
DEL4		287.95

1IEA [60].
[2]IPCC [61].

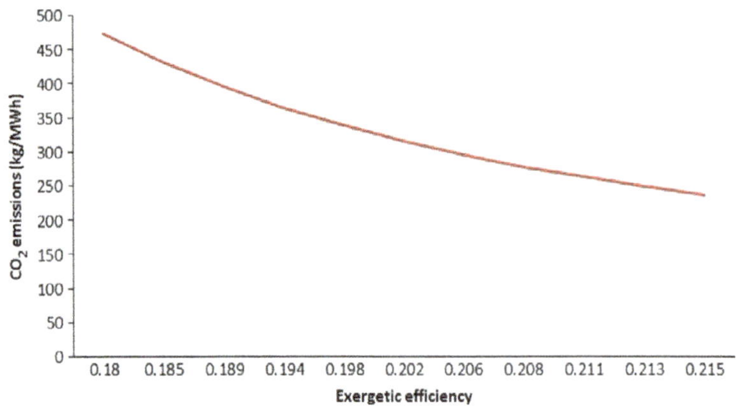

Figure 5. Variation of CO_2 emissions with exergetic efficiency.

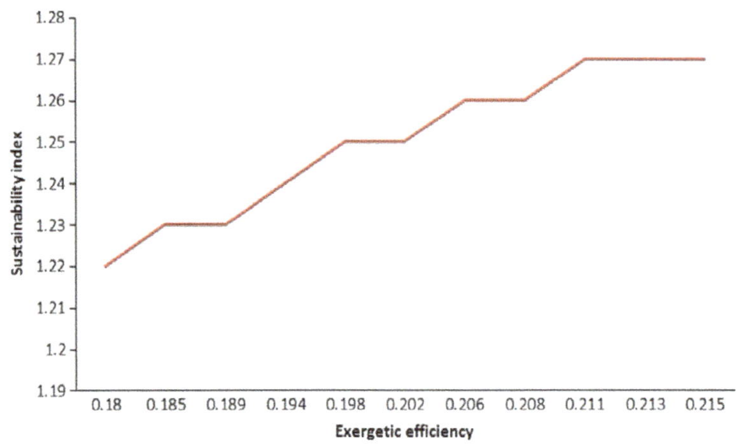

Figure 6. Variation of sustainability index with exergetic efficiency.

concluded that CO_2 emissions from the selected GT plants is not relatively too high.

Figure 5 shows the variation of CO_2 emissions with exergetic efficiency for plant AES1 (PB204) as sample for analysis. Figure 5 illustrates that increasing exergetic efficiency results in CO_2 emission reduction. The increase of exergetic efficiency is related to reduction of ambient inlet air temperature into the compressor.

Figure 6 shows the effect of increasing exergy efficiency of GT power plant on the SI of the whole system. The efficiency of the system is directly linked to the entire system. However, it is apparent that the overall exergy destruction of the cycle decreases, while the SI increases with decreasing CIT.

Further consideration of the results for CO_2 emissions, the effect of GTIT on cost of environmental impact and carbon dioxide emissions was investigated. The variations of both CO_2 emissions and cost of environmental impact

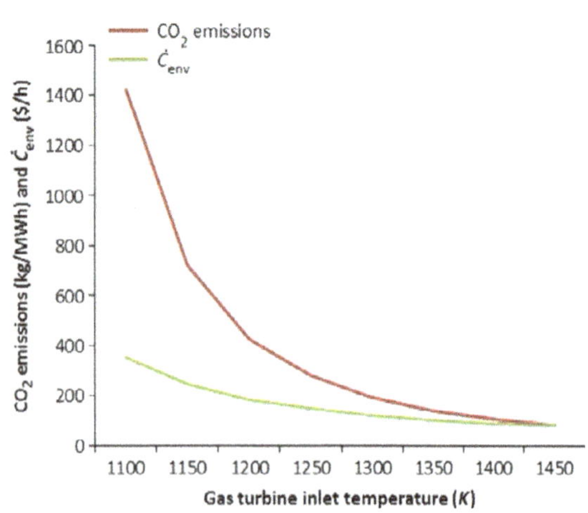

Figure 7. Variation of CO_2 emissions and cost rate of environmental impact with gas turbine inlet temperature.

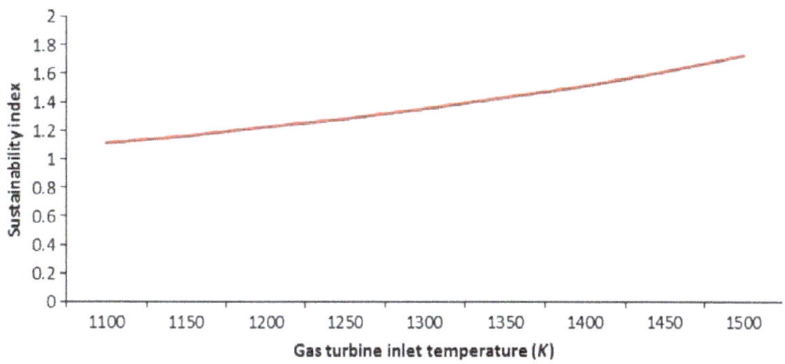

Figure 8. Variation of sustainability index with gas turbine inlet temperature.

for the gas power plant with GTIT are shown in Figure 7. CO_2 emissions and cost of environmental impact are seen to decrease with increasing GTIT. Although the mass flow rate of gases through the CC increases with GTIT, so does the net output power. The variation of GTIT with SI is shown in Figure 8. The SI is seen to increase with GTIT. Since the compressor PR is constant the exergy flow entering the CC is fixed. From the exergy balance for the CC, it is seen that increasing the GTIT increases the outlet exergy flow and exergy destruction in the CC decreases.

From above discussion, exergy efficiency, exergy destruction, environmental impact, and sustainability are apparently linked in such system, and thus supporting the utility of exergoenvironomic analysis.

From this study, it is seen that improvement of power plant's efficiency is twofold. By improving the most inefficient components (e.g., reducing the compressor inlet air temperature) of the system and utilizing the minimum adequate fuel flow rate ensuring maximum burn. The reduction in wasted unburned fuel and the reduction in overall system inefficiencies results in net CO_2 emissions reduction.

Validation of computer simulation code

The model developed is validated by the actual data that were taken from operational GT power plants in Nigeria. The operating parameters were taken from 2005 to 2010. Average parameters recorded within the period under review are set as baseline for comparison with the calculated results. The parameters considered in GT engine during simulation are inlet temperature of the air compressor, the mass flow rate of fuel and TIT.

The schematic flow diagram of a simple GT power plant that depicts these power plants is shown in Figure 1. The average operating data for the selected GT power plants are presented in Table 3. The results of thermodynamic properties of the cycle form the modeling part and the power plant data are illustrated in Tables 8–10.

The comparison of simulation results and the actual data from the power plants shows that the average difference in the simulation results and the actual data varies from 2.32% to 7.60%. The maximum difference is about 13.87% for mass flow rate of fuel in GT unit GT10 in Delta power plant. The minimum difference is about 0.089% for compressor outlet temperature in unit PB209 in AES power plant. This validates the correct performance of the developed simulation code to model the GT power plants in Nigeria, as the results of the simulation values are close to the actual operating data of the plants considered in this study.

Conclusions and Recommendations

In the present study, exergy, thermoeconomic, and thermoenvironomic analyses were performed for 11 selected GT power plants in Nigeria.

The results from the exergy analysis show that the CC is the most significant exergy destructor in the selected power plants, which is due to the chemical reaction and the large temperature difference between the burners and working fluid. Moreover, the results show that an increase in the TIT (GTIT) leads to an increase in GT exergy efficiency due to a rise in the output power of the turbine and a decrease in the CC losses.

The results from the thermoeconomic analysis, in common with those from the exergy analysis, show that the CC has the greatest cost of exergy destruction compared to other components. In addition, the results show that by increasing the TIT (GTIT) the GT cost of exergy destruction can be decreased. The finding solidifies the concept that the exergy loss in the CC is associated with the large temperature difference between the flame and the working fluid. Reducing this temperature difference reduces the exergy loss. Furthermore, cooling compressor inlet air allows the compression of more air per cycle, effectively increasing the GT capacity. The results of this study revealed that an increase in the TIT of about 200 K can lead to a reduction

Table 8. Results between the power plant data and simulation code (Afam power plant).

Data	Unit	AF1 (GT17)			AF3 (GT18)			AF3 (GT19)			AF4 (GT20)		
		Measured data	Simulation code	Difference (%)	Measured data	Simulation code	Difference (%)	Measured data	Simulation code	Difference (%)	Measured data	Simulation code	Difference (%)
T_2	K	593.73	616.79	3.88	595.82	625.13	4.92	610.90	618.88	1.31	618.32	619.95	0.26
T_4	K	712.73	714.60	0.26	723.75	747.45	3.27	770.07	755.12	1.94	807.32	764.87	5.26
m_f	kg/sec	6.50	5.95	8.46	6.4	6.56	2.5	8.1	7.46	7.90	8.4	7.6	9.52

Table 9. Results between the power plant data and simulation code (Delta power plant).

Data	Unit	DEL1 (GT9)			DEL2 (GT10)			DEL3 (GT18)			DEL4 (GT20)		
		Measured data	Simulation code	Difference (%)	Measured data	Simulation code	Difference (%)	Measured data	Simulation code	Difference (%)	Measured data	Simulation code	Difference (%)
T_2	K	613.73	646.80	5.38	619.07	647.38	4.57	634.32	643.89	1.51	630.32	645.63	2.43
T_4	K	710.56	738.35	3.91	707.48	738.40	4.37	730.15	758.26	3.85	705.07	758.70	7.61
m_f	kg/sec	3.08	2.68	12.99	3.10	2.67	13.87	8.15	7.54	7.48	8.13	7.54	7.26

of about 29% in the cost of exergy destruction. Therefore, TIT is the best option to improve cycle losses.

From exergy costing analysis, the unit cost of electricity produced in the selected power plants varies from cents 1.99/kWh (₦3.16/kWh) to cents 5.65/kWh (₦8.98/kWh).

The results of thermoenvironomic analysis show that CO_2 emissions in kg/MWh of electricity generated for the selected GT power plants vary from 100.18 to 408.78 kg CO_2/MWh. The cost rate of environmental impact varies from 40.18 to 276.97 \$/h. The study further shows that an increase in exergetic efficiency of GT plant results in a reduction in CO_2 emissions. The increase of exergetic efficiency is related to reduction of ambient inlet air temperature into the compressor. This implies that improvement of a system's efficiency is twofold. By improving the most inefficient components (e.g., reducing the compressor inlet air temperature) of the system and utilizing the minimum adequate fuel flow rate ensuring maximum burn. The reduction in wasted unburned fuel and the reduction in overall system inefficiencies results in net CO_2 emissions reduction. Moreover, CO_2 emissions and cost of environmental impact decrease with increasing GTIT. The SI is seen to increase with GTIT.

In summary, from the study and thermodynamic analysis of selected GT power plants, it was concluded that bigger η_{AC} and η_{GT} guarantee less exergy destruction in compressor and turbine as well as less net cycle fuel consumption and operating cost. Moreover, increasing GTIT also decreases the exergy destruction in CC, saves fuel consumption, and minimizes CO_2 emissions as well.

Recommendations to improve performance of the selected power plants

The average efficiency of GT plants in the Nigerian energy utility sector over the past two decades was in the range 27–30% [43]. In order to address increasing electricity demand, cost of electricity generation, and concern for environmental safety in the selected power plants, it is imperative to install power plants based on advanced technologies which are more energy efficient, environmentally acceptable, and economically viable. Results of this research provide insight into the economic and environmental impacts from exergy analysis perspective of the selected GT power plants in Nigeria. Based on the results of this research work, the following possible economical methods and technologies to improve performance of the selected GT power plants are hereby recommended:

1. The results of this study revealed that the CC has the largest irreversibility and cost of exergy destruction. This large exergy loss can be reduced in the selected power plants by addition of spray water and preheating of the reactants in the CC.

Table 10. Results between the power plant data and simulation code (AES power plant).

		AES1 (PB204)			AES2 (PB209)			AES3 (PB210)		
Data	Unit	Measured data	Simulation code	Difference (%)	Measured data	Simulation code	Difference (%)	Measured data	Simulation code	Difference (%)
T_2	K	622.31	629.59	1.17	627.48	628.04	0.089	636.28	628.98	1.15
T_4	K	750	775.02	3.34	755	776.51	2.86	827.05	816.53	1.27
m_f	kg/sec	2.58	2.45	5.04	2.54	2.47	2.76	2.81	2.63	6.41

2. Application of advanced clearance control schemes and sealing technologies. GTs are constructed with cases around the blades to contain and control the working fluid. Every molecule of working fluid that the blade does not extract work from as it passes by is called "leakage" which also reduces turbine efficiency. The problem of leakage is common in the simple GT plants in Nigeria. The possible method to control and limit the amount of leakages in turbines is through advanced clearance control schemes and sealing technologies.
3. Retrofitting with a GT air inlet cooling system (evaporative cooling, inlet fogging, or inlet chilling method) is a useful option for increasing power output of the selected power plants. This helps to increase the density of the inlet air to compressor [62].
4. The compressor airfoils of older turbines tend to be rougher than a newer model simply because of longer exposure to the environment. In addition, the compressor of older models consumes a larger fraction of the power produced by the turbine section. Therefore, improving the performance of the compressor will have a proportionately greater impact on total engine performance. Application of coatings to GT compressor blades (the "cold end" of the machine) would improve the selected GT engines performance. Compressor blade coatings provide smoother, more aerodynamic surfaces, which increase compressor efficiency. In addition, smoother surfaces tend to resist fouling because there are fewer "nooks and crannies" where dirt particles can attach. Coatings are designed to resist corrosion, which can be a significant source of performance degradation, particularly if a turbine is located near saltwater. As AES Barge GT plant is located on lagoon, compressor coating technology would improve the plant performance significantly.
5. Another option for improving the selected GT plants performance is to apply ceramic coatings to internal components. Thermal barrier coatings (TBCs) are applied to hot section parts in advanced GTs. As some of the selected GTs are over 25 years in operation, TBCs can be applied to the hot sections of the selected GTs. The TBCs provide an insulating barrier between the hot combustion gases and the metal parts. TBCs will provide longer parts life at the same firing temperature, or will allow the user to increase firing temperature while maintaining the original design life of the hot section.

Acknowledgments

The authors appreciate the Management of AES power plant, Afam power plant, and Delta power plant for providing the data used in this study.

Conflict of Interest

None declared.

References

1. Park, S. R., A. K. Pandey, V. V. Tyagi, and S. K. Tyagi. 2014. Energy and exergy analysis of typical renewable energy systems. Renew. Sustain. Energy Rev. 30:105–123.
2. Mousafarash, A., and P. Ahmadi. 2014. Exergy and exergo-economic based analysis of a gas turbine power generation system. Pp. 97–108 *in* I. Dincer, A. Midilli and H. Kucuk, eds. Progress in sustainable energy technologies vol II. Springer International Publishing, Switzerland.
3. Mitrovic´, D., D. Zivkovic, and M. S. Lakovic´. 2010. Energy and exergy analysis of a 348.5 MW steam power plant. Energy Sources Part A 32:1016–1027.
4. Aljundi, I. H. 2009. Energy and exergy analysis of a steam power plant in Jordan. Appl. Therm. Eng. 29:324–328.
5. Silveira, J. L., C. E. Tuna, W. Q. Lamas, and I. A. Castro Villela. 2010. A contribution for thermoeconomic modelling: a methodology proposal. Appl. Therm. Eng. 30:1734–1740.
6. Suresh, M. V. J. J, K. S. Reddy, and K. K. Ajit. 2006. Pp. 1–6 *in* Energy and exergy based thermodynamic analysis of a 62.5 MWe coal-based thermal power plant – a case study. International Conference on Energy and Environment, Malaysia.
7. Ebadi, M. J., and M. Gorji-Bandpy. 2005. Exergetic analysis of gas turbine plants. Int. J. Exergy 2:31–39.

8. Edgerton, R. H. 1992. P. 1992 *in* Available energy and environmental economics. D.C. Heath, Toronto, Canada.
9. Dincer, I., and M. A. Rosen. 2013. Pp. 257–276 *in* Exergy, energy, environmental and sustainable development. Elsevier, Oxford, U.K.
10. Moran, M. J., and E. Sciubba. 1994. Exergy analysis: principles and practice. J. Eng. Gas Turbines Power 116:285–290.
11. Rosen, M. A., I. Dincer, and M. Kanoglu. 2008. Role of exergy in increasing efficiency and sustainability and reducing environmental impact. Energy Pol. 36:128–137.
12. Ofodu, J. C., and D. P. S. Abam. 2002. Exergy analysis of Afam thermal power plant. NSE Tech. Trans. 37:14–28.
13. Abam, D. P. S., and N. N. Moses. 2011. Computer simulation of a gas turbine performance. Global J. Res. Eng. 11:37–44.
14. Ganapathy, T. K., N. A. Murugesan, and R. P. Gakkar. 2009. Exergy analysis of operating lignite fired thermal power plants. J. Eng. Sci. Technol. Rev. 2:123–130.
15. Rajkumar, S. C., and K. Ashok. 2009. Energy and exergy analysis of non-reheat thermal power plant. Proceedings of International Conference of Energy and Environmental, ISSN: 2070-3740
16. Balkrishna, M. C. 2009. Energy and exergy analysis of a captive steam power plant. Proceedings of International Conference of Energy and Environmental, ISSN: 2070 – 3740
17. Fagbenle, R. L., A. B. C. Oguaka, and O. T. Olakoyejo. 2007. A thermodynamic analysis of a biogas-fired integrated gasification steam injected gas turbine (BIG/STIG) plant. Elsevier Appl. Therm. Eng. 27:2220–2225.
18. Ehsana, A., and M. Z. Yilmazoglub. 2011. Design and exergy analysis of a thermal power plant using different types of Turkish lignite. Int. J. Thermodyn. 14:125–133.
19. Ighodaro, O. O., and B. A. Aburime. 2011. Exergetic appraisal of Delta IV power station, Ughelli. J. Emerg. Trends Eng. Appl. Sci. 2:216–218.
20. Rashidi, M. M., A. Aghagoli, and M. Ali. 2014. Thermodynamic analysis of a steam power plant with double reheat and feed water heaters. Adv. Mech. Eng. 2014:1–12.
21. Dincer, I., and M. A. Rosen. 2003. Thermoeconomic analysis of power plants: an application to a coal-fired electrical generating station. Energy Convers. Manage. 44:2743–2761.
22. Connelly, L., and C. P. Koshland. 1997. Two aspects of consumption: using an energy based measure of degradation to advance the theory and implementation of industrial ecology. Resour. Conserv. Recycl. 19:199–217.
23. Dincer, I., and M. A. Rosen. 1999. Exergy analysis of waste emissions. Int. J. Energy Res. 23:1153–1163.
24. Buchgeister, J. 2010. Exergoenvironmental analysis – a new approach to support the design for environment of chemical processes? Chem. Eng. Technol. 33:593–602.
25. Ahmadi, P., and I. Dincer. 2010. Exergo-environmental analysis and optimization of a cogeneration plant system using Multimodal Genetic Algorithm (MGA). Energy 35:5161–5172.
26. Amrollahi, Z., I. Ertesvag, and O. Bolland. 2011. Thermodynamic analysis on post-combustion CO_2 capture of natural-gas fired power plant. Int. J. Greenhouse Gas Control 5:422–426.
27. Barzegar Avval, H., P. Ahmadi, A. Ghaffarizadeh, and M. H. Saidi. 2011. 'Thermoeconomic environmental multiobjective optimization of a gas turbine power plant with preheater using evolutionary algorithm. Int. J. Energy Res. 35:389–403.
28. Ahmadi, P., M. A. Rosen, and I. Dincer. 2011. Greenhouse gas emission and exergo-environmental analyses of a trigeneration energy system. Int. J. Greenhouse Gas Control 5:1540–1549.
29. Memon, A. G., K. Harijan, M. A. Uqaili, and R. A. Memon. 2013. Thermo-environmental and economic analysis of simple and regenerative gas turbine cycles with regression modeling and optimization. Energy Convers. Manage. 76:852–864.
30. Seyyedi, S. M., H. Ajam, and S. Farahat. 2011. Thermoenvironomic optimization of gas turbine cycles with air preheat. Proc. Inst. Mech. Eng. Part A: J. Power and Energy 225:12–23.
31. Mert, M. S., O. F. Dilmaç, S. Özkan, F. Karaca, and E. Bolat. 2012. Exergoeconomic analysis of a cogeneration plant in an iron and steel factory. Energy 46:78–84.
32. Gorji-Bandpy, M., and H. Goodarzian. 2011. Exergoeconomic optimization of gas turbine power plants operating parameters using genetic algorithms: a case study. Therm. Sci. 15:43–54.
33. Sahoo, P. K. 2008. Exergoeconomic analysis and optimization of a cogeneration system using evolutionary programming. Appl. Therm. Eng. 28:1580–1588.
34. Fellah, G. M., F. A. Mgherbi, and S. M. Aboghres. 2010. Exergoeconomic analysis for unit Gt14 of South Tripoli gas turbine power plant. Jordan J. Mech. Ind. Eng. 4:507–516.
35. Regulagadda, P., I. Dincer, and G. F. Naterer. 2010. Exergy analysis of a thermal power plant with measured boiler and turbine losses. Appl. Therm. Eng. 30:970–976.
36. Ahmad, P., A. H. Barzegar, A. Ghaffarizadeh, and M. H. Saidi. 2010. Thermoeconomic- environmental multi-objective optimization of a gas turbine power plant with preheater using evolutionary algorithm. Int. J. Energy Res. 35:389–403.

37. Tsatsatronis, G. 1993. Thermoeconomic analysis and optimization of energy systems. Prog. Energy Combust. Sci. 19:227–257.
38. Abam, F. I., I. U. Ugot, and D. I. Igbong. 2012a. Effect of operating variables on exergetic efficiency of an active gas turbine power plant. J. Emerg. Trends Eng. Appl. Sci. 3:131–136.
39. Rivero, R., M. Garcia, and J. Urquiza. 2004. Simulation, exergy analysis and application of diabatic distillation to a tertiary amyl methyl ether production unit of a crude oil refinery. Energy 29:467–489.
40. Hammond, G. P. 2004. Engineering sustainability: thermodynamics, energy systems, and the environment. Int. J. Energy Res. 28:613–639.
41. Van Gool, W. 1997. Energy policy fairy tales and factualities. Pp. 93–105 *in* O. D. D. Soares, A. Martins da Cruz, G. Costa Pereira, I. M. R. T. Soares and A. J. P. S. Reis, eds. Innovation and technology-strategies and policies. Kluwer, Dordrecht.
42. Van Gool, W. 1992. Exergy analysis of industrial processes. Energy 17:791–803.
43. Abam, F. I., I. U. Ugot, and D. I. Igbong. 2012b. Performance analysis and components irreversiblities of a (25 MW) gas turbine power plant modeled with a spray cooler. Am. J. Eng. Appl. Sci. 5:35–41.
44. Siahaya, Y. 2009. Thermoeconomic analysis and optimization of gas turbine power plant. Pp. 1–10. *in* Proceedings of the International Conference on Fluid and Thermal Energy Conversion 2009 – FTEC 2009, Tongyeong, South Korea, 7–10 December.
45. Ahmadi, P., and I. Dincer. 2011. Thermodynamic and exergoenvironmental analyses, and multi-objective optimization of a gas turbine power plant. Appl. Therm. Eng. 31:14–15.
46. Bejan, A., G. Tsatsaronis, and M. Moran. 1996. Thermal design and optimization. John Wiley & Sons Inc, New York, NY.
47. Moran, M. J. 1982. Availability analysis: a guide to efficient energy use. Prentice Hall, Englewood Cliffs, NJ.
48. Kim, S., S. Oh, Y. Kwon, and H. Kwak. 1998. Exergoeconomic analysis of thermal systems. Energy 2:393–406.
49. Kwon, Y., H. Kwak, and S. Oh. 2001. Exergoeconomic analysis of gas turbine cogeneration systems. Exergy Int. J. 1:31–40.
50. Tsatsaronis, G., and M. Winhold. 1984. Thermoeconomic analysis of power plants, EPRI final report AP – 3651. Palo Alto, CA.
51. Ameri, M., P. Ahmadi, and A. Hamidi. 2009. Energy, exergy and exergoeconomic analysis of a steam power plant: a case study. Int. J. Energy Res. 33:499–512.
52. Valero, A., M. A. Lozano, L. Serra, G. Tsatsaronis, J. Pisa, C. Frangopoulos, et al. 1994. CGAM problem: definition and conventional solution. Energy 19:279–286.
53. Ameri, M., P. Ahmadi, and S. Khanmohammadi. 2007. Exergy analysis of a 420 MW combined cycle power plant. Int. J. Energy Res. 32:175–183.
54. Moran, M. J., and G. Tsatsaronis. 2000. Engineering thermodynamic. Pp. 1–109 *in* F. Kreith, ed. The CRC handbook of thermal engineering. CRC Press LLC, Boca Raton, FL.
55. Gülder, O. L. 1986. Flame temperature estimation of conventional and future jet fuels. J. Eng. Gas Turbines Power 108:376–380.
56. Lazzaretto, A., and A. Toffolo. 2004. Energy, economy and environment as objectives in multi – criterion optimization of thermal systems design. Energy 29:1139–1157.
57. Altayib, K. 2011. Energy, exergy and exergoeconomic analyses of gas-turbine based systems. M.Sc. thesis, University of Ontario Institute of Technology.
58. Chandramohan, S. 2008. First and second law analysis of organic rankine cycle. Ph.D. thesis, Mississippi State University.
59. Kotas, T. J. 1995. The exergy method of thermal plant analysis. Krieger Publishing Company, Malabar, FL.
60. International Energy Agency (IEA). 2012. Energy statistics and balances of non-OECD countries and energy statistics of OECD countries.
61. Inter-governmental Panel on Climate Change. 1996. IPCC second assessment climate change 1995. Volumes 1–3. Cambridge University Press, Bracknell, U.K.
62. Oyedepo, S. O., R. O. Fagbenle, S. S. Adefila, and S. A. Adavbiele. 2014. Performance evaluation and economic analysis of a gas turbine power plant in Nigeria. Energy Convers. Manage. 79:431–440.

13

Techno-economic analysis of hybrid PV–diesel–battery and PV–wind–diesel–battery power systems for mobile BTS: the way forward for rural development

Lanre Olatomiwa[1,2,3], Saad Mekhilef[1,2], A. S. N. Huda[1,2] & Kamilu Sanusi[4]

[1]Power Electronics and Renewable Energy Research Laboratory (PEARL), University of Malaya, 50603 Kuala Lumpur, Malaysia
[2]Department of Electrical Engineering, Faculty of Engineering, University of Malaya, 50603 Kuala Lumpur, Malaysia
[3]Department of Electrical & Electronic Engineering, Federal University of Technology, PMB 65, Minna, Nigeria
[4]Department of Electrical Engineering, Waziri Umaru Federal Polytechnics, Birnin-Kabbi, Nigeria

Keywords
HOMER, hybrid renewable energy system (HRES), mobile base station, Nigeria, techno-economic

Correspondence
Lanre Olatomiwa, Department of Electrical & Electronic Engineering, Federal University of Technology, PMB 65 Minna, Nigeria.

E-mail: olatomiwa.l@futminna.edu.ng

Funding Information
This study was supported by the Ministry of Higher Education, Malaysia, and the Bright Spark Unit of University of Malaya, Malaysia, under grant no. UMRG project RP015D-13AET.

Abstract

In recent times, hybrid renewable energy systems are increasingly being utilized to provide electricity in remote areas especially where the grid extension is considered too expensive. This study presents the results of techno-economic analysis of hybrid system comprising of solar and wind energy for powering a specific remote mobile base transceiver station (BTS) in Nigeria. All the necessary modeling, simulation, and techno-economic evaluation are carried out using the assessment software package HOMER (Hybrid Optimization Model for Electric Renewable). Two best optimal system configurations namely PV–diesel–battery and PV–wind–diesel–battery systems are compared with the conventional stand-alone diesel generator (DG) system. Findings indicated that PV array (10 kW) – DG (5.5 kW) – battery (64 units Trojan L16P) is the most economically viable option with the total net present cost of $69,811 and per unit cost of electricity of $0.409. The sensitivity analysis is also carried out to find the effects of probable variation in solar radiation, wind speed, and diesel price in the optimal system configurations. Finally, the environmental benefit of hybrid systems over the conventional stand-alone diesel system is described. The obtained results show that the hybrid PV–diesel–battery system provides a reduction in CO_2 emissions of about 16.4 tons per year as compared to the stand-alone DG system.

Introduction

Nigeria, a developing nation with population of about 174.5 million is the most populous country in Africa, whereas more than 50% of its population live in urban areas [1]. The current status of electric power infrastructure in Nigeria is playing a negative role in the growth of mobile telecommunications in terms of network coverage and tremendous impact on the operation cost of running the system due to nonavailability of grid power supply. On average, power availability from the national grid across Nigeria's electrified communities is less than 5 h in a day [2]. It was reported that about 11,692 mobile sites are connected to national grid, out of which only 9% have up to 6 h grid outage/day, 10% have 6–12 h outage/day, and 81% have more than 12 h outage per day, whereas about 12,560 sites are completely off-grid [2]. These uncertainties in power availability have compelled telecom operators in Nigeria to the use of diesel generator (DG) for proving uninterrupted power supply to the telecom equipment in order to meet the needs of its numerous subscribers. Powering these sites with generators requires significant maintenance, fuel costs and also increases environmental pollution. Annually, more than

500 million liters of diesel are consumed by telecom sites in Nigeria, which produce about 1.3 million metric tonnes of resulting CO_2 [3]. Owning to the deregulation of diesel prices and for reducing carbon emission, it has become an imperative issue for telecom operators in the country to evaluate all alternative options in order to improve network operation and reduce energy cost.

Several efforts have been made in some countries around the world to optimize energy cost by converting indoor base transceiver station (BTS) into outdoor one in order to eliminate the use of air conditioner [4], installing energy efficient equipment [5] and the use of green energy sources to power telecom sites [6]. Research and development in renewable energy have shown excellent potential as a form of contribution to the improvement of conventional power generation systems. It has been observed and verified that the green energy sources reduce the level of diesel usage, thereby lessen carbon emission and bring in expansion of telecom infrastructure to off-grid areas at reasonable cost.

With the availability of several forms of renewable energy sources such as solar, wind, biomass, fuel cell, and geothermal, etc., there is a possibility of vast expansion of cost-effective and reliable telecommunications in rural and remote areas [7]. Renewable technologies offer clean, sustainable, environmental friendly, reliable as well as low operating and maintenance cost of power supply [8–10]. Among these energy sources, solar and wind have proven their abilities to meet the rapidly growing energy demand and contribute substantially to global climate protection effort (reduction in greenhouse gas emission) [11–14]. However, standalone solar and wind system alone cannot provide reliable power supply due to the intermittent nature of the resources because of changes in atmospheric conditions [15]. Therefore, in order to meet the continuous typical load demand of a mobile base station during varying atmospheric conditions, different energy sources need to be integrated for extended usage of alternative energy.

A hybrid renewable energy system (HRES) is defined as a cluster of different renewable energy sources operating in a coordinated and independent manner [16]. Several studies on the feasibility, techno-economic analysis and performance of HRES have be reported in the literatures [17–21], These literatures have found the system to exhibit higher degree of reliability as well as lower cost of energy production as compare to system consisting only one energy source. The combination of different energy sources also allows improving the system efficiency and reduces the energy storage requirements compared to systems with single renewable energy source [21]. With the complementary nature of solar and wind energy resources, hybrid solar/wind power generation systems with storage banks offer a highly reliable source of power [22]. The use of the stand-alone solar-wind with diesel backup hybrid system for the power supply of remote areas gives an economically attractive alternative for mobile telecom sector over the use of conventional standalone DG [23]. A HRES with DG offers several advantages including reduction in operating cost by decreasing operating time of the generator, reduction in environmental pollution, and lower maintenance costs, etc. The inclusion of battery storage reduces the number of start/stop cycle of the generator and thus minimizing fuel consumption considerably [15].

This study therefore, presents the results of technical and economic feasibility of employing HRES to power a remote location mobile telecom BTS in Nigeria. The study aimed to determine the viability of hybrid PV–diesel–battery and PV–wind–diesel–battery power systems as well as selecting the most cost-effective and environmental friendly configuration for the site. Hybrid optimization model for electric renewable (HOMER), one of the most widely used optimization tool for renewable energy systems was employed to carry out the techno-economic analysis [24, 25]. The selection of the suitable configuration was based on net present cost (NPC), levelized cost of energy (LCOE), and renewable energy fraction.

The rest of the study is organized as follows: Assessment and analysis presents the brief description of HOMER software, location, load profile, and solar and wind resources of the study area. Configuration of the proposed system discusses the configuration of the proposed system as well as basic mathematical models and cost summaries of the require hybrid system components. Results and Discussion presents the optimization and sensitivity analysis results. Finally, concluding remarks are included in Conclusions.

Assessment and Analysis

HOMER software

HOMER, a renewable-energy-based system optimization tool developed by National Renewable Energy Laboratory (NREL), USA was used to model the hybrid system [25]. This is a general-purpose system design tool that facilitates the design of electric power system either for off-grid or grid-connected application. HOMER sizes up the system specifications such as the load profile, wind resources, solar resources, diesel price, system control parameter, constraint parameter as well as components technical and economics details. HOMER performs numbers of hourly simulation to ensure the best possible matching between the load and the supply in order to design an optimal system. Thereafter, it creates a list of feasible system configurations sorted according to cost effectiveness and presents the optimal configuration based on the lowest NPC. HOMER also performs the sensitivity analysis to explore the effect of changes of different factors on the different system configurations [26]. Climatic data (wind speed and solar irradiation), site load profile, hybrid system component's technical details and costs, system control as well as system constraints serve as input to the software.

Site location

The mobile telecom base station considered for this study is located at Doka-Sharia nature reserve, in Kaduna state, Northern region of Nigeria with latitude (9° 47′N) and longitude (7° 20′E). The weather in the Northern part of Nigeria can be characterized as hot and humid, and there are two major seasons namely rainy and dry seasons. Each season remains about 6 months. People living in this study location usually dependent on wood for cooking, gas lamps for lightings and the electrical equipment in most of the cases are powered by DGs.

Site load profile

Typically, a conventional BTS site load consists of BTS equipment load (base station RBS and microwaves) as well as air conditioner and lighting loads. Figure 1 shows a typical off-grid PV–wind–diesel powered mobile base station. A second-generation GSM mobile base station consisting of one number BTS with three transceivers was considered in this study. The maximum power consumption of air conditioner is considered as 1.8 kW, whereas the BTS equipment's power and miscellaneous loads (battery charging load and lighting) are 1500 W and 480 W, respectively [27]. Normally, the full load will be the constant BTS equipment load, miscellaneous load, and the air conditioner running power, but when the temperature goes low, the air conditioner can be configured to switch off automatically, thereby minimizing the energy consumption [28]. Although, measured hourly load data for the considered site are not available, it is, however, important to note that, the stated rating of BTS equipment (base station RBS and microwaves) will not be loaded to full capacity at all times due to the variability in operating hour of the equipment which often depends on the level of usage by the mobile subscribers. This load is therefore classified into three categories according to the level of usage, namely; low hour, medium hour, and busy hour load. Based on the first category, it is assumed that the subscriber utilized only 10% of the total BTS load, this usually occurred during 12 AM to 6 AM in the morning, whereas in the second category is considered 30% usage, this occurred between 7 AM and 10 AM in the morning and between 6 PM and 11 PM in the night. Finally the third category is taking as 50% and usually occurs during the business hours of the day (10 AM to 6 PM). It should also be noted, that the air conditional is considered to only operate during the business hour (sunshine hour) of the day when high power is expected to be demanded from the BTS equipment, whereas the miscellaneous loads (lighting) is operated only in the night hours (6 PM to 6 AM). Based on this variation, a day-to-day random variability of 5% and hour-to-hour random variability of 10% was specified in HOMER so as not to underestimate the peak load the proposed system can serve. The hourly load profile of the site for a typical day (1 January) is shown in Figure 2 from where scaled average energy consumption per day, scaled daily peak demand, and daily average demand are found to be approximately 37 kWh, 3.3 kW, and 1.53 kW, respectively. The mismatch in the peak value and value shown in Figure 2 is due to the scaling process and the random variability introduced in HOMER to make the load pattern unique. In addition, monthly averages of AC primary load for the considered BTS site is shown in Figure 3.

Figure 1. Off-grid hybrid PV–wind–diesel powered mobile base station.

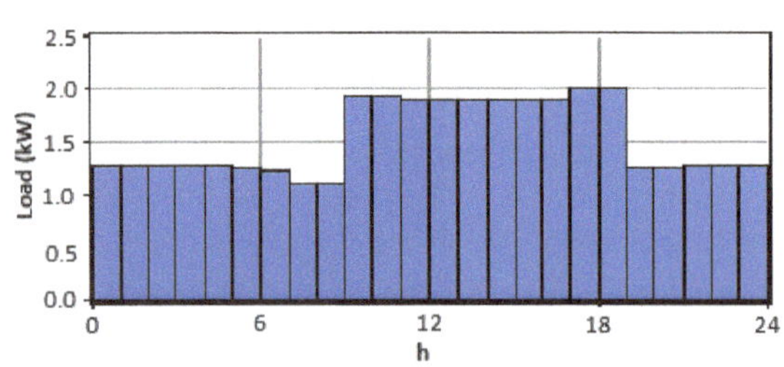

Figure 2. Daily load profile of the site [25].

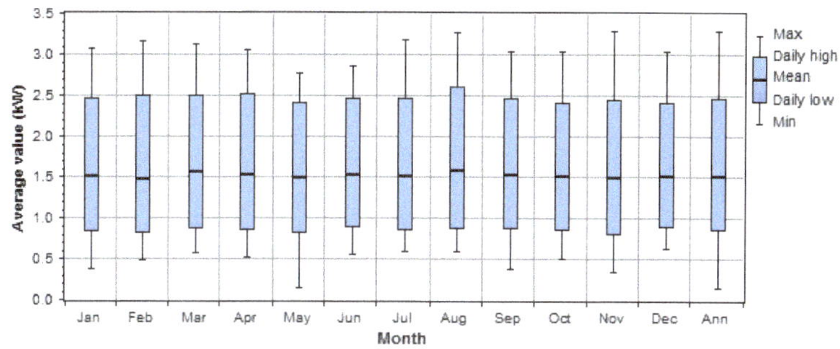

Figure 3. Annual energy load profile of the site [25].

From this figure, it can be observed that the maximum load ranges between 2.8 kW and 3.6 kW monthly with lowest in May and highest in August and the daily peak it seen to vary from 2.4 kW to 2.7 kW, whereas the average load is almost the same in all the months. All this variation is due to the above mentioned random variability factor introduced in HOMER to cater for the variation that may occur in the load each day of the year.

Solar energy

Nigeria is endowed with abundant renewable energy resources due to its coordinate position. It lies within a high sunshine belt and thus has enormous solar energy potential. Monthly mean daily solar radiation data averaged over a period of 22 years (1983–2005) were obtained for the site from NASA surface meteorology and solar energy database [29]. Scaling this data to consider the long-term annual average, HOMER introduces a clearness index from the latitude and longitude of the site considered. Figure 4 shows the monthly variation in solar radiation with the clearness index. These data serve as input to the HOMER for the analysis. Solar radiation is seen to be well distributed in the site with average monthly solar radiation of 5.8 kWh/m^2 per day and average daily sunshine of 6 h. In addition, Figure 5 shows synthesized daily solar radiation distribution over a year. From this figure it can be seen that solar energy resources is well distributed in the site, with highest value close to 8.0 kWh/m^2 per day in some days of the year, whereas the minimum value occurs only during the peak rainy season months (July and August) in the country. The average value close to 6.0 kWh/m^2 occurs most days in the year. From this analysis, it can be seen that the considered site has enormous potentials for PV applications.

Wind resource

Wind energy is available with an annual average speed of about 2.0 m/sec in the coastal region and 4.0 m/sec in the far northern region of the country [30]. With an air density of 1.1 kg/m^3, the intensity of wind energy perpendicular to the wind direction ranges between 44 W/m^2 at the coastal areas and 352 W/m^2 at the far northern region of the country [31]. In order to estimate the expected power output of the wind turbine in the proposed hybrid system, monthly mean data pertaining the wind resources of the site are obtained from NASA surface meteorology and solar energy database [29]. According to NASA, the data were recorded over a 10-year period from 1996 to 2005 at an anemometer height of 10 m. Figure 6 shows the monthly wind speed variation in the site where the

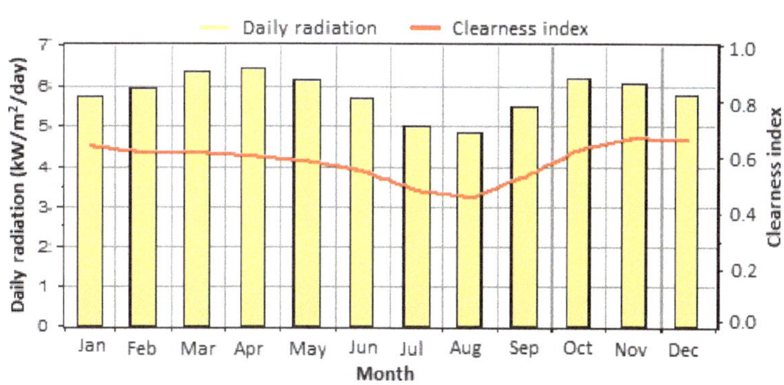

Figure 4. Average monthly solar radiation and clearness index profile at the selected site [25].

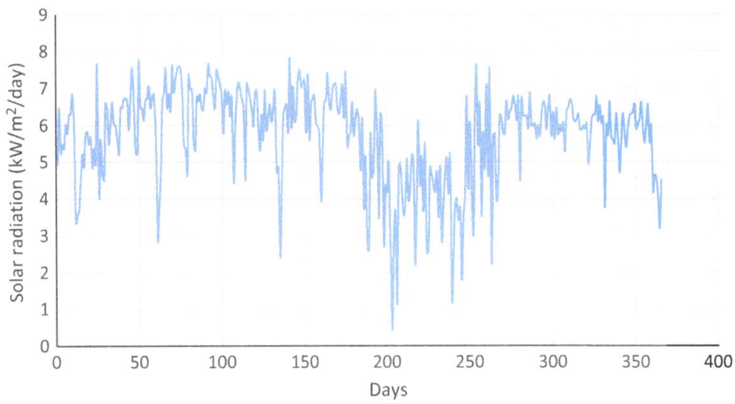

Figure 5. Daily distribution of solar radiation over a year.

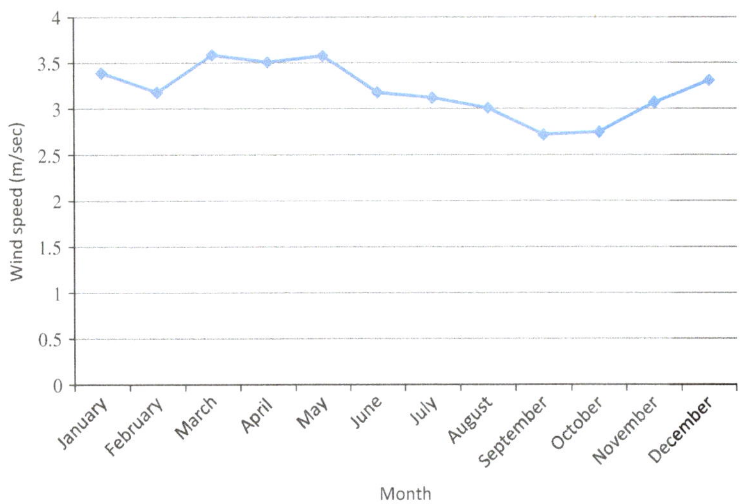

Figure 6. Average monthly wind speed of the site.

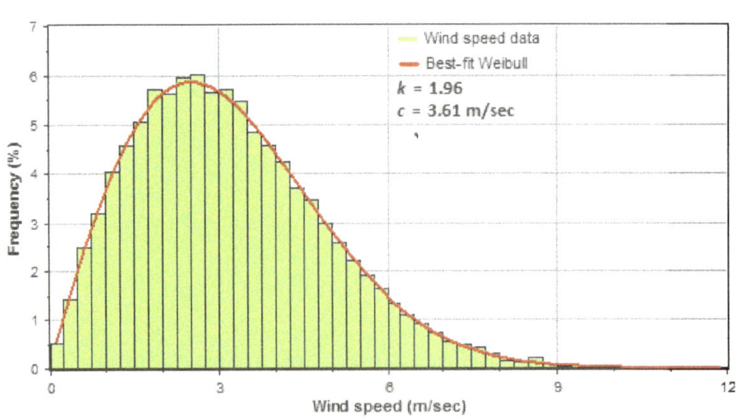

Figure 7. Probability distribution function of wind speed data synthesized by HOMER [25].

average wind speed is found to be 3.2 m/sec. In addition, the wind speed variation over a period at the site is described by the probability density function (PDF) shown in Figure 7. This distribution is generated by HOMER from the wind resource data input. Weibull shape factor (k), which is a measure of wind speed distribution over a year obtained by MATLAB curve fitting toolbox is given as 1.96, the autocorrelation factor, which shows the randomness of the wind speed is found to be 0.85, and the diurnal pattern, which is a measure of wind speed variation over a day is given as 0.25 [26]. From this figure it can be seen that the most probable wind speed range 2.5–3.5 m/sec occurs approximately 60% of time, whereas high value of wind speed above this occurs only for limited period of time. This shows the site to be unsuitable for wind power production, unless with application of low speed wind turbine due to the low wind speed regime at the site.

Configuration of the Proposed System

In this study, solar- and wind-based hybrid system is proposed for the site based on the justification from previous authors as earlier discussed in Introduction. A typical hybrid power system consisting of PV, wind, DG, and battery is shown in Figure 8. The solar PV, wind, and DG components are combined to harness the output power of the system as well as to compensate for the unpredictable variation in the climatic condition. The converter is added to maintain the flow of energy between the AC and DC components, whereas the battery is employed as a backup in order to ensure uninterrupted power and to maintain the desired power quality at the load point. In such a system, the battery bank absorbs energy when the renewable energy output exceeds the load and discharges energy when the load exceeds the renewable output. The use of these three energy sources in parallel with battery storage provides a smooth and uninterrupted power supply. Hence makes the hybrid system more reliable and efficient. The initial choice of the components size is based on the site load profile. The brief description of the main components of the proposed hybrid system is summarized in the following section;

Basic mathematical model and economic details of system components

A more detailed mathematical model of the components of the proposed hybrid system (Solar PV, wind turbines, DG, batteries, and converter) and the control strategy can be found in previous literatures [32–34]. Nonetheless, a brief summary of each mathematical models and economic parameters is presented as follows:

PV model

A photovoltaic system employs solar panels composed of a number of solar cells in series/parallel to generate power from solar PV. The power output of the PV panel is given by the following equation [32]:

$$P_{\text{pv-out}} = P_{\text{pv-rated}} \times (G/G_{\text{ref}}) \times [1 + K_T(T_c - T_{\text{ref}})], \quad (1)$$

where $P_{\text{pv-out}}$ is the power output of the PV cell, $P_{\text{pv-rated}}$ is the PV rated power at reference condition, G is solar radiation (W/m^2), G_{ref} is the solar radiation at standard temperature condition (G_{ref} = 1000 W/m^2), T_{ref} is cell temperature as reference conditions (T_{ref} = 25°C), and K_T is temperature coefficient of the PV module. The cell temperature $T_c = T_{\text{amb}} + (0.0256 \times G)$, where T_{amb} is ambient temperature. In this study, capital cost and replacement cost for 1 kW of PV array are taken as $2500 and $2000, respectively. In this case, six different types of PV arrays are considered to get the optimal size (0, 2, 5, 7, 8, 9, and 10 kW). The lifetime of PV arrays was taken as 20 years and no tracking system was considered. Derating factor of the PV modules was assumed as 90%.

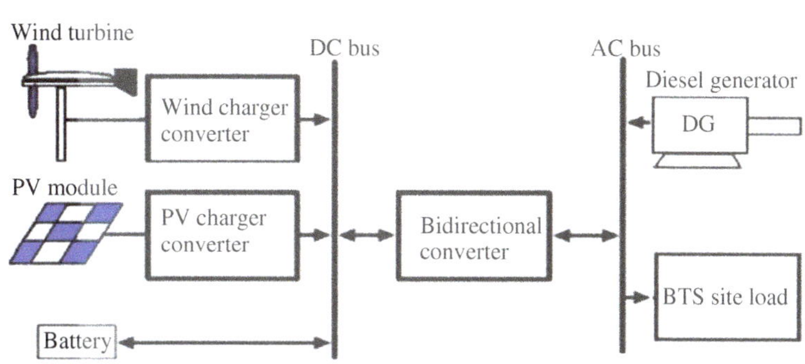

Figure 8. Typical hybrid PV/wind/diesel/battery configuration.

Wind turbine model

Wind energy is transformed into mechanical power through wind turbine and then converted into electrical power. The mechanical power over an area A is given by the following equation [34]:

$$P_m = \frac{1}{2} \times \rho \times A \times V^3, \quad (2)$$

where ρ is the air density (1.225 kg/m^3), and V the wind speed (m/sec). The electrical power of the wind energy conversion system is given by the following equation:

$$P_w = \frac{1}{2} \times \rho \times C_e \times A \times V^3 \times 10^{-3}, \quad (3)$$

where C_e the maximum power extraction efficiency of the wind generator and other electrical components connected to the generator. A BWCXL.1 model wind turbine with a rated capacity of 1 kW is considered in this study. The initial cost of one unit is considered as $6760. Replacement and operational maintenance costs were assumed as $4595 and $25/year, respectively. In order to find an optimal size, five different wind turbine options were analyzed: 0, 1, 2, 3, 4, and 5 turbines. The operational lifetime of a turbine is considered as 20 years [35].

Diesel generator model

The renewable energy systems have intermittent output characteristics and are integrated with conventional power sources to deliver a steady power output. In various HRES, DG acts as steady source of power. The DG systems are designed to supply the load and also charge the battery energy storage system, if the renewable energy source along with battery is unable to supply the load. Proper energy balance is required for optimum system operation as the consumption of fuel is proportional to the power being supplied by the DG. The fuel consumption of the DG, F_G (L/h) is modeled as dependent on the DG output power as [33]:

$$F_G = B_G \times P_{G\text{-rated}} + A_G \times P_{G\text{-out}}, \quad (4)$$

where $P_{G\text{-rated}}$ is the nominal power of the DG, $P_{G\text{-out}}$ is the output power, whereas A_G and B_G represents the coefficients of fuel consumption curve as defined by the user (L/kWh).

The initial capital cost of the DG is assumed $230/kW. Replacement and operational costs were assumed $230/kW and $0.5/h, respectively. The operating lifetime was also considered 15,000 h [35]. Currently, per liter price of diesel in Nigeria is $1.1. In this study, 0, 5.5, and 7.5 kW sizes of DGs were also considered for simulation. The maximum load ratio was set at 10%.

Battery energy storage system

Battery is a storage device essential for storing electrical energy for maximum utilization of intermittent renewable resources. The lead-acid battery which is often used in HRES is complex, nonlinear device controlling operational states of the system. Modeling of lead-acid batteries for real time analysis of HRES must account for the dependence of battery parameters on (a) state of charge, (b) battery storage capacity, (c) rate of charge/discharge, (d) ambient temperature, (e) life and other internal phenomenon, such as gassing, double layer effect, self-discharge, heating loss, and diffusion. The battery storage capacity is given as [36]:

$$C_{Wh} = (E_L \times AD)/(\eta_{inv} \times \eta_{Batt} \times DoD), \quad (5)$$

where E_L is the average daily load energy (kWh/day), AD is daily autonomy of the battery, DoD is battery depth of discharge, whereas η_{inv} and η_{Batt}, respectively, represent the inverter and battery efficiency. A Trojan L16P model battery with 6 V, 360 Ah, and 2.16 kWh is selected in the proposed system. The initial cost of one unit is considered as $174. Replacement and operational maintenance costs were assumed as $174 and $5/year, respectively. In order to find an optimal configuration, the battery bank was assumed to consist of any number of batteries (0, 16, 32, 48, 64, and 80 units). Each battery string contains eight batteries and the lifetime throughput of each battery is estimated to be 1075 kWh.

Power converter

The proposed hybrid system consist of both AC and DC system, hence a power converter is needed to maintain the flow of power between the system and the load. Both capital and replacement cost of the converter used in this work are taken as $329/kW. The operational and maintenance cost is assumed $10/year. The efficiency of the converter is assumed to be 90% and the lifetime is considered as 10 years [35]. Different sizes of converters (0, 2, 4, 6, 8, and 10 kW) are considered during analysis using HOMER.

The overall summary of technical and economic parameters of the hybrid system component are presented in Table 1.

Sensitivity variables

Sensitivity analysis helps in exploring the effect of the changes in the available resource and economic condition. This analysis shows the range of the variables for which it makes sense to include the renewable energy in the

Table 1. Technical and economic parameters of the hybrid system component.

Component parameters	Values	Component parameters	Values
Wind turbine		Diesel generator	
Brand/model	Bergey wind	Brand/model	Cummins
Rated power (kW)	1	Rated power (kW)	5.5
Cut-in speed (m/sec)	3	Rated current (A)	22.6
Cut-off speed (m/sec)	14	Minimum load ratio (%)	10
Initial capital cost ($/kW)	6760	Fuel curve ratio (L/h per kW)	0.33
Replacement cost ($/kW)	4595	Fuel curve intercept (L/h per kW)	0.05
O & M cost ($/year)	25	Fuel price ($/L)	1.1
Operational lifetime (year)	20	Initial capital cost ($/kW)	230
Solar PV		Replacement cost ($/kW)	230
Rated power (Wp)	250	O & M cost ($/h)	0.5
Max power (Wp)	250	Operational lifetime (h)	15,000
Max voltage (V)	31.1	Battery	
Short circuit current (A)	8.28	Brand/model	Trojan Ll6p
Derating factor (%)	90	Nominal voltage (V)	6
Initial capital cost ($/kW)	2500	Nominal capacity (Ah)	360
Replacement cost ($/W)	2500	Maximum charging current (A)	11
O & M cost ($/year)	–	Round-trip efficiency (%)	85
Operational lifetime (year)	20	Minimum SOC (%)	30
Power converter		Initial capital cost ($/unit)	174
Rated power (kW)	4	Replacement cost ($/unit)	174
Efficiency (%)	90	O & M cost ($/unit/year)	5
Initial capital cost ($/kW)	329	Operational lifetime (kWh/unit)	1075
Replacement cost ($/kW)	329		
O & M cost ($/year)	10		
Operational lifetime (year)	10		

Table 2. Sensitivity parameter values

Sensitivity variables Solar radiation (kWh/m^2 per day)	Wind speed (m/sec)	Diesel price ($/L)
5.0	2.0	0.7
5.5	2.5	0.8
5.8	3.0	0.9
6.0	3.2	1.0
6.5	3.5	1.1
		1.2
		1.3
		1.4
		1.5

system design. Three sensitivity parameters were considered in this study, namely: solar radiation, wind speed, and diesel price. The sensitivity values entered for each of these parameters are presented in Table 2.

System control parameters, economic inputs and constraints

System control parameters defines how HOMER model the operation of battery and generator in the hybrid system, economic inputs play an integral role in the entire HOMER simulation and optimization process, whereas the constraints are set of predefined conditions which systems must fulfill. These parameters are further discussed below;

Simulation time set

The simulation time step is the time step that HOMER uses to simulate the operation of each system configuration. The time for simulation can be entered from several hours down to 1 min. In the proposed system, the duration of simulation is set to 60 min.

Dispatch strategy

A dispatch strategy is a set of rules that controls the operations of the DG and the battery bank. Two types of dispatch strategies are available in HOMER, namely; cycle charging and load following [37]. Under the load-following strategy, a generator produces only enough power to serve the load and does not charge the battery bank. On the other hand, under the cycle charging strategy, the generator operates at its maximum rated capacity to serve the load and charges the battery bank using excess power. The type of dispatch strategy which will be optimal for the proposed system depends on various characteristics including the sizes of the generators and battery bank, the price of fuel, the O&M cost of the generators, the amount of renewable power output in the system, and the nature of the renewable resources. In the proposed system, both the strategies are considered which means HOMER will simulate each system using both dispatch strategies and finally, will determine the optimal configuration.

Economics inputs

In HOMER, the system life-cycle cost is represented by total NPC. The NPC is comprised of initial capital cost, replacement cost, annual operating, and maintenance cost as well as fuel costs. Thus is given by the following equation [26]:

$$C_{NPC} = \frac{TAC}{CRF(i, N)}, \tag{6}$$

where TAC is the total annualized cost ($/year) and CRF is the capital recovery factor which is a function of annual real interest rate (i), and the project lifetime (N in years). CRF is given by the following equation [26]:

$$CRF(i, N) = \frac{i(1+i)^N}{(1+i)^N - 1}. \tag{7}$$

In this study, the project lifetime is considered to be 25 years and the annual interest rate is fixed at 6%. HOMER aimed to minimize the total NPC and cost of electricity (COE) by finding the optimal system configuration. It should be noted that all economic factors considered in HOMER are calculated in constant dollar (US $) terms.

Constraints

Constraints are preconditions which systems must fulfill; otherwise HOMER neglects those systems that do not content the defined constraints. In the proposed system, the maximum renewable fraction is considered as a range from 0 to 100% and the maximum unserved energy is assumed 0%, whereas the different values (0, 4, 6, and 10%) are taken as the maximum annual capacity shortage. The maximum allowable value of the capacity shortage fraction, which is the total capacity shortage divided by the total annual electric load.

Results and Discussion

An hourly time series simulation was performed by HOMER for every possible system configuration on a yearly basis in order to evaluate the operational charac-

Table 3. Performance parameters of different system configurations

Parameter	Configurations		
	PV/diesel/battery hybrid system (best optimal)	PV/wind/diesel/ battery hybrid system (second best optimal)	Stand-alone diesel system (third best optimal)
PV capacity (kW)	10	10	–
Wind turbine capacity (kW)	–	1	–
Diesel generator capacity (kW)	5.5	5.5	5.5
Battery capacity (no)	64	48	32
Converter capacity (kW)	4	4	4
Dispatch strategy	CC	CC	CC
Total capital cost ($)	44,393	48,369	13,825
Total NPC ($)	69,811	76,023	245,855
Total annual capital cost ($/year)	3473	3784	1081
Total annual replacement ($/year)	1383	1623	2664
Total O&M cost ($/year)	720	731	9151
Total fuel cost ($/year)	249	243	6477
Total annual cost ($/year)	5461	5947	19,232
Operating cost ($/year)	1988	2163	18,151
Cost of energy (COE) ($/year)	0.409	0.445	1.440
PV production (kWh/year)	17,241	17,241	–
Wind production (kWh/year)	–	775	–
Diesel production (kWh/year)	654	611	17,823
Total electricity production (kWh/year)	17,896	18,627	17,823
AC primary load serve (kWh/year)	13,359	13,359	13,359
Renewable fraction (%)	95.1	95.4	0
Capacity shortage (kW)	0	0	0
Unmet load (kWh/year)	0	0	0
Excess electricity (%)	9.6	13.5	0
Fuel consumption (L/year)	224	221	5888

teristics such as, annual electricity production, annual load served, excess electricity, renewable fraction, etc. The renewable energy sources and the DG were evaluated in order to determine the feasibility of the system to meet load demand. HOMER searches for optimum system configuration and component sizes that meet the load requirement at the lowest NPC and then presents the results of the simulation in terms of optimal systems and sensitivity analysis. The optimal results are categorized based on the sensitivity variables chosen. The following subsections discuss the results of optimum configuration.

Optimization results

The result panel shows the list of configurations available based on the input data earlier introduced to HOMER and ordered according to the lowest NPC. In this section, first best two optimal hybrid system configurations will be compared with the conventional stand-alone diesel/battery system. The performance parameters of PV, wind, DG, battery and converter, respectively, for PV/diesel/battery, PV/wind/diesel/battery and stand-alone diesel systems configuration is presented in Table 3.

PV/diesel/battery configuration

The best optimal combination at 5.8 kWh/m^2 per day average global solar radiation, 3.2 m/sec annual average wind speed and $1.1/L diesel price comprised of 10 kW PV array, 5.5 kW DG, 64 units Trojan-L16P battery and a 4 kW converter. There is no inclusion of wind turbine in this configuration due to low intensity of wind speed at the site. This system has a total NPC of $69,811, COE of $0.409/kWh, and a renewable fraction of 95.1%. The NPC for the PV, diesel, and battery bank were 39.7%, 19.9%, and 37.2% of total NPC, respectively. The total NPC for the stand-alone diesel system is $245,855 which is 252% higher than PV/diesel/battery configuration. Therefore, renewable-energy-based hybrid system is considered as a cost-effective solution in comparison to the diesel standalone system.

The monthly mean electric production from PV array and DG is presented in Figure 9. It can be seen that the site can adequately rely on the RE source (solar) to power the BTS loads due to high presence of solar radiation at the site and the DG only contributes to meet up the load demand during the months of low solar irradiance (i.e., June, July August and September) as PV module provides

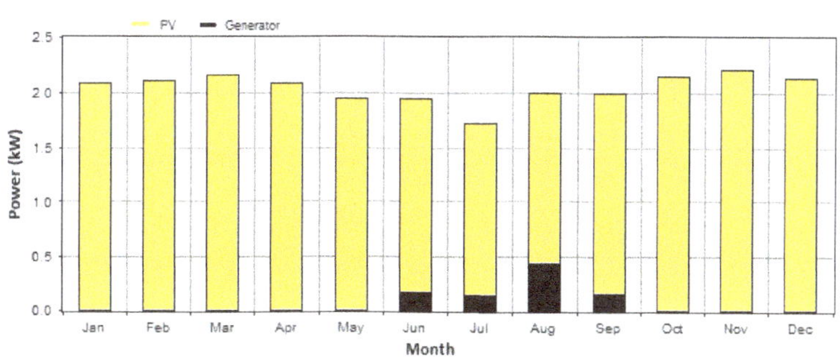

Figure 9. Monthly average electricity production of PV/diesel/battery hybrid system [25].

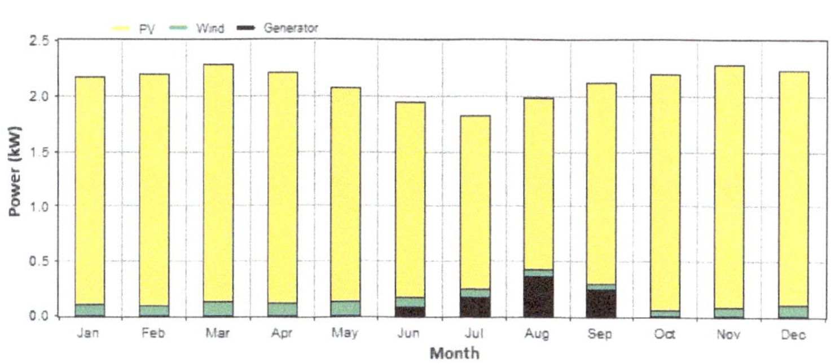

Figure 10. Monthly average electricity production of PV/wind/diesel/battery hybrid system [25].

95% of the total energy production during the year. The DG operates for just 138 h/year (capacity factor 1.36%), to produce 564 kWh/year of energy with 224 L of fuel. Limited hour of operation of DG will reduce the operating cost and also contribute to reduce CO_2 emission. Increasing the size of the PV panel and the addition of more string of battery can also compensate the shortage of electrical production and thereby optimize the usage of excess electricity and increase the overall system autonomy. The percentage excess electricity production for this hybrid system if found to be 9.6%.

PV/wind/diesel/battery

The second best optimal combination at 5.8 kWh/m^2 per day average global solar radiation, 3.2 m/sec annual average wind speed and \$1.1/L diesel price comprised of 10 kW PV array, 1 BWCXL.1 wind turbine, 5.5 kW DG, 48 units Trojan L16P battery and a 4 kW converter. The COE of this system is \$0.445/kWh with renewable fraction of 95.4%. The initial capital cost and total NPC of the above system were found to be \$48,369 and \$76,023, respectively. On the other hand, NPC of the stand-alone diesel system is more than 223% higher than this system. Therefore, this system configuration is still considered as a cost-effective solution in comparison with diesel-alone system. It should be noted that the total cost of the PV array in this system is nearly 36% of the total NPC, out of which the initial capital cost occupies about 90%. The cost of 48 units Trojan L16P battery accounts for 30% of the system total NPC, followed by the cost of DG and wind turbine, which takes up approximately 19% and 11%, respectively. The least cost item is the converter, which is 3% of the total NPC.

The monthly average electric production from the hybrid system is shown in Figure 10. Solar energy provides almost 93% of the total annual production followed by wind turbine and then diesel, accounting 4% and 3%,

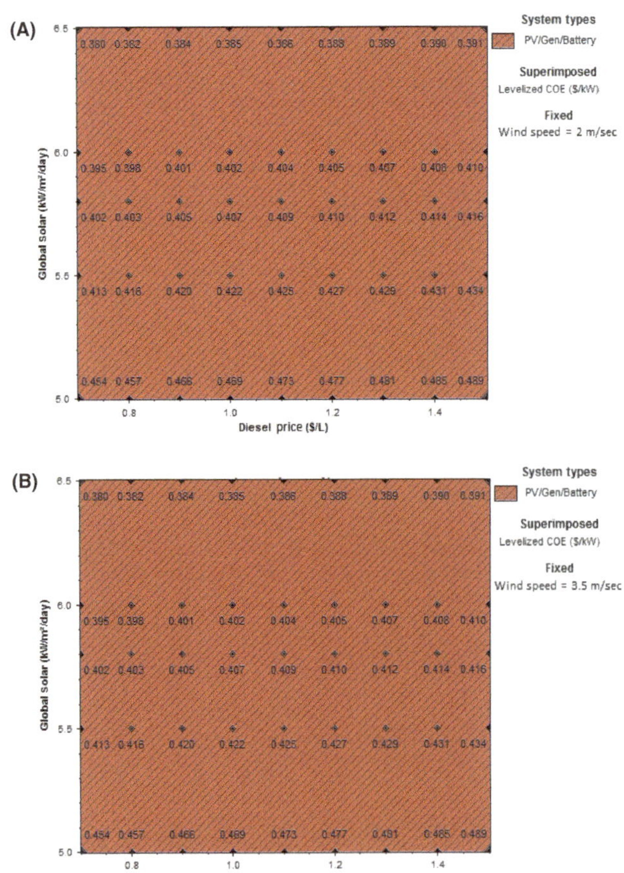

Figure 11. Optimal system types graph for varying wind speed values: (A) 2.0 m/sec, (B) 3.5 m/sec [25].

respectively. As the PV system will operate more compared to the wind turbine, which means the PV array serves the base load. The electricity production from the wind turbine is limited due to the lower wind speed at the site, but the DG contributes to fulfill the demand during the period of low solar irradiance. The DG operates for 155 h/year (capacity factor 1.27%) during the month of June, July, August and September to produce 611 kWh/year with 221 liters of fuel. On the other hand, the stand-alone diesel system consumes a total of 5888 L/year and run for 3255 h to produce 17,823 kWh/year of electricity. The excess electricity of this hybrid system is 13.5%. The excess electricity produced can be used to supply the dump load (water pumping, street lighting, heating, or cooling load), it can as well be sold to the grid.

Sensitivity analysis

For each sensitivity variable presented in sensitivity variables, HOMER simulates the entire system with respect to the search space. The feasible systems are ranked based on increasing NPC, whereas the software eliminates all infeasible combinations. Figures 11, 12, and 13 shows the Optimal System Type (OST) sensitivity results for three different scenarios, namely; variation in wind speed and solar radiation with constant diesel price, variation in wind speed and diesel price with constant solar radiation and finally, the variation in solar radiation and diesel price with constant wind speed.

Varying average wind speed value

In this sensitivity analysis, OSTs are investigated for five different values of wind speed, as presented in Table 2. Optimal system types graph for wind speed range 2.0 and 3.5 m/sec are shown in Figure 11. It is obvious from this figures that for the specified ranges of wind speed values is no output power from the wind turbine system irrespective of the variation in solar radiation and diesel price. Hence, HOMER does not consider wind turbine to be feasible at the considered site, therefore it is more economical to exclude wind turbine from the system and

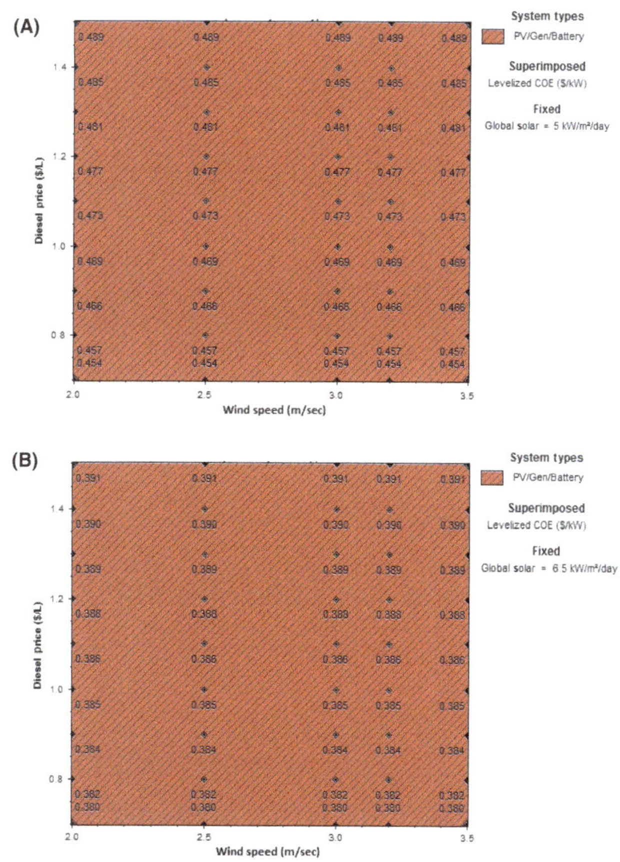

Figure 12. Optimal system types for varying solar radiation values: (A) 5 kWh/m^2 per day, (B) 6.5 kWh/m^2 per day [25].

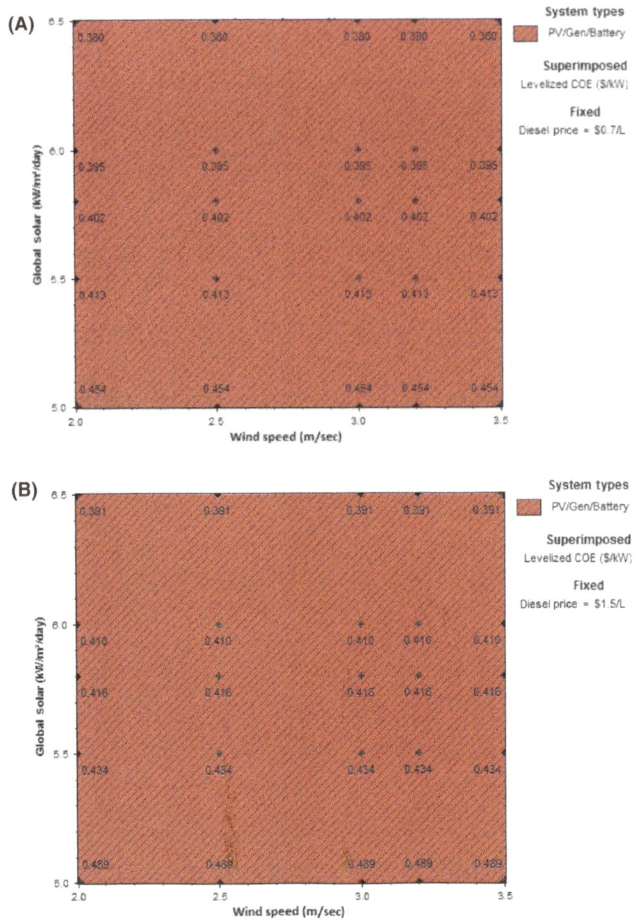

Figure 13. Optimal system types for varying diesel price: (A) $0.7/L, (B) $1.5/L [25].

consider PV/diesel/battery configuration to be the best optimal. It can also be observed that the superimposed LCOE remains the same for the two cases. It is further observed from these figures that the superimposed LCOE reduces as the global solar radiation increases, whereas the reversed is the case for the increase in diesel price.

Varying average solar radiation value

In this analysis, OSTs are tested for five different value of solar radiation as specified in Table 2. Optimal system types graph for two samples of these scenarios are shown in Figure 12. At all values of solar radiation irrespective of the variation in wind speed and diesel price, PV/diesel/battery configuration is still considered more feasible and economical than any other system configurations. However, increasing in solar radiation, reduces the LCOE, whereas wind speed variation has no effect on the superimposed LCOE.

Table 4. Pollutants emission in different system configurations.

Emissions (kg/year)	Configurations		
	PV/diesel/ battery	PV/wind/ diesel/battery	Diesel only
Carbon dioxide	591	582	15,505
Carbon monoxide	1.46	1.44	38.3
Unburned hydrocarbons	0.162	0.159	4.24
Particulate matter	0.11	0.108	2.89
Sulfur dioxide	1.19	1.17	31.1
Nitrogen oxides	13	12.8	341
Total	606.9	597.7	15,922

Varying diesel price

In this analysis, OSTs are investigated for nine different diesel prices as contained in Table 2 OSTs graph for these different diesel prices (lowest and highest) are shown in Figure 13. For both the lower and higher diesel price, same PV/diesel/battery hybrid system shows to be more feasible and economical other system configuration. However, reduction in the price of diesel reduces the LCOE.

Environmental effects

Renewable energy sources are one of the most efficient and effective clean and sustainable energy solutions in a country. Table 4 shows the quantity of pollutants emission (Carbon dioxide, carbon monoxide, unburned hydrocarbons, sulfur dioxide, nitrogen oxides, etc.) in the first and second best optimal hybrid systems compared with the conventional-diesel-based configuration. The conventional-diesel-based power generation system produces 15,505 kg/year of CO_2 in the site. On the other hand, the hybrid PV/diesel/battery and PV/wind/diesel/battery systems produce a total of 591 and 582 kg/year of CO_2, respectively. Therefore, a total of 96.1% and 96.6% of CO_2 emission can be reduced if the first and second best optimal systems, respectively, are considered. The percentage of reduction of pollutants in the system is related to the penetration of renewable energy in the system. Other pollutants also reduce in the hybrid system compared to the conventional diesel system. Therefore, in addition to the PV/diesel/battery hybrid system configuration being more economically viable option, the system also can help to abate prevalent global warming which usually occurred as a result of CO_2 emission into the environment.

Conclusions

Techno-economic analysis of hybrid power system for a remote telecommunication mobile base station (BTS) using HOMER, hybrid system optimization tools is presented in this study. The study indicates that the site, Doka-Sharia rural area of Kaduna State in Nigeria blessed with considerable annual average global solar radiation (5.8 kWh/m^2 per day) and average wind speed (3.2 m/sec) is a potential site for installation of PV/diesel/battery hybrid system. The simulation results from HOMER show that the most economically feasible configuration for the site load is composed of 10 kW PV array, 5.5 kW DG and 64 units of batteries of which each has a nominal voltage of 6 V and capacity of 360 Ah. The configuration has a renewable penetration of about 95%. The study shows that the conventional standalone diesel configuration is not economically feasible due to the high cost of diesel, which might have significant effect on the operating cost of the mobile telecom operation and also environmental impact due to high CO_2 emission from the system. Also, from the sensitivity analysis results, it could be concluded that the selected site is not a potential site for wind turbine applications due to low wind speed regime at the site. Therefore, utilizing renewable-energy-based hybrid system in conjunction with conventional DG for powering BTS sites will not just only decrease the operating hours of the DG leading to reduction in operating cost of the telecom company which has direct impact on the low-income rural telecoms subscriber, but as well reduces diesel consumption, thereby bring low greenhouse gas emission to the environment.

This present proposal can bring about improvement in the standard of living and increase in the economic activities among the rural dweller in the location where the BTS is situated, as the excess electricity generated from such hybrid system can serve to supply dump loads (water pumping and street lighting, etc.) as a form of social responsibility of on the part of the telecom company to the area. Finally, the results of the findings from the sensitivity analysis carried out, as well as information gained from the entire study can be applied in the design, execution, or development of HRES for any applications in other locations in the country having similar geographical coordinate as the site considered in this study.

Acknowledgments

The authors thank the Ministry of Higher Education, Malaysia, and the Bright Spark Unit of University of Malaya, Malaysia, for providing the enabling environment and financial support under the grant no. UMRG project RP015D-13AET, to successfully carry out this research.

Conflict of Interest

None declared.

References

1. Central Intelligence Agency. 2014. The World Factbook: Afirca-Nigeria. Available at http://www.cia.gov/library/publications/the-world-factbook/geos/ni.html (accessed 30 October 2014).
2. GSMA. 2013. Powering Telecoms: West Africa Market Analysis. London.
3. GSMA. 2013. Market research and Analysis. London.
4. Roy, S. N. 2008. *Energy logic: a road map to reducing energy consumption in telecom munications networks*. in *Telecommunications Energy Conference*, INTELEC 2008. IEEE 30th International. 2008. IEEE.
5. Koutitas, G., and P. Demestichas. 2010. A review of energy efficiency in telecommunication networks. Telfor J. 2:2–7.
6. Hashimoto, S., T. Yachi, and T. Tani. 2004. *A new stand-alone hybrid power system with wind generator and photovoltaic modules for a radio base station*. in

Telecommunications Energy Conference, INTELEC 2004. 26th Annual International. 2004. IEEE.

7. Nema, P., R. Nema, and S. Rangnekar. 2010. Minimization of green house gases emission by using hybrid energy system for telephony base station site application. Renew. Sustainable Energy Rev. 14:1635–1639.
8. Daud, A.-K., and M. S. Ismail. 2012. Design of isolated hybrid systems minimizing costs and pollutant emissions. Renew. Energy 44:215–224.
9. Hrayshat, E. S. 2009. Techno-economic analysis of autonomous hybrid photovoltaic-diesel-battery system. Energy Sustainable Dev. 13:143–150.
10. Hessami, M.-A., H. Campbell, and C. Sanguinetti. 2011. A feasibility study of hybrid wind power systems for remote communities. Energy Pol. 39:877–886.
11. Dihrab, S. S., and K. Sopian. 2010. Electricity generation of hybrid PV/wind systems in Iraq. Renew. Energy 35:1303–1307.
12. Dursun, B., C. Gokcol, I. Umut, E. Ucar, and S. Kocabey. 2013. Techno-economic evaluation of a hybrid PV—Wind power generation system. Int. J. Green Energy 10:117–136.
13. Nandi, S. K., and H. R. Ghosh. 2010. Techno-economical analysis of off-grid hybrid systems at Kutubdia Island, Bangladesh. Energy Policy 38:976–980.
14. Shaahid, S. M., and I. El-Amin. 2009. Techno-economic evaluation of off-grid hybrid photovoltaic–diesel–battery power systems for rural electrification in Saudi Arabia—A way forward for sustainable development. Renew. Sustainable Energy Rev. 13:625–633.
15. Jacobus, H., et al. 2011. Evaluating the impact of adding energy storage on the performance of a hybrid power system. Energy Convers. Manage. 52:2604–2610.
16. Wang, C. 2006. *Modeling and control of hybrid wind/ photovoltaic/fuel cell distributed generation systems.* Montana State University Bozeman.
17. Neves, D., C. A. Silva, and S. Connors. 2014. Design and implementation of hybrid renewable energy systems on micro-communities: a review on case studies. Renew. Sustainable Energy Rev. 31:935–946.
18. Nema, P., R. K. Nema, and S. Rangnekar. 2009. A current and future state of art development of hybrid energy system using wind and PV-solar: a review. Renew. Sustainable Energy Rev. 13:2096–2103.
19. Adaramola, M., O. Oyewola, and S. Paul. 2012. Technical and economic assessment of hybrid energy systems in South-West Nigeria. Energy Explor. Exploit. 30:533–552.
20. Rehman, S., and L. M. Al-Hadhrami. 2010. Study of a solar PV–diesel–battery hybrid power system for a remotely located population near Rafha. Saudi Arabia. Energy 35:4986–4995.
21. Nandi, S. K., and H. R. Ghosh. 2010. Prospect of wind–PV-battery hybrid power system as an alternative to grid extension in Bangladesh. Energy 35:3040–3047.
22. Zhou, W., C. Lou, Z. Li, L. Lu, and H. Yang. 2010. Current status of research on optimum sizing of stand-alone hybrid solar–wind power generation systems. Appl. Energy 87:380–389.
23. Kusakana, K., and H. J. Vermaak. 2013. Hybrid renewable power systems for mobile telephony base stations in developing countries. Renewable Energy 51:419–425.
24. Connolly, D., H. Lund, B. V. Mathiesen, and M. A. Leahy. 2010. A review of computer tools for analysing the integration of renewable energy into various energy systems. Appl. Energy 87:1059–1082.
25. NREL. 2009. *HOMER*. Available at http://www.homerenergy.com.
26. Lambert, T., P. Gilman, and P. Lilienthal. 2006. Micropower system modeling with HOMER. Integration Alternative Sources Energy 1:379–385.
27. Naikodi, A. *Solar-wind hybrid power for rural Indian cell sites.* in IEEE International Energy Conference and Exhibition (EnergyCon). 2011. Manama.
28. Diamantoulakis, P. D., and G. K. Karagiannidis. 2013. On the design of an optimal hybrid energy system for base transceiver stations. J. Green Eng. 3:127–146.
29. NASA/SSE. 2013. *Surface meteorology and energy.* Available at http://eosweb.larc.nasa.gov/sse.
30. Uzoma, C. C., C. E. Nnaji, C. N. Ibeto, C. G. Okpara, O. O. Nwoke, and I. O. Obi. 2011. Renewable energy penetration in Nigeria: a study of the South-East zone. Continental. J. Environ. Sci. 5:1–5.
31. Sambo, A. S. 2009. Strategic Developments in Renewable Energy in Nigeria. International Association for Energy Economics. Ohio, United State. 3rd Quarter: pp.15–19.
32. Elhadidy, M. 2002. Performance evaluation of hybrid (wind/solar/diesel) power systems. Renew. Energy 26:401–413.
33. Dufo-López, R., and J. L. Bernal-Agustín. 2008. Multi-objective design of PV–wind–diesel–hydrogen–battery systems. Renew. Energy 33:2559–2572.
34. Mahmoud, M. M., and I. H. Ibrik. 2006. Techno-economic feasibility of energy supply of remote villages in Palestine by PV-systems, diesel generators and electric grid. Renew. Sustainable Energy Rev. 10:128–138.
35. Olatomiwa, L., S. Mekhilef, and A. Huda. 2014. Optimal sizing of hybrid energy system for a remote telecom tower: A case study in Nigeria. Pp. 243–247 *in* IEEE Conference on Energy Conversion (CENCON).
36. Erdinc, O., and M. Uzunoglu. 2012. Optimum design of hybrid renewable energy systems: overview of different approaches. Renew. Sustainable Energy Rev. 16: 1412–1425.
37. Farret, F. A., and M. G. Simões. 2006. Integration of alternative sources of energy. John Wiley & Sons, Hoboken, New Jersey.

14

The remarkable impact of renewable energy generation in Sicily onto electricity price formation in Italy

Francesco Meneguzzo[1], Rosaria Ciriminna[2], Lorenzo Albanese[1] & Mario Pagliaro[2]

[1]Institute of Biometeorology, CNR, via Caproni 8, 50145 Firenze, Italy
[2]Institute for the Study of Nanostructured Materials, CNR, via U. La Malfa 153, 90146 Palermo, Italy

Keywords
Energy end use, merit order effect, photovoltaics, Sicily, wind energy

Correspondence
Francesco Meneguzzo, Institute of Biometerology, CNR, via Caproni 8, 50145 Firenze FI, Italy. E-mail: f.meneguzzo@ibimet.cnr.it
and
Mario Pagliaro, Institute for the Study of Nanostructured Materials, CNR, via U. La Malfa 153, 90146 Palermo PA, Italy. E-mail: mario.pagliaro@cnr.it

Funding Information
No funding information provided.

Abstract

During the first half of 2015, for the first time, the zonal electricity price in Sicily decreased below than the national wholesale price in Italy. Showing the unique pattern of electricity consumption in Italy's largest region at different time scales, we identify the effectiveness of the impact of renewable power generation on utility-scale in Sicily upon the whole of Italy's electricity market. Increasing the electrification of the energy end uses, as it is happening despite prolonged reduction in electricity demand, will lead to further benefits for power consumers throughout the whole country.

Introduction

Sicily is the largest and the most solar irradiated among Italy's regions. For comparison, the global horizontal irradiation yearly total of the second largest region (Lombardia) is 1100 kWh/m^2 whereas Sicily has an average value of 1800 kWh/m^2 [1].

While photovoltaic (PV) power generation was negligible till 2008–2009, the deployment of the Feed-In-Tariff (FiT) incentive scheme in Italy between 2006 and 2013 caused an impressive surge in the PV installed power nationwide as well as in Sicily, jumping from few tenths of MW to 1270 MW during 3 years (2011–2013) [2].

Even without the FiT subsidies (practically terminated on June 2013), in 2014 Sicily saw the installation of another 130 MW bringing the overall power to 1400 MW [3], with owners of the new PV systems benefiting of net metering, namely self-consuming the energy produced by their PV installations. This accelerated deployment of solar electricity took place almost concomitantly to massive adoption of wind energy as Sicily is also the most ventilated region of Italy, so that the installed wind power rose from zero in 2005 to about 2000 MW at the end of 2014 [3].

A preliminary estimate suggests that the PV power deployed in Sicily rose further up to more than 1400 MW during 2015, while the combined PV and wind power generation exceeded 2 TWh [4].

This rapid change, along with significant economic interests behind the incentives that reward only the electricity produced and actually fed into the grid, pushed the State owner of the Italian grid to deploy massive investments to widen and improve Sicily's high voltage grid [5]. The new "Sorgente-Rizziconi" connection between Sicily and the Italian peninsula [6], now almost completed, is a 380 kV line that by the end of 2016 will bring the interconnection capacity from the existing 1000–3000 MW, relieving the bottleneck that historically caused higher electricity prices in Sicily compared to the rest of the country.

From the viewpoint of electric interconnection, indeed, Sicily is an isolated island. For example, in 2014 the existing 1000 MW line was congested for 90% of the year's time [3], limiting, for example, the possibility to export the surplus of renewable energy during holydays and weekends.

We have recently shown why and how the growing penetration of PV renewable power generation in Italy (18 GW as of late 2014) has caused a substantial fall in the price of electricity in the Italian wholesale electricity market (IPEX) [7]. Now we show how the significant PV and wind power generation has changed the electricity price formation in Sicily affecting the national price (PUN) to such an extent to cause a significant decrease in the national electricity price.

In other words, the Sicily zone is substantially a one-way feeder into the Italian electricity market, not only on the side of locally generated power but also as a price steering element, since the zonal price affects the PUN in the weighted average leading to its formation. We also develop a model linking the daily zonal power price in Sicily to the demand not met by PV and wind energy.

The study is concluded suggesting a few practical ways to achieve further economic (and environmental) benefits for electricity consumers not only in Sicily, but also throughout the whole country.

Italy's and Sicily's Electricity Markets

As mentioned above, until recently wholesale electricity prices in Sicily have been constantly higher than in the rest of the country. The monthly average was more than 40% higher than the average national prices in Italy in 2008 and 2009, exceeding the PUN by more than €20/MWh for more than 50% of the months between January 2008 and December 2014 (Fig. 1). Following the spikes in 2008–2009, the Italian energy regulator (AEEG) launched an investigation into Sicilian prices whose outcomes, on August 2009, were transferred to the antitrust regulator (ICA) [8].

According to two following ICA investigations into Sicily wholesale electricity prices launched on January 2010, capacity withholding (mainly economically) was used to raise Sicily's zonal price, in turn affecting the national wholesale purchasing price. Eventually, the two major utilities agreed, the first, to limit the price of power produced in Sicily (a bid cap) to €190 per MWh in 2011, and the other to set prices based on market [9].

Nevertheless, it was only in February 2015 that the Sicily's zonal price fell below €10/MWh over the PUN, never exceeding that threshold again during the subsequent 13-months (until February 2016), even falling *below* the PUN by €1.51/MWh on July 2015 (Fig. 1).

The trends of absolute monthly power prices, PUN and Sicily's zonal price share few key features, such as the spikes in 2008 and the sustained downward trend after 2012 (Fig. 2). Again, the closing gap between Sicily's and Italy's prices becomes apparent since early 2015.

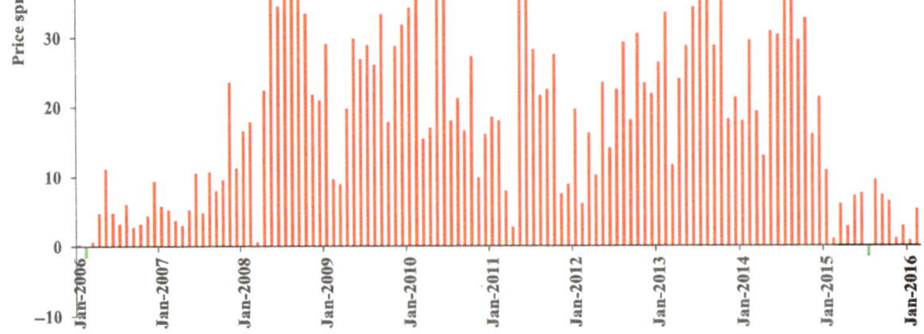

Figure 1. Spread between monthly average Sicily's zonal power price and Italy's PUN. Positive spread in red, negative in green. Source: Energy Markets Authority (GME).

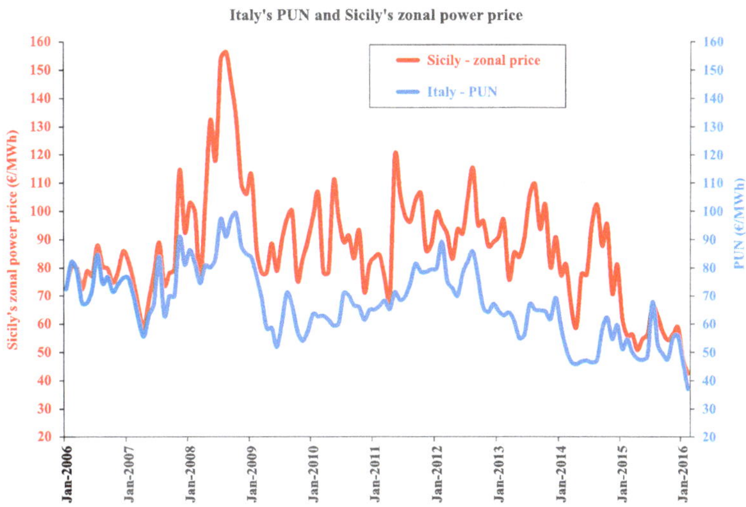

Figure 2. Yearly average prices on the day ahead market in Italy (PUN) and in the Sicily's zone. Source: Energy Markets Authority (GME).

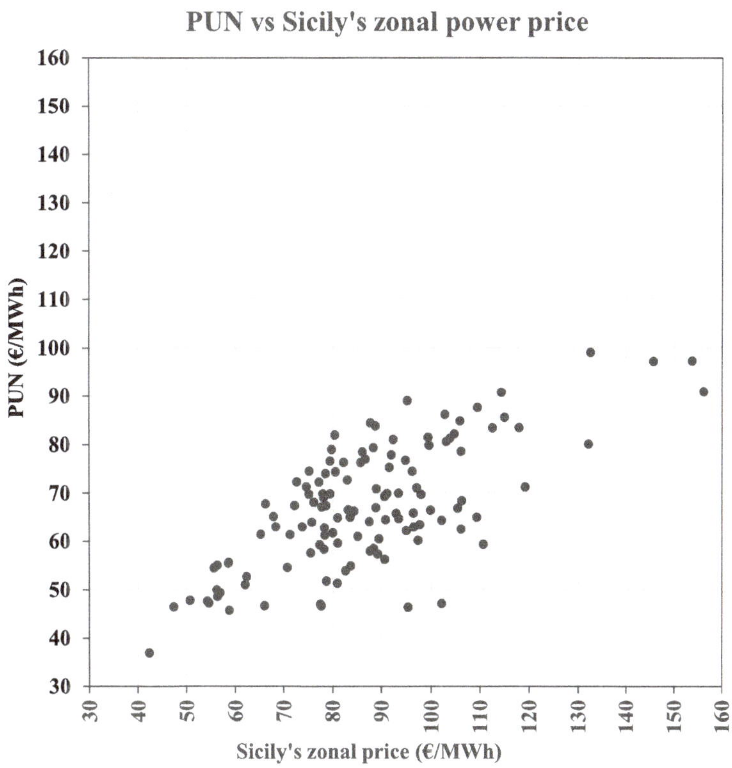

Figure 3. Sicily's zonal power price explains 50% of the variance of the national wholesale price (PUN).

The linear correlation between the PUN and the Sicily's zonal price is especially noticeable, since as much as 50% of the variance of the former is explained by the latter, as shown in Figure 3, a value which is not so distant from the correlation figure found for the peak PUN (wholesale power price during peak hours) model using an autoregressive term, without which the correlation strength fell below 60% [7].

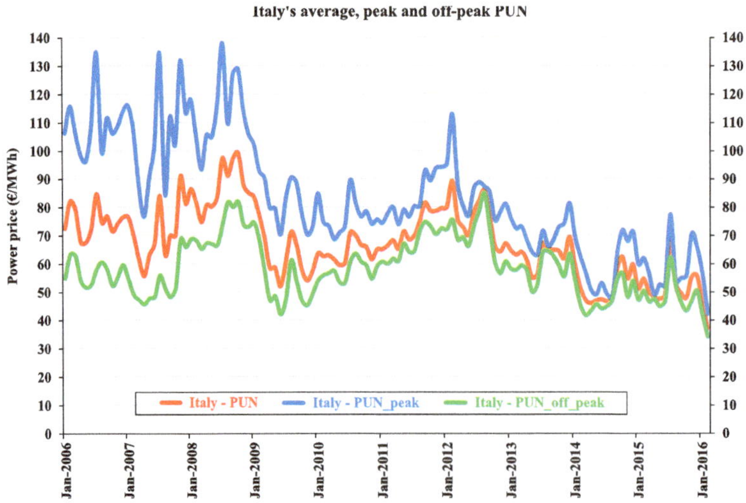

Figure 4. Italy's average, peak and off-peak wholesale power price (PUN). Source: GME.

Power prices formed in the Sicily's zone obviously reflect the impact of the changing prices in raw fuels such as oil and natural gas needed to feed conventional generation plants, and changes in the regional power demand. The closing gap between peak PUN (when most of the demand occurs) and off-peak national prices shown in Figure 4 may be attributed to the growing national generation from renewable sources, mainly PV due to its straightforward impact upon the peak diurnal demand in the merit-order price formation scheme [7].

In the year 2015, while the overall contribution of the renewable energy sources (hydroelectric, wind and PV) to the domestic power generation fell to 33.2% from 37.9% in 2014, mainly due mainly to significantly lower (−24.9%) hydropower generation, the PV production increased by as much as 13%, totaling almost 25 TWh, that is more than 55% of the hydropower and exceeding 9% of the domestic generation (8% in 2014) [10].

It was previously shown that the PV generation affects the peak PUN more than the other renewable sources [7], therefore contributing to close the gap between peak and off-peak PUN. Figure 5 shows the yearly series of PV and wind power generation in Italy where, after the sharp increase from 2010 to 2013, a slower yet sustained upward trend persisted until early 2016.

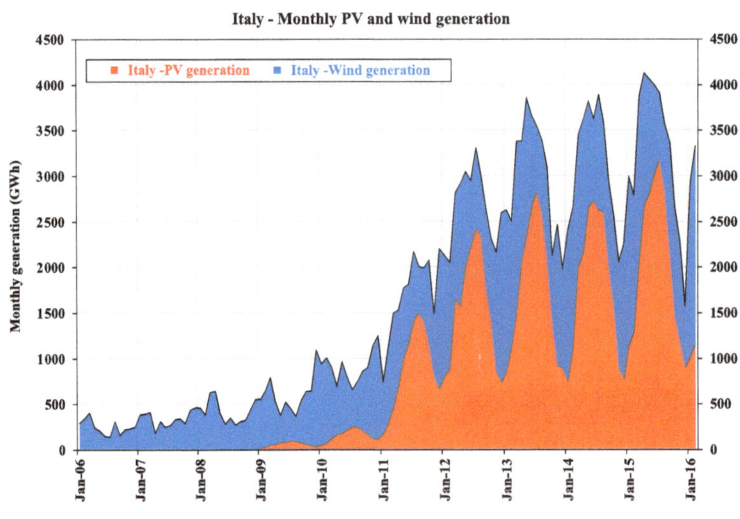

Figure 5. PV and wind power generation in Italy. Source: Terna.

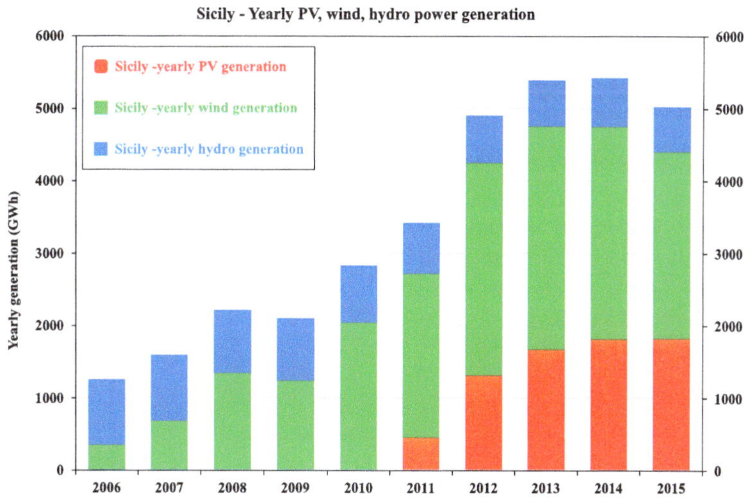

Figure 6. PV, wind, and hydro power generation in Sicily. Source: Terna.

A reputed market research company concluded that, thanks to the renewable energy boom, Italy's electricity market now faces a dramatic overcapacity, with the largest utility having announced closures of several plants for at least 11 GW of capacity, and the fifth largest producer exiting the market with sale of its thermo-electric assets [11].

In this context, in 2013 – following a sharp increase the year before – the amount of renewable energy produced in Sicily exceeded for the first time the threshold of 5 TWh, namely almost 25% of the about 23 TWh produced and fed into the grid [2]. Such a threshold was exceeded in each of the subsequent 2 years, despite a noticeable climatically-forced drop of wind and hydro contributions in 2015, which were partially offset by the slowly increasing PV generation (Fig. 6) [3, 4].

Following the early drop occurring as a consequence of the economic crisis during 2008–2009, both Italy's and Sicily's power demands underwent sustained fall starting in 2012. Yet, while the former halted in 2015, demand of electricity in Sicily nosedived, hitting the 11-years lowest point in February 2016 (Fig. 7). No wonder, therefore, that the historically lowest price was observed in the same month, in its turn helping to explain the partly unexpected record low PUN as a result of the steering effect of Sicily's zonal power price (Fig. 2).

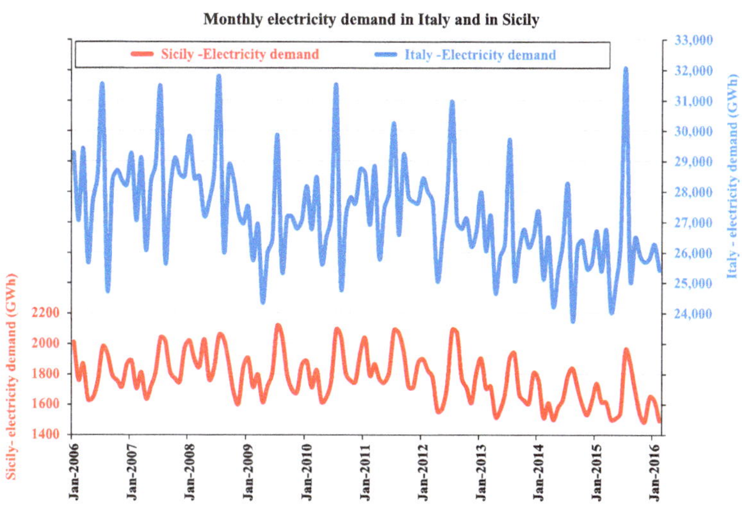

Figure 7. Monthly power demand in Italy and in Sicily. Source: Terna.

In view of the above-mentioned drastic regulatory intervention upon the formation of the zonal power price in Sicily, defining a statistical model linking the monthly Sicily's regional power price with local demand, renewable generation and conventional fuel price, as was done for the Italian market as a whole [7], de facto is impracticable. Nevertheless, there is room for a modeling effort on a daily basis over periods encompassing different seasons, that is, different compositions of the power generation mix, with particular reference to the PV share of the overall generation mix. This is the subject of the next Section.

Impact of Solar and Wind Power Generation

Aiming to show the remarkable impact of solar and wind electric generation in Sicily (and in Italy), we focus on the zonal and national prices of power in the course of the first 6 months of 2015. Prices on the day-ahead market for both Sicily and Italy were collected from GME [12], the manager of the Italian energy markets.

As expected, during the first half of 2015, different climatic conditions affected Sicily, leading to strong changes of solar PV and wind power generation superimposed to

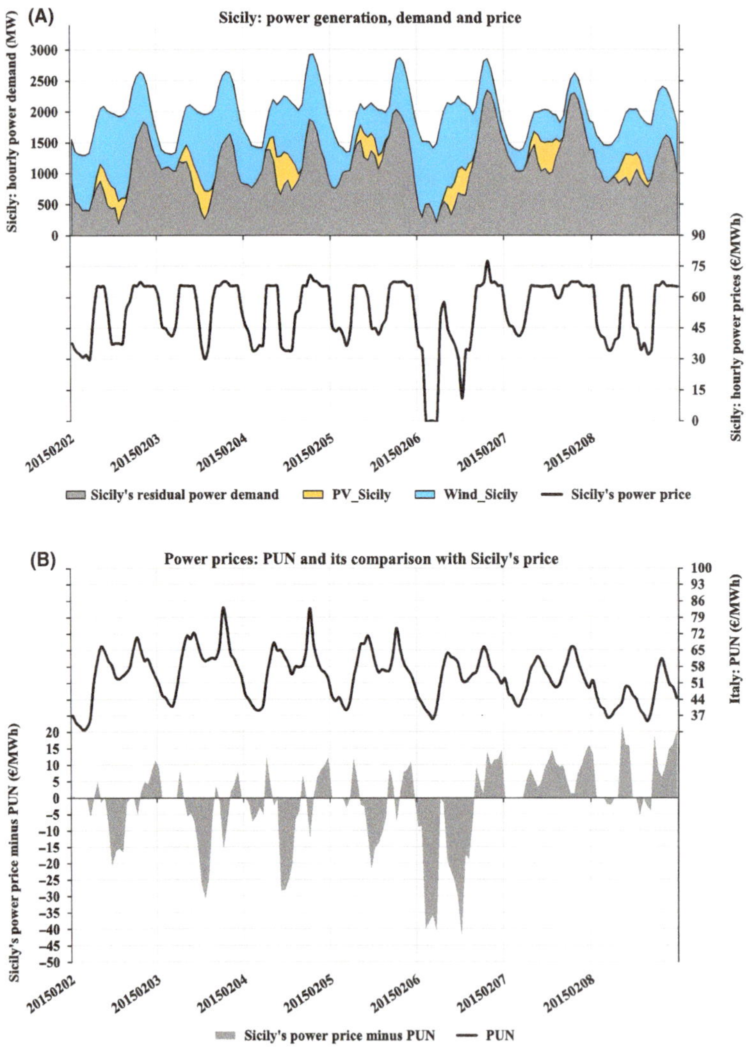

Figure 8. (A) Solar PV and wind hourly power generation superimposed to the residual power demand in Sicily, along with the respective power zonal price in the first week of February 2015; (B) national wholesale power price (PUN) along with the difference between Sicily's power price and PUN in the first week of February 2015. (C) Same as (A) but for the last week of June 2015; (D) same as (B) but for the last week of June 2015.

the daily cycle. Looking at the first week of February (Monday to Sunday) when wind generation was far higher than solar PV generation, Figure 8A shows that the residual electricity demand, that is, demand resulting after subtracting from the total the renewable (solar PV and wind) power generation, is correlated with the zonal power prices on an hourly basis. In other words, price grows along with residual demand, both during day and night, not only during working days but also during the weekend.

A further comparison between the values of Sicily's zonal power price and the national wholesale price during the same week (Fig. 8B) shows that low residual demand in Sicily leads to zonal prices lower than PUN, thereby lowering the national electricity bill.

In Italy, the national power price is formed in the day-ahead market, which works as a periodic multi-unit price auction [13]. The GME uses a merit-order principle to construct and aggregate the supply and demand curves. If electricity flows through the grid violate the transmission constraints (as it happens every day), the market is split into zonal areas: North, Center North, Center South, South, Sardinia and Sicily. Power-generating companies receive the zonal prices, whereas buyers pay the *Prezzo Unico Nazionale* (PUN) which is the average of the zonal prices, weighted by the zonal electricity consumption levels.

Similar results concerning residual electricity demand and zonal hourly power prices were observed during the

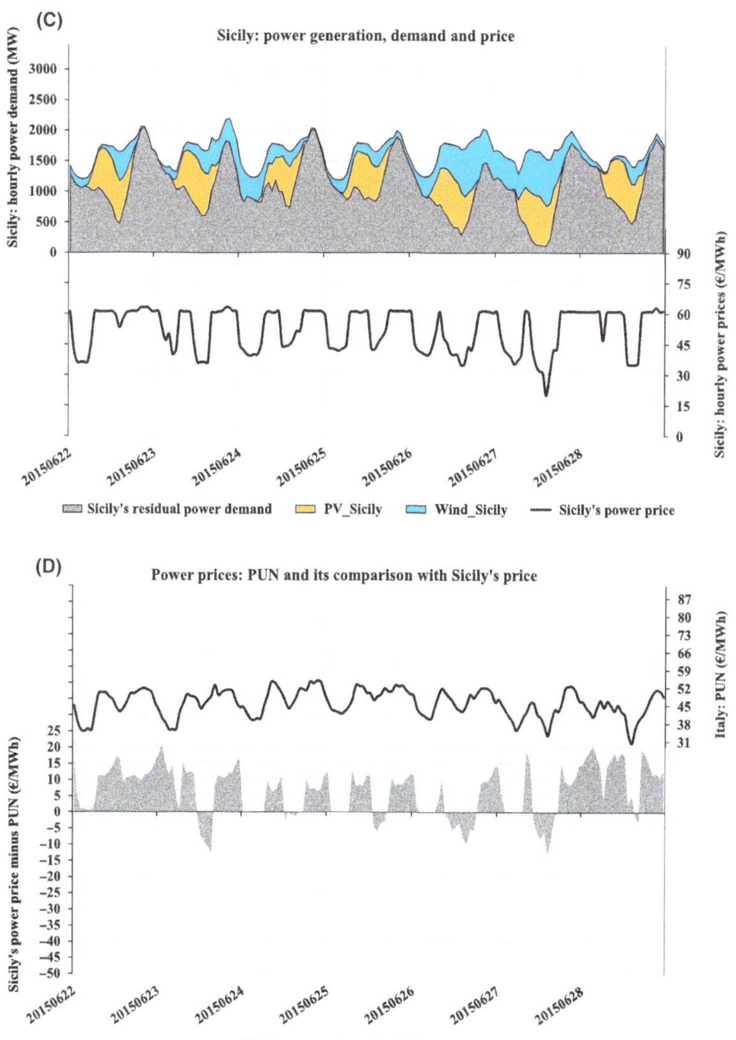

Figure 8. *(Continued)*

last week of June 2015 (Monday to Sunday), when solar PV generation was generally higher than wind-derived generation (Fig. 8C).

However, the substantially lower PUN values due to widespread solar PV generation across Italy, led to negative values of the difference between Sicily's zonal price and PUN only when the zonal residual demand fell below about 500 MW (Fig. 8D).

With reference to the same 2 weeks of 2015 in February and June, comparison between the hourly overall electricity demand in Sicily and Italy shows striking differences (Fig. 9). While the morning and evening peak demands are indeed comparable at the national level, in Sicily the evening peak is much larger both in February (Fig. 9A) and in June (Fig. 9B).

Moreover, the decrease in the electricity demand in Sicily during the weekend is much smaller than in Italy. Both these outcomes show evidence of the general lack of widespread significant industrial activities in Sicily, where three large, energy-intensive oil refineries and two petrochemical parks are installed in northern (Milazzo) and southern Sicily (Gela, Augusta and Priolo Gargallo), working in the continuous mode, independently of regular or weekend days of the week.

Widening the analysis to the whole period 1 January 2015 through 30 June 2015, the relationship between the

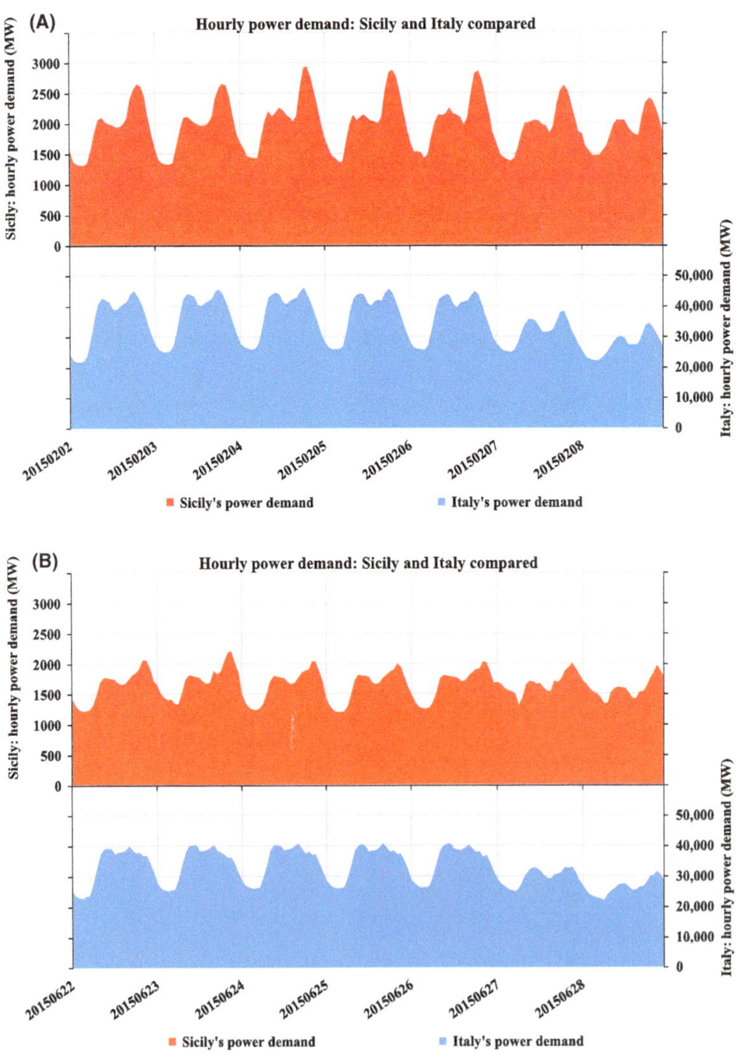

Figure 9. (A) Hourly series of power demand in Sicily and Italy in the first week of February, 2015. (B) Same as (A) but for the last week of June, 2015.

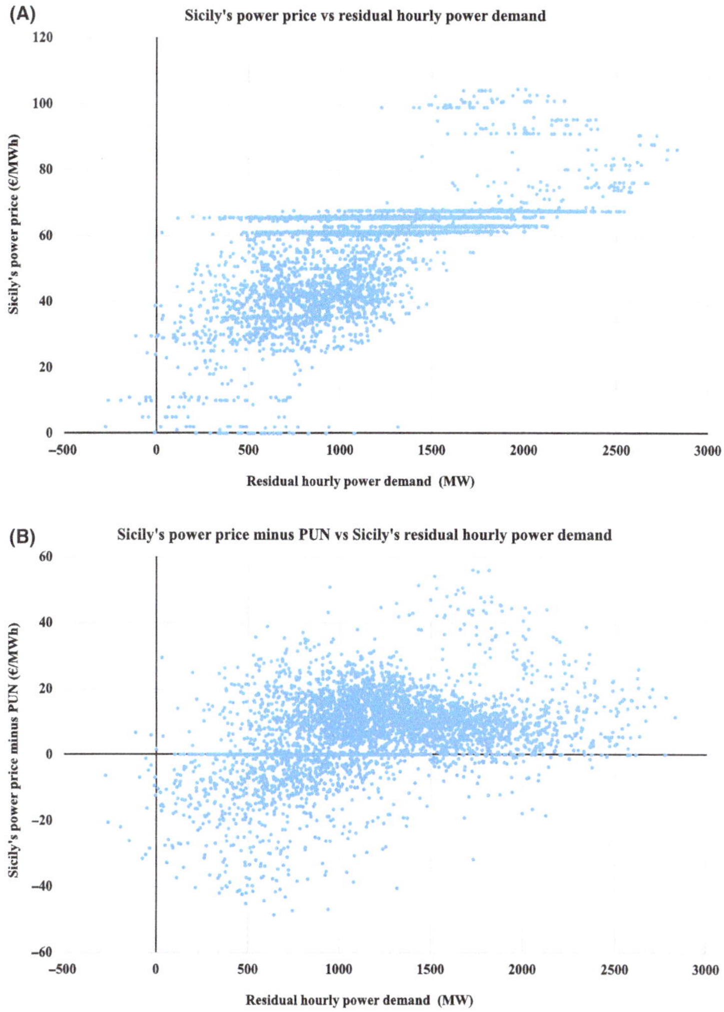

Figure 10. (A) Power price in the Sicily's zone against zonal residual electricity demand; (B) difference between Sicily's zone power price and national wholesale power price (PUN) against zonal residual electricity demand.

Sicily's zonal price and the zonal residual electricity demand, that is the remaining demand after subtraction of solar PV and wind generation, becomes apparent from the plots shown in Figure 10.

Most Sicily's zonal prices fall below the €40/MWh threshold when the residual demand lies below 500 MW (Fig. 10a). Similarly, during the hours in which the residual demand of electricity in Sicily is not generated by renewable sources lies below the same 500 MW threshold, the zonal prices lie below the PUN (Fig. 10B).

In order to outline a more homogeneous statistics, the power prices in the Sicily's zone were averaged across the values observed in six classes of the residual demand.

Figure 11 shows the results along with the standard deviation associated to the average price in each class. Best fit to this curve is a monotonic, logarithmic increase in the zonal price with the residual demand (eq. 1, where price is in €/MWh and the residual hourly demand is in units of MWh) explaining about 98.7% of the sample variance:

$$\text{Sicily's hourly zonal price} = 21.6\ \ln(\text{Residual hourly demand}) - 93.1. \quad (1)$$

Electrification of Energy End Uses

It is now well understood that the beneficial effects of PV generation onto wholesale electricity price formation rapidly increase with electricity demand [7]. Hence, to allow further reduction in the cost of electricity in both

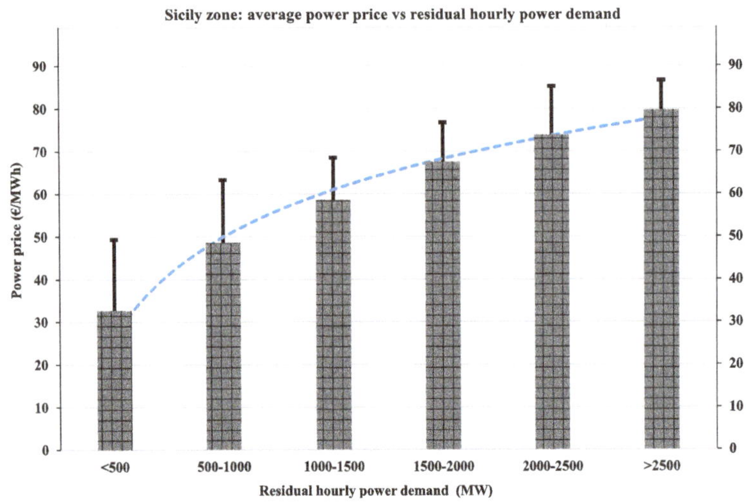

Figure 11. Average power price in the Sicily's zone in classes of the residual electricity demand.

Sicily and in Italy, it is important that the power demand in Sicily returns to grow.

Nevertheless, the contrary occurred both in Sicily and in Italy since 2007, as shown in Figure 7, down to unprecedented low consumption, which weakens the efficiency of solar PV generation concerning its ability to lower the power prices according to the merit-order price formation scheme.

In brief, policy makers in Sicily should boost the electrification of the energy end uses [14], as it happened recently with the new train connection linking the two main cities of the island (Palermo and Catania), as well as replacing obsolete heating systems based on burning natural gas, with much more efficient heat pumps in both residential, commercial, industrial, and public buildings.

Furthermore, new legislation should be passed and aimed to promote the use of pumping storage as an ideally suited option to store the excess of renewable PV and wind energy, using the associated significant (600 MW) hydroelectric capacity installed in Sicily [15].

Of course, electrification means also the adoption of storage systems on a large scale: because of its poor interconnection with Italy's grid, Sicily cannot rely on neighboring regions to balance short-term fluctuations in its grid.

For that reason, the owner of Italy's grid (Terna) has built the new "Sorgente-Rizziconi" and Sicily–Malta cable connection lines, and is adding new storage capacity in Sicily (using lithium-ion batteries) to make the existing grid more flexible, and thus more stable [5].

Conclusions and Perspectives

The analysis of the wholesale electricity prices in Sicily and in Italy reported in this study shows large, positive changes in Sicily's and Italy's electricity markets, due to significant penetration of new renewable energy sources, namely solar PV and wind energy. In 2015, the historical spread between Sicily's electricity zonal price and the national price (the PUN) eventually vanished, occasionally assuming negative values.

Furthermore, the zonal price has been shown to increase with the electric demand not met by renewable energy, according, as we find out, to a log-normal equation.

The forthcoming inauguration of the new 3000 MW high voltage connection between Sicily and the rest of Italy will only enhance such impact that so far has been restricted by the easily saturated 1000 MW existing connection.

Consistently lower values of the electricity national price (the PUN) will be achieved, especially during the sunniest months of the year when the electricity peak demand that has recently surpassed winter's demand (56,883 MW as of 8 July 2015) is met in large part by concomitant PV generation (40% of the above-mentioned peak of July 2015) [16].

Electrification of energy uses is slowly, but inevitably, taking place along with the global solar energy revolution [17].

Railway transportation in Italy, including Sicily, is knowing a renaissance with new high speed trains, new tram and new underground lines. The use of heat pumps in place of conventional heaters to produce low temperature heat is similarly undergoing a slow yet sustained growth [18], starting from large commercial buildings. The ongoing boom of electric bikes and motorbikes is prodromal to the forthcoming diffusion of electric cars [19].

Under these conditions, the reshaping of Sicily's and Italy's electricity market will continue. Investing in the electrification of energy user needs will translate into further economic (and environmental) benefits not only for Sicily's but also for all Italy's electricity consumers.

Acknowledgments

This article is dedicated to Daniele Tringali, eminent engineer at AMG Energia, whose daily engineering and management work does citizen's good in the city of Palermo.

Conflict of Interest

None declared.

References

1. European Commission, Joint Research Centre, Institute for Energy and Transport. Photovoltaic Geographical Information System (PVGIS). Available at http://re.jrc.ec.europa.eu/pvgis/ (accessed 20 April 2016).
2. Palmisano, G., M. Pagliaro, F. Meneguzzo, and R. Ciriminna. 2014. Sicily's solar report 2014. Simplicissimus Book Farm, Catania, Italy. ISBN 9788868859985.
3. Pagliaro, M., R. Ciriminna, F. Meneguzzo, and L. Albanese. 2015. Sicily's solar report 2015, Simplicissimus Book Farm, Catania, Italy. ISBN 9788869094231.
4. Meneguzzo, F., M. Pecoraino, L. Albanese, and M. Pagliaro. 2016. Sicily's solar report 2016. Simplicissimus Book Farm, Catania, Italy. ISBN 9788892500037.
5. Bastioli, C., M. Dal Fante, and P. Cristofori. Terna strategic plan 2015–2019. Available at http://download.terna.it/terna/0000/0086/99.pdf (accessed 20 April 2016).
6. Ries, J., L. Gaudard, and F. Romerio. 2016. Interconnecting an isolated electricity system to the European market: The case of Malta. Utilities Policy (In Press).
7. Meneguzzo, F., F. Zabini, R. Ciriminna, and M. Pagliaro. 2014. Assessment of the minimum value of photovoltaic electricity in Italy. Energy Science and Engineering 2:94–105.
8. Noce, A. Abuse of dominant position by energy incumbents. The Italian experience, Energy Community Competition Network Meeting, Athens. Available at https://www.energy-community.org/pls/portal/docs/2106181.PDF (accessed 4 June 2013).
9. Reuters. 2010. Enel to cap Sicily power price – antitrust watchdog. Available at http://fr.reuters.com/article/idUKLDE6781GW20100809 (accessed 20 April 2016).
10. Terna. Available at https://www.terna.it/it-it/sistemaelettrico/dispacciamento/datieservizio/rapportomensile.aspx (accessed 20 April 2016).
11. Rossetto, N. An oversized electricity system for Italy. Available at http://www.ispionline.it/it/energy-watch/oversized-electricity-system-italy-12135#sthash.exajQhg8.dpuf (accessed 22 January 2015).
12. Energy Markets Authority (GME). Available at http://www.mercatoelettrico.org/en/Statistiche/ME/DatiSintesi.aspx (accessed 20 April 2016).
13. Petrella, A., and S. Sapio. 2010. No PUN intended: a time series analysis of the Italian day-ahead electricity prices, EUI RSCAS, 2010/03; Loyola de Palacio Programme on Energy Policy, Bruxelles.
14. Edmonds, J., T. Wilson, M. Wise, and J. Weyant. 2006. Electrification of the economy and CO_2 emissions mitigation. Environmental Economics and Policy Studies 7:175–203.
15. Artizzu, G. 2013. Mercato elettrico: il PUN in altalena e lo zampino delle fonti rinnovabili. Available at http://www.qualenergia.it/articoli/20130414-mercato-elettrico-il-PUN-in-altalena-e-lo-zampino-delle-energie-rinnovabili (accessed 28 March 2016).
16. Terna. Flegetonte spinge le rinnovabili e il fabbisogno energetico. 56.883 MW: nuovo record assoluto dei consumi elettrici in Italia, press release, Rome, July 8, 2015. Available at https://www.terna.it/ViewDocumenti/tabid/1095/docid/33932/docType/TCAT-CS/language/it-IT/Default.aspx (accessed 28 March 2016).
17. Meneguzzo, F., R. Ciriminna, L. Albanese, and M. Pagliaro. 2015. The great solar boom: a global perspective into the far reaching impact of an unexpected energy revolution. Energy Science and Engineering 3:300–309.
18. Franci (REF-E), T. 2014. Efficienza energetica: evoluzione della domanda e tendenze di sviluppo tecnologico. Presentazione dei risultati della ricerca di mercato, *Nuova etichettatura energetica per gli impianti*, Congresso nazionale Domotecnica, Torino, 19 September 2014.
19. Albanese, L., R. Ciriminna, F. Meneguzzo, and M. Pagliaro. 2015. The impact of electric vehicles on the power market. Energy Science and Engineering 3:300–309.

15

Shale gas in North America and Europe

Michael H. Stephenson

British Geological Survey, Keyworth, Nottingham NG12 5GG, United Kingdom

Keywords
Economics, environment, hydraulic fracturing, shale gas

Correspondence
Michael H. Stephenson, British Geological Survey, Keyworth, Nottingham NG12 5GG, United Kingdom.

E-mail: mhste@bgs.ac.uk

Funding Information
No funding information provided.

Abstract

According to the U.S. Energy Information Administration, shale gas will provide half of the United States' domestic gas by 2035. The United States has already moved from being one of the world's largest importers of gas to being self-sufficient in less than a decade, bringing hundreds of thousands of jobs and attracting back companies that long ago left America in search of cheap manufacturing costs. But the increase in shale gas extraction has also had an environmental cost. There is clear scientific evidence of leaking shale gas wells and induced earthquakes, and in some areas a population increasingly turning against the industry. The technology of horizontal drilling and hydraulic fracturing that was developed in the United States is now being tried outside the United States, including in Europe, Argentina, and China. There are clear reasons why shale gas might be attractive to Europe. It may offer security of energy supply to some countries particularly dependent on Russian gas; it could stimulate growth and jobs; and it could supply a cleaner fuel than coal in power stations. However, prospective shale often underlies areas of high population density in Europe, and moreover, populations that are unfamiliar with onshore gas operations. The main challenge in Europe therefore is not mainly technological but for the industry to achieve a "social license" and for Government and regulations to be manifestly protecting the public and property.

Introduction

Shale is a fine grained, dark colored sedimentary rock that often contains natural gas (methane) as well as other gases. Its origins lie in mud deposited in sea and lake beds. Most of the mud is made up of stable minerals that are the result of advanced weathering of older rocks, but it also contains (often more than 10% by weight) organic matter that comes from plants growing on nearby land areas, as well as algae and plankton that live in the water column [1]. It is this organic material that, through heating and pressure supplied by deep burial under other later sediments, is converted to oil and gas through a complex series of chemical reactions. The temperature required is between 60°C and 120°C, with gas being formed at the high end of this range, and oil at the low end. Thus, shale can contain oil (known as "shale oil") in certain geological circumstances and gas in others. Shale whose thermal history lies outside the 60–120°C range may not contain any oil or gas [2].

The mineral material that makes up the bulk of shale is very fine and very tightly packed with the result that oil and gas created within the shale cannot readily move within the rock. Unless natural fractures are present, shale will tend to retain its hydrocarbons. This low permeability is the root of the idea of unconventional hydrocarbons, so-called because the oil and gas industry has to resort to new unconventional methods to extract oil and gas. The main advance in the last few decades has been hydraulic fracturing from long horizontal wells that target deep shale layers. Although hydraulic fracturing has been used for decades throughout the world [3], the extent to which the technique is being used now is unprecedented. About

33,000 gas wells are drilled in the United States every year and about 90% of these are fracked [4]. Rarely has a new technology produced such controversy. Cheap energy, energy security, economic growth, and climate amelioration are claims made by shale gas' advocates. Its detractors regard shale gas as a cause of environmental degradation, and also doubt whether its economic benefits are real or long lasting.

An important distinction to bear in mind in any discussion of shale gas is that between resources and reserves. The first concept is the total amount of gas that in geological terms constitutes material of potential value. Reserve, a smaller amount, is the proportion of the resource that is economical to produce, within environmental and social limits. The "recovery factor" connects the two concepts in that it is the percentage of the resources that can become reserves.

Only Canada and the United States have a commercial shale gas business and proven reserves, and so consideration of the effects of shale gas needs to concentrate in those countries.

North America

The main shale areas

The Barnett shale (Fig. 1) was the first shale to be exploited on an industrial scale. It underlies 28,000 square miles in north-central Texas, including the city of Fort Worth. The core producing area is the Newark East gas field covering 500 square miles, with over 2400 producing wells and 2.7 tcf (trillion cubic feet) of proven reserves.

The Marcellus shale is an older shale covering over 75,000 square miles of Ontario, New York, Pennsylvania, Ohio, West Virginia, Maryland, and Virginia with the best shale running in a belt from the northeast in New York, through northeast Pennsylvania into southwest Pennsylvania. The core area – where the formation exceeds 50 feet thick and has the right maturity – covers 50,000 square miles making it the largest potential shale region in North America.

The Eagle Ford shale trends across Texas from the Mexican border up into East Texas, in an area roughly

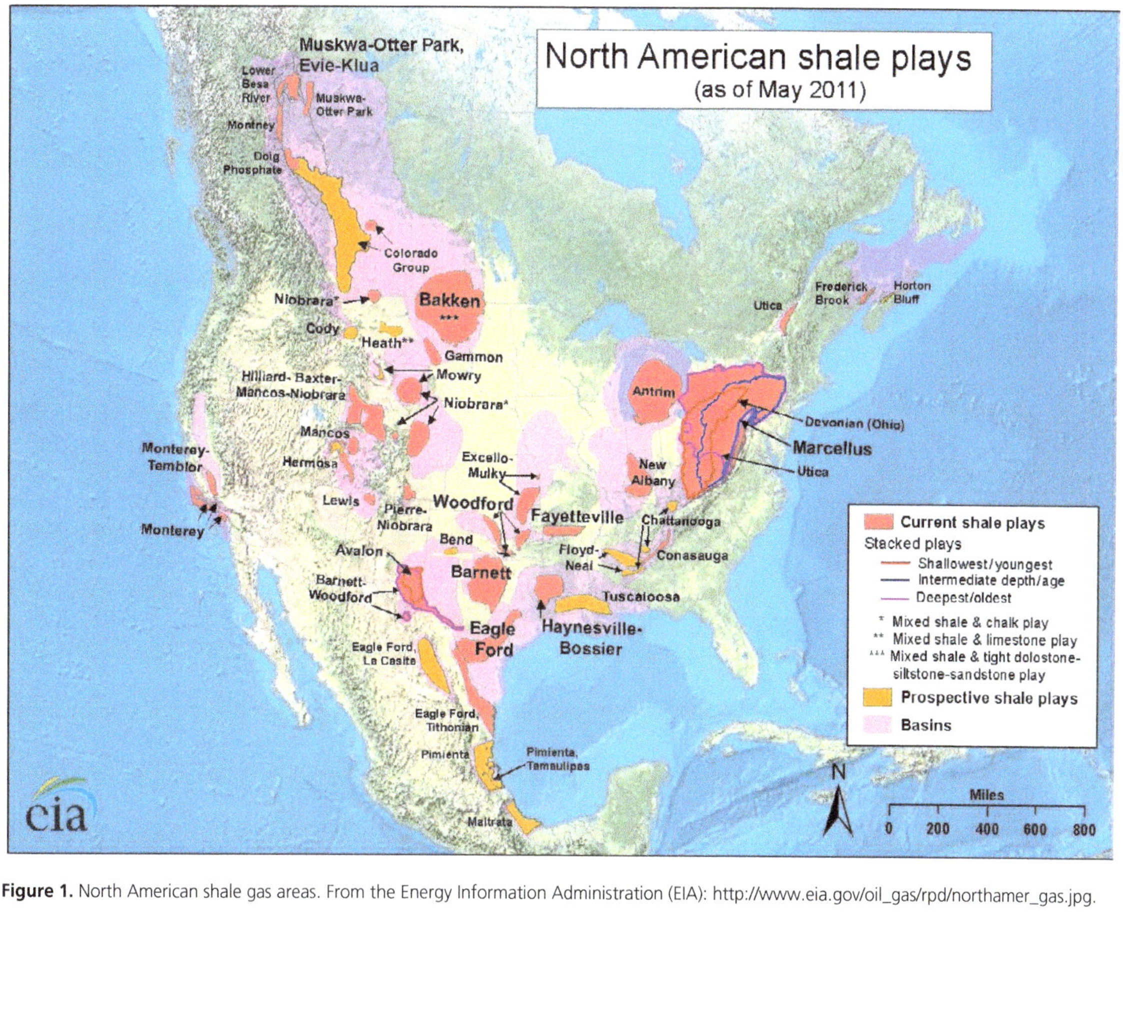

Figure 1. North American shale gas areas. From the Energy Information Administration (EIA): http://www.eia.gov/oil_gas/rpd/northamer_gas.jpg.

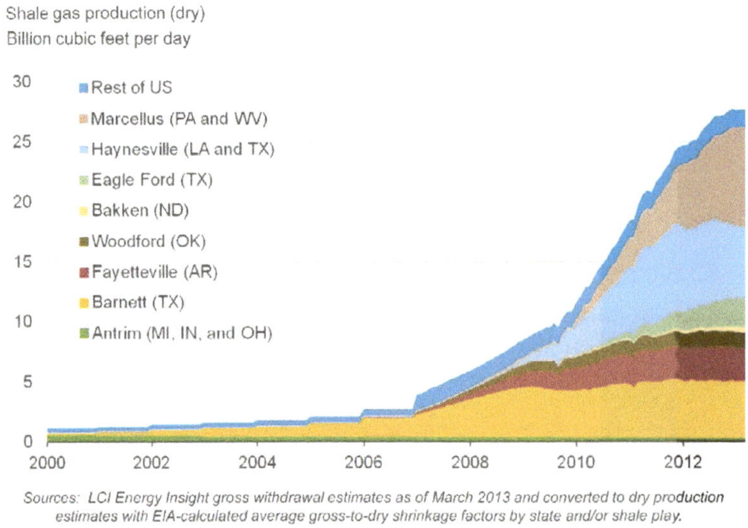

Figure 2. The relative importance of different U.S. shale layers. From Newell 2011, presentation to the OECD, Paris.

50 miles wide and 400 miles long. Here, the shale has an average thickness of 250 feet. The Haynesville shale underlies an area of about 9000 square miles of east Texas and west Louisiana averaging about 200–300 feet thick in the main productive area. The shale is very deep in the area occurring 10,500–13,000 feet below the surface, and produces mainly gas. This makes it quite expensive to drill. The Haynesville shale was one of the fastest growing shale gas businesses in the United States until about 2012 when gas prices fell and it became more difficult to operate profitably because of the depth. Many companies shifted to more profitable shales where they could get oil as well as gas, for example, the Eagle Ford shale.

The Fayetteville shale is roughly the same age as the Barnett shale and underlies much of northern Arkansas and adjacent states. It produces natural gas in the central portion of the Arkoma basin in Van Buren and Cleburn counties where the shale is at a few hundred to 7000 feet below the surface.

In Canada, two shales are important: the Muskwa (Horn River) shale and the Montney shale. The Muskwa shale was first described in outcrop on the banks of the Horn River, a tributary of the Mackenzie River, in the Northwest Territories. In eastern British Colombia, it ranges in depth from 6300 to 10,200 feet, averaging 8000 feet for the prospective area. The Montney shale in northwestern Alberta covers an area of ~1900 square miles at a depth ranging from 3000 to 9000 feet, averaging 6000 feet for the prospective area.

Commercial development of North American shale began in north east Texas in the Barnett shale where the increase in wells drilled is particularly evident between 1997 and 2009 (see [3]). The success in the Barnett shale lead quickly

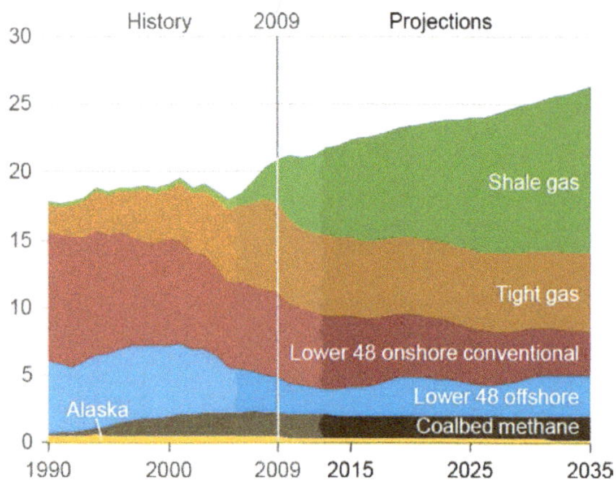

Figure 3. U.S. natural gas production, 1990–2035 (trillion cubic feet per year). From the EIA's 2011 Annual Energy Outlook.

to the development of the Woodford, Fayetteville, Haynesville, Marcellus, and Eagle Ford shales, and also shale in Canada. The relative importance of the different U.S. shale layers in terms of production is shown in Figure 2.

Economic effects of shale gas

The Energy Information Administration's 2011 Annual Energy Outlook [5] showed the recent history of shale gas and a projection to 2035 (Fig. 3), predicting that shale gas would make up almost half of U.S. natural gas production by 2035. The EIA Annual Energy Outlook of 2014 [6] indicated that the abundance of cheap shale gas is affecting the way that electricity is generated in the United States, and the balance of imports and exports of natural gas. Electricity generated by coal will remain the same or will slightly decline while gas-fired electricity generation will markedly increase. The EIA Annual Energy Outlook of 2014 also forecast that exports of U.S. natural gas might grow, both by pipelines to Mexico and Canada, but also as liquefied natural gas (LNG). The United States has already moved from being one of the world's largest importers of gas to being self-sufficient in less than a decade [3].

Shale gas has contributed to American energy security. American conventional natural gas production peaked in the early 2000s, and the widely discussed solution to the shortfall was to import LNG from conventional natural gas fields in the Middle East, Australia, and Russia. In 2006, there were four operating onshore LNG import terminals in the United States and more were expected to be built [3]. Since that time shale gas production has reduced U.S. natural gas imports to a level not seen since 1994. The EIA Annual Energy Outlook of 2012 [7] suggests that the United States will become a net exporter of gas in the form of LNG in 2016. More than 2 billion cubic feet per year may be being exported by 2019 [7].

The EIA Annual Energy Outlook of 2014 [6] predicts a jump in U.S. manufacturing powered by cheaper fuel in the form of shale gas. The main industries affected will be energy intensive bulk chemicals and primary metals, both of which provide products used by the mining and other downstream industries such as fabricated metals and machinery. The bulk chemicals industry is also a major user of natural gas and, increasingly, hydrocarbon gas liquids which are often produced with gas from shale. One of these is ethane which is used to make ethylene which in turn has many industrial products including PVC, polystyrene, latex, detergent, and vinyl. The Dow Chemical Company announced in 2012 that it would build a $1.7 billion ethylene and propylene production facility in Texas using ethane from local shale.

The low price of U.S. fuel now means that manufacturing costs are decreasing and companies that once left the United States to find cheaper production abroad are "reshoring" or returning production to the United States, which could result in 1 million more manufacturing jobs as companies build new factories [8]. Shale gas created 600,000 jobs in the United States as a whole up to 2010 and is predicted to provide 1 million jobs by 2025 [9] and 1.6 million by 2035 [10]. In Pennsylvania alone, the home of the huge Marcellus shale, the shale gas business is predicted to bring $18 billion extra business, nearly $2 billion local tax and 200,000 jobs by 2020 [11]. The U.S. Bureau of Labor Statistics [12] observed that between 2007 and 2012 employment in Pennsylvania's oil and natural gas industry (mainly shale gas) increased by 15,114 (259.3%), also that while the state's average annual pay increased by $5158 (11.9%), to $48,397 in 2012, wages in Pennsylvania's oil and natural gas industry rose by $22,104 (36.3%), to $82,974 in 2012. For the United Kingdom recently, Ernst and Young [13] predicted 64,500 jobs at peak for a U.K. shale gas industry of 4000 wells.

One result of the substitution of gas for coal in electricity generation is that for the years 2006–2011, CO_2 from fossil fuel consumption declined by 430 million tons (or 7.7%) in the United States [14]; although U.S. energy-related CO_2 emissions increased from 5267 million tons in 2012 to 5396 million tons in 2013 [6].

Environment

Concerns over degradation of the environment in North America center on three main topics: groundwater contamination by stray methane, uncontrolled emissions of methane to the atmosphere (fugitive emissions), and earthquakes directly or indirectly related to hydraulic fracturing. Perhaps the area that has caused most concern has been the first. At least two peer-reviewed papers representing studies of groundwater in Pennsylvania close to hydraulic fracturing of shale have reported contamination of shallow water wells [15, 16].

Osborn et al. [16] described high concentrations of methane from water wells close to active hydraulic fracturing sites and used the chemical composition of the methane to show that it was thermogenic rather than biogenic, and so likely to have come from shale rather than shallow biological action. The paper was criticized (e.g., Davies [17]) for the rather small amount of supporting data, and for the lack of baseline data on how much methane naturally occurs in water wells in Pennsylvania (see, e.g., [18]). However, Jackson et al. [15] working on the same area in Pennsylvania showed statistically significant evidence from a larger number of data points that water wells within 1 km of hydraulic fracturing wells contained high levels of stray methane. Molofsky et al. [18] and Molofsky et al. [19] showed that the methane – though clearly thermogenic shale gas – probably emanated from shale layers above the hydraulically fractured layer (the Marcellus shale) suggesting that hydraulic fracturing was not the direct cause of stray gas, but that leaking or faulty production wells were [15]. However, large studies of areas undergoing intense hydraulic fracturing elsewhere, such as the Arkansas Fayetteville shale, have found very low concentrations of only biogenic methane in water wells, suggesting that this leakage is a local problem for Pennsylvania [20]. Recent tentative evidence from study of a groundwater supply contamination incident in Pennsylvania showed that additives probably derived from drilling or hydraulic fracturing fluid were present in groundwater [21].

However, concerns over additives used in hydraulic fracturing fluid mainly center on them reaching the environment from spills at the surface or in transport, from illegal dumping of wastewater, or from damage to the liners of wastewater impoundment dams [22]. Most of the hydraulic fracturing fluid is water but the nontoxic sand proppant is also an important addition. On average 0.17% of the volume of the hydraulic fracturing fluid is made up of other additives [23]. These usually include inhibitors to prevent scale on the walls of the well, acid to help initiate fractures, and biocide to kill bacteria that can produce acids that lead to corrosion.

In a recent survey [24], 81 common additives were identified and categorized according to their functions. Of these, 17 chemicals were considered to present potential treatment challenges but most of the additives were considered "nontoxic or of low toxicity" with three "classified as Category 2" oral toxins according to standards in the "Globally Harmonized System of Classification and Labeling of Chemicals" (see [24]). Stringfellow et al. [24] noted that toxicity information was not located for 30 of the additives and recommended that further research and assessment was needed.

Richardson et al. [25] found that just under half of a survey of U.S. states permitting hydraulic fracturing currently require some form of fracturing fluid disclosure but the detail varies across states. Environmental groups (and some in industry) have suggested that states should insist on disclosure whether the federal government asks for it or not. Many companies voluntarily disclose information on the web database FracFocus (http://fracfocus.org/).

Howarth et al. [26] studied the methane emissions from the surface installations of several shale gas wells finding high levels of methane release into the atmosphere as a result of two routine processes: flow back and drill out. Flow back is the liquid that returns to the well and then to the surface following hydraulic fracturing. Typically it carries methane. The practice of leaving the flow back water in open tanks at the surface (ready for disposal or reuse) allows methane to escape directly into the atmosphere. After drill out (where plugs in the well that partitioned parts of the well for hydraulic fracturing are removed), methane also escapes from the well to the atmosphere. Howarth et al. [26] published data that showed a very high level of methane leakage of between 3% and 8% of a well's production over its lifetime. Given that methane is a potent greenhouse gas, [26] considered that such levels of emissions cast doubt on the claims that shale gas could be a low carbon fuel, for example, in comparison with coal. Howarth's paper has been contested and recent work [27] suggested that the high figures for emissions quoted in the paper were unrepresentative of the industry as a whole. Recent work from a larger number of shale gas drilling sites [28] suggests that the leakage rate is about half of 1% of gas production.

Evidence of earthquakes caused directly by hydraulic fracturing is extremely rare in the United States, though in Canada there are well-documented cases [29]. However the associated activity of deep geological disposal of flow back water is increasingly associated with earthquakes. Frohlich [30] showed that flow back disposal wells in the Barnett shale area in northeast Texas were responsible for several small earthquakes and concluded that the wells that were near faults, and particularly faults with stress, were the most likely to trigger earthquakes. The few studies like these suggest that permanent underground disposal of waste water is much more likely to produce earthquakes than hydraulic fracturing mainly because the water injected is permanent rather than temporary and because very large volumes can build up underground.

Europe

The EIA [31] indicated that much of the world's shale gas is probably outside of North America – in Europe, China, Argentina, and Mexico, though there has been very little development in these areas. The way that this resource is seen, and its potential for commercialization, depends largely on the individual circumstances of the country with regard to energy prices, energy supply security, and environmental concerns.

European shale areas

The most prospective areas in Europe include the eastern Baltic, north-eastern Ukraine, western Ukraine, the Balkans, central Poland, northern Germany, parts of southern Norway and Sweden, Netherlands, northern France, and northern England. European countries with the most advanced shale gas development, in terms of exploration, drilling, and licenses are Poland and the United Kingdom.

Poland

Poland appears to have a substantial shale gas resource [31]. The Polish Geological Institute [32] estimated shale gas resources in the range 346–768 billion cubic meters.

Poland also has gas supply security concerns, and as a member state of the European Union is part of ambitious plans to reduce Europe-wide greenhouse gas emissions through the EU Emissions Trading Scheme (ETS). Almost 70% of Poland's natural gas is imported (mainly from Russia) and almost 90% of its electricity is generated using local Polish coal [33]. There is therefore a clear policy need for Poland to not only switch away from coal to generate electricity, but also to lower dependency on Russian supplies. Shale gas is therefore being considered very seriously by the Government of Poland and much of the country is licensed for gas exploration. The concession areas of exploration cover 37,000 square kilometers, which constitutes 11% of the country's territory and extend as a wide belt from Pomerania (northeastern Poland), through Mazowsze and Podlasie (central Poland), to Lubelszczyna (southeastern Poland) [33]. By the end of 2012, companies had drilled 33 exploration wells and fracked 11 of those, but results from drilling and testing appear to have been mixed.

The United Kingdom

The United Kingdom has three main prospective areas, the north of England to the east and west of the Pennine Hills, the central lowlands of Scotland, and the Weald in south-eastern England. The largest resource is within the Bowland–Hodder shale unit (containing 1300 tcf) in the north of England; the central lowlands contain 80.3 tcf of gas, and 6.0 billion barrels of oil; and the Weald contains 4.4 billion barrels of oil [2, 34, 35].

Western European countries tend to have more diverse sources of energy supply than those in eastern Europe. The United Kingdom sources are coal, gas, and nuclear. British gas power stations use Norwegian gas and Qatari LNG. Since the peak of production from the North Sea, Britain has become a net gas importer and as British North Sea oil and gas continue to decline, dependency on imports will increase. Although British oil imports come mainly from Norway, other sources are in less stable countries and there is a concern that world events like the "Arab spring" that began in 2010, might affect Britain's ability to source its energy. Thus, security of supply is a significant issue.

The United Kingdom has also committed to greenhouse gas reductions of 80% by 2050. Much of this reduction could be supplied by nuclear, but the cost of nuclear new build and decommissioning power stations – and the public dislike of them – has made their development more and more difficult. Another of the United Kingdom's plans to reduce emissions while still using fossil fuels in power stations – carbon capture and storage – has been slow to start. The technology, which has only been proven on a small scale, is expensive and difficult to commercialize without a high price for carbon emissions per ton through the ETS [36]. For all these reasons the United Kingdom

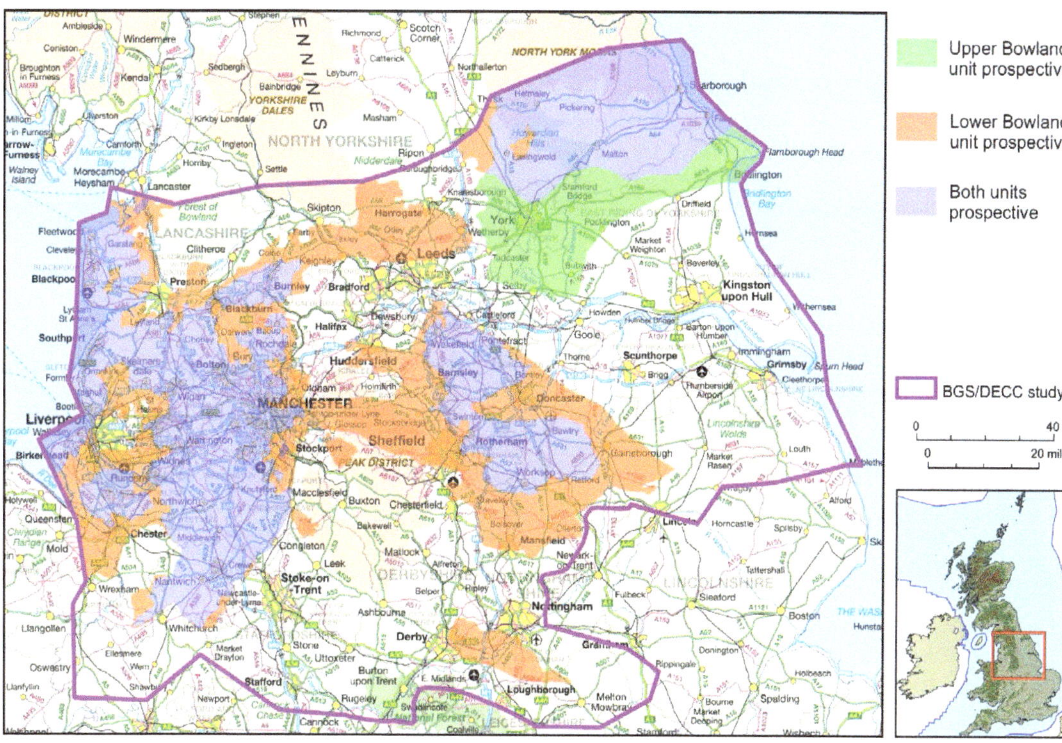

Figure 4. The two main prospective layers of shale in the North of England and their distribution with respect to population and infrastructure. From Andrews [2].

government has shown interest in the development of shale gas and is offering support, including tax incentives. As of early 2014, a percentage of revenues from production will be available to local people and local councils will receive an increase in the rates from operations.

Environmental concerns in Europe

The most detailed estimate for shale gas thus far in Europe has been that carried out in the north of the United Kingdom to assess the size of the resource in the Bowland–Hodder shale layer [2]. Among the findings of the report were a substantial resource of 1300 tcf, very thick shale layers underlying parts of the north of England either side of the Pennine hills – and crucially a concentration of prospective shale under areas with relatively high population density (Fig. 4; [2]). The cities of Manchester, Liverpool, Blackpool, Leeds, York, and Sheffield are all underlain by prospective shale as are high-value agricultural land and National Parks. The coincidence of populated areas and shale is not entirely a matter of chance since the Bowland–Hodder shale layer is related to the Coal Measures that geologically overly it. These coal deposits fueled the 19th Century industrial revolution that encouraged human settlement.

This coincidence highlights the central problem in Europe – that shale gas resources often underlie dense populations and populations that are also unused to oil and gas drilling, unlike in the United States and Canada. This is the main reason for the overall lack of support such that there are moratoria on drilling in several countries (e.g., France, Netherlands, Czech Republic, and Bulgaria); and in many others, even with substantial resources, there is very slow progress in drilling.

In the United Kingdom there have also been some very public problems in early drilling, for example, the first attempts at hydraulic fracturing of the Bowland–Hodder shale layer near Blackpool in the northwest resulted in a number of small earthquakes, including two above 1.0 on the ML scale [37]. Following these in May 2011, hydraulic fracturing was suspended by the U.K. Government and an intensive study of hydraulic fracturing in the area was commissioned by the U.K. Department of Energy and Climate Change (DECC) and carried out by an independent panel of experts in seismology, induced seismicity, and hydraulic fracturing. Among the recommendations were that: (1) hydraulic fracturing should invariably include a smaller preinjection and monitoring stage before the main injection; (2) hydraulic fracture growth and direction should be monitored during operations; (3) future hydraulic fracturing operations in Lancashire should be subject to an effective monitoring system that can provide automatic locations and magnitudes of any seismic events in near real time. Perhaps most important, the report recommended that operations should be halted and remedial action instituted, if seismic events of magnitude 0.5 ML or above are detected. This has become known as the traffic light system [37].

In the United Kingdom hydraulic fracturing and shale gas did not come to mass public attention, however, until protests started at a drilling site in Balcombe in the southeast of England in July 2013. Ironically the well was being drilled for a conventional accumulation of hydrocarbons (not shale) but nevertheless the site captured media attention. A large number of protesters tried to stop the drilling. This and a heavy police presence made operations very difficult so the company involved eventually suspended drilling in August 2013.

An advantage of the concerned public reaction has been a more cautious approach to shale gas development (both in terms of environment and regulation) than in North America. In the United Kingdom and Poland studies of geological conditions before hydraulic fracturing (known as baselines) have been initiated. The purpose of these is to establish if changes have occurred that could be unequivocally attributed to hydraulic fracturing.

The most important of these baselines is the British Geological Survey (BGS) Groundwater Baseline Survey which looks at the natural amounts of methane in British groundwater in areas most likely to be developed for shale gas (Fig. 5). BGS has also calculated the underground vertical separation between aquifers that carry potable water and prospective shale gas layers in a series of aquifer-shale separation maps intended for the guidance of planners, policy makers and shale gas exploration companies.

As was seen at Balcombe in the United Kingdom, even with a well-developed regulatory and licensing system for oil and gas and with exploration licenses and planning permissions in place, protests by local people can make operations difficult, or impossible. Sociologists and economists refer to general acceptance on the part of the public that a new technology can and should proceed, as *social license*. This has proved difficult to obtain for shale gas companies operating in some parts of Europe.

The nature of the extraction technique of hydraulic fracturing and the fact that it operates deep underground in an environment that for many people is unfamiliar may mean that more effort may have to be made in educating and informing the affected public. In modern democracies, the trend is toward ever increasing information and data availability often through the internet. Many people want to know precisely about the activities that might go on under their land or houses. This trend is often at odds, however, with the culture of oil and gas companies which will tend not to provide information about operations – often because they do not think it is

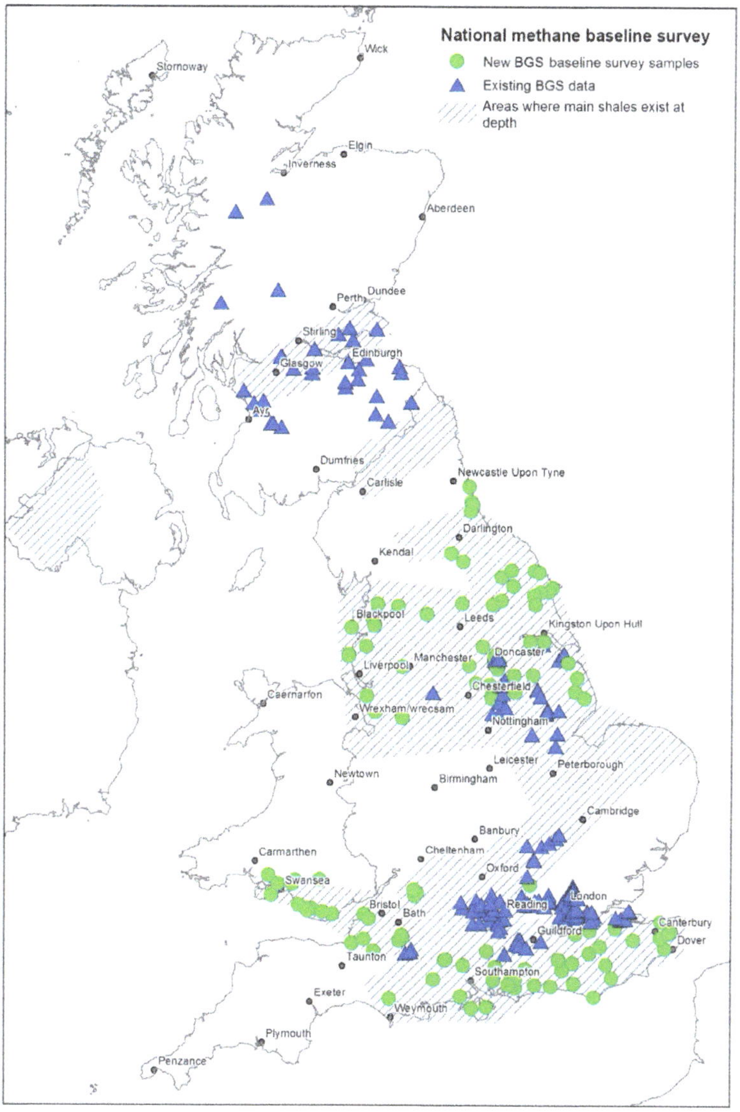

Figure 5. A survey of methane in groundwater for Britain will establish what the natural methane levels are before any shale gas extraction. British Geological Survey © NERC 2014. Contains Ordnance Survey data © Crown Copyright and database rights 2014.

necessary, or because there are matters of commercial advantage at stake. For example, a company may not want to reveal the precise contents of its hydraulic fracturing fluid because it believes that this fluid gives the company a particular commercial advantage over other companies. In the United States, the so-called "Halliburton loophole" has excused companies from disclosing all the materials that they use in their hydraulic fracturing fluids. However this is changing as companies realize that a crucial part of obtaining a social license is to be fully open in the activities and materials involved in hydraulic fracturing.

It may be however that to get social license for unfamiliar extraction techniques or any subsurface usage (e.g., deep geological radioactive disposal) may require a more radical approach. Recently it has been proposed that full disclosure of materials and activities should be accompanied by monitoring and measurement of a wide range of environmental variables, for example groundwater quality and air quality – and that the data so derived should be made open and available in real time for anyone to see [38]. The kind of monitoring involved could include seismic and groundwater monitoring, down hole sensors, and other down hole techniques such as electrical resistivity tomography, as as remote sensing. In the United States and Canada, dedicated websites already provide this in-depth information to allow the public to peruse local energy extraction activities and their effects on the environment.

A good example is the Alberta Government's Oil Sands Information Portal. Oil sands have only recently been considered to be part of the world's oil reserves, as higher oil prices and new technology enable profitable extraction and processing. The oil sands of Athabasca are among the most controversial in the world in relation to the environmental effects of their exploitation. Environmentalists' concerns relate to CO_2 emissions of the process by which they extracted, other non-CO_2 air pollutant emissions, the use of water in processing, the effect of exploitation on ground and surface water, and the long-term effects on landscape, wildlife and ecology. The oil sands are also a huge resource of oil, but many Albertans and other Canadians are uneasy about the effects of mining on the environment. The Oil Sands Information Portal provides environmental information on the web, as far as possible in real time, to show the effects (if any) of the processes of extraction on the environment including information on the mines and tailings ponds, and air and water quality. Another example dealing with shale gas and the environment (albeit on a smaller scale) includes Colorado Water Watch (CWW), a real-time groundwater monitoring pilot program for shale gas sites.

Conclusions

The technology of shale gas is well developed and has made a huge impact in the United States and is making an impact in Canada. Whether shale gas extraction makes an impact elsewhere in the world depends largely on the applicability of technology developed in U.S. shale layers to shale layers outside the United States, and perhaps more importantly on the public attitude to shale gas.

The technology developed in the Barnett shale was quickly adapted for widely different shales in the United States This was helped by a generally entrepreneurial oil and gas culture, by subsurface ownership rules favorable to rapid development, and by a buoyant gas market – at least in the early years of shale gas. The variety of shale types that have been successfully exploited already within the United States is probably comparable to the amount of variety that drilling companies will encounter elsewhere in the world and so it is not unreasonable to imagine that U.S. techniques will quickly adapt to new conditions.

However, as noted before, many prospective areas in Europe, for example, are densely populated and so shale gas extraction will require very high levels of environmental assurance to gain a social license. This is probably a more intractable problem than that of technology and may need radical solutions, for example, increasing levels of transparency in operations, including full public disclosure of data relating to extraction.

The cost of gaining a social license will likely be high in Europe, and seen against the falling prices of oil and gas, may make commercial investment less attractive and more challenging for industry both in Europe and the United States

Acknowledgments

The author publishes with the permission of the Executive Director of the British Geological Survey.

Conflict of Interest

None declared.

References

1. Könitzer, S. F., S. J. Davies, M. H. Stephenson, and M. J. Leng. 2014. Depositional controls on mudstone lithofacies in a basinal setting: implications for the delivery of sedimentary organic matter. J. Sediment. Res. 84:198–214.
2. Andrews, I. J. 2013. The Carboniferous Bowland Shale gas study: geology and resource estimation. British Geological Survey for Department of Energy and Climate Change, London, U.K.
3. Wang, Q., X. Chen, A. N. Jha, and H. Rogers. 2014. Natural gas from shale formation – the evolution, evidences and challenges of shale gas revolution in United States. Renew. Sustain. Energy Rev. 30:1–28.
4. Prud'homme, A. 2014. Hydrofracking: what everyone needs to know. Oxford Univ. Press, Oxford, U.K.
5. Energy Information Administration (EIA). 2011. Annual energy outlook.
6. Energy Information Administration (EIA). 2014. Annual energy outlook 2014, early release.
7. Energy Information Administration (EIA). 2012. Annual energy outlook.
8. The Economist. 2013. Available at: http://www.economist.com/news/special-report/21569570-growing-number-american-companies-are-moving-their-manufacturing-back-united (accessed 24 September 2015).
9. Price Waterhouse Coopers. 2011. Shale gas: a renaissance in US manufacturing. December 2011.
10. IHS. 2011. The economic and employment contributions of shale gas in the United States complete report and methodology. Available at: www.ihs.com/EconomicContributionofShaleGasintheUS (accessed 24 September 2015).
11. Considine, T. J., R. Watson, and S. Blumsack. 2010. The economic impacts of the Pennsylvania Marcellus shale

natural gas play: an update. Report of Pennsylvania State University, College of Earth and Mineral Sciences.
12. US Bureau of Labor Statistics. 2014. Monthly Labor Review. The Marcellus Shale gas boom in Pennsylvania: employment and wage trends. Available at: http://www.bls.gov/opub/mlr/2014/article/the-marcellus-shale-gas-boom-in-pennsylvania.htm (accessed 24 September 2015).
13. Ernst & Young. 2014. Getting ready for UK shale gas: supply chain and skills requirements and opportunities. Ernst & Young LLP, 1 More London Place, London, SE1 2AF.
14. IEA. 2011. World energy outlook. Paris.
15. Jackson, R. B., A. Vengosh, T. H. Darrah, N. R. Warner, A. Down, R. J. Poreda, et al. 2013. Increased stray gas abundance in a subset of drinking water wells near Marcellus shale gas extraction. Proc. Natl. Acad. Sci. USA 110:11250–11255.
16. Osborn, S. G., A. Vengosh, N. R. Warner, and R. B. Jackson. 2011. Methane contamination of drinking water accompanying gas-well drilling and hydraulic fracturing. Proc. Natl. Acad. Sci. USA 108:8172–8176.
17. Davies, R. J. 2011. Methane contamination of drinking water caused by hydraulic fracturing remains unproven. Proc. Natl Acad. Sci. 108:E871.
18. Molofsky, L. J., J. A. Connor, A. S. Wylie, T. Wagner, and S. K. Farhat. 2013. Evaluation of methane sources in groundwater in northeastern Pennsylvania. Ground Water 51:333–349.
19. Molofsky, L. J., J. A. Connor, S. K. Farhat, A. S. Wylie, and T. Wagner. 2011. Methane in Pennsylvania water wells unrelated to Marcellus shale fracturing. Oil Gas J. 54–67.
20. Warner, N. R., T. M. Kresse, P. D. Hays, A. Down, J. D. Karr, R. B. Jackson, et al. 2013. Geochemical and isotopic variations in shallow groundwater in areas of the Fayetteville Shale development, north-central Arkansas. Appl. Geochem. 35:207–220.
21. Llewellyn, G. T., F. Dorman, J. L. Westland, D. Yoxtheimer, P. Grieve, T. Sowers, et al. 2015. Evaluating a groundwater supply contamination incident attributed to Marcellus shale gas development. Proc. Natl. Acad. Sci. USA 112:6325–6330. doi: 10.1073/pnas.1420279112
22. Bamberger, M., and R. E. Oswald. 2012. Impacts of gas drilling on human and animal health. New Solut. 22:51–77.
23. Royal Society and Royal Academy of Engineering. 2012. Shale gas extraction in the UK: a review of hydraulic fracturing DES2597.
24. Stringfellow, W. T., J. K. Domen, M. K. Camarillo, W. L. Sandelin, and S. Borglin. 2014. Physical, chemical, and biological characteristics of compounds used in hydraulic fracturing. J. Hazard. Mater. 275:37–54.
25. Richardson, N., M. Gottlieb, A. Krupnick, and H. Wiseman. 2013. The state of state shale gas regulation. Report for Resources for the Future, Washington, DC.
26. Howarth, R., R. Santoro, and A. Ingraffea. 2011. Methane and the greenhouse-gas footprint of natural gas from shale formations. Clim. Change. 106:679–690.
27. McKay, D. J. C., and T. J. Stone. 2013. Potential greenhouse gas emissions associated with shale gas extraction and use. Report for the UK Department of Energy and Climate Change.
28. Allen, D. T., V. M. Torres, J. Thomas, D. W. Sullivan, M. Harrison, Al Hendler, et al. 2013. Measurements of methane emissions at natural gas production sites in the United States. Proc. Natl. Acad. Sci. USA 110:17768–17773.
29. British Colombia Oil and Gas Commission Report. 2012. Investigation of observed seismicity in the Horn River Basin. https://www.bcogc.ca/node/8046/download (accessed 24 September 2015).
30. Frohlich, C. 2012. Two-year survey comparing earthquake activity and injection-well locations in the Barnett Shale, Texas. Proc. Natl. Acad. Sci. USA 109:13934–13938.
31. Energy Information Administration (EIA). 2013. Technically recoverable shale oil and shale gas resources: an assessment of 137 shale formations in 41 countries outside the United States.
32. Polish Geological Institute – National Research Institute (PGI). 2012. Assessment of shale gas and shale oil resources of the lower Paleozoic Baltic-Podlasie-Lublin Basin in Poland. Available at: http://www.pgi.gov.pl/en/dokumenty-in-edycja/cat_view/294-aktualnoci-2012/297-zasoby-gazu.html (accessed 24 September 2015).
33. Shale gas Information Platform (SHIP). 2013. Shale gas in Poland. Available at: http://www.shale-gas-information-platform.org/de/areas/die-debatte/shale-gas-in-poland.html (accessed 24 September 2015).
34. Andrews, I. J. 2014. The Jurassic shales of the Weald Basin: geology and shale oil and shale gas resource estimation. British Geological Survey for Department of Energy and Climate Change, London, U.K.
35. Monaghan, A. A. 2014. The Carboniferous shales of the Midland Valley of Scotland: geology and resource estimation. British Geological Survey for Department of Energy and Climate Change, London, U.K.
36. Stephenson, M. H. 2013. Returning carbon to nature: coal, carbon capture, and storage. Elsevier, Amsterdam, The Netherlands.
37. Green, C. A., P. Styles, and B. J. Baptie. 2012. Preese Hall shale gas fracturing: review and recommendations for induced seismic mitigation. Department of Energy and Climate Change, London. Available at: http://og.decc.gov.uk/assets/og/ep/onshore/5075-preese-hall-shale-gas-fracturing-review.pdf (accessed 24 September 2015).
38. Stephenson, M. H. 2015. Shale gas and fracking: the science behind the controversy. Elsevier, Amsterdam, The Netherlands.

16

Study of thermal behavior of deoiled karanja seed cake biomass: thermogravimetric analysis and pyrolysis kinetics

Radhakumari Muktham[1,2], Andrew S. Ball[2], Suresh K. Bhargava[2] & Satyavathi Bankupalli[1]

[1]Chemical Engineering Division, CSIR – Indian Institute of Chemical Technology, Hyderabad 500007, India
[2]School of Applied Sciences, Royal Melbourne Institute of Technology, Melbourne 3083, Australia

Keywords
Biochar, karanja cake, kinetics, pyrolysis, TGA

Correspondence
Satyavathi Bankupalli, Chemical Engineering Division, CSIR – Indian Institute of Chemical Technology, Hyderabad, 500007, India.

E-mail: satya@iict.res.in

Funding Information
This study was supported by the CSIR-IICT.

Abstract

Karanja is a medium sized evergreen tree which has minor economic importance in India. The nonedible seed kernel contains 27–30% oil that is used for biodiesel production, leaving the remaining nonedible seed cake as a waste product. The aim of the present work was to obtain kinetic parameters in relation to technological parameters in nonedible seed cake biomass pyrolysis conversion process to bio-oil and biochar. Effects of heating rate on karanja seed cake slow pyrolysis behavior and kinetic parameters were investigated at heating rates of 5, 10, and 20°C/min using thermogravimetric analysis (TGA). Thermogravimetric experiments showed the onset and offset temperatures of the devolatilization step shifted toward the high-temperature range, and the activation energy values increased with increasing heating rate. In the present study, isoconversional method was applied for the pyrolysis of karanja seed cake biomass by TGA and the activation energies (118–124 kJ/mol) and the pre-exponential factors obtained using progressive conversion. Proximate–ultimate analyses, energy value, surface structure, and Fourier transform infrared spectra of the biomass processed under conditions were reported. The pyrolysis resulted in upgradation of the energy value of seed cake biomass from 18.1 to 24.5 MJ/kg; importantly with high carbon and low oxygen contents. The approach represents a novel method for the upgrading of karanja seed cake that has significant commercial potential.

Introduction

With the growing concerns about fossil fuel exhaustion and environmental problems such as global warming, efforts are underway to decrease CO_2 emission rates and to develop alternative energies which substitute for limited fossil fuels. Biodegradable nontoxic biomass having low emission profiles compared to petroleum-based fuels (petrol and diesel) is now widely recognizing as a potential energy source.

Biomass pyrolysis is a thermal decomposition of biomass taking place in the absence of oxidizing agent with products biochar, bio-oil, and gases such as carbon dioxide, carbon monoxide, hydrogen, and methane. Pyrolysis is a promising technology for the production of bio-oil and biochar. Bio-oil, also known as pyrolytic oil, can be upgraded to light hydrocarbons which contain low levels of aromatics with the absence of sulfur compared to petroleum-based fuels. Biochar, on the other hand, is an ecofriendly carbon-rich product from the pyrolysis of renewable feedstocks like nonedible seed cakes after the extraction of oil and has been widely used in agriculture as a soil amendment and for improving soil fertility due to its carbon sequestration ability [1]. Biochar is also reported to promote nitrogen fixation thereby decreasing the emission of N_2O [2] and other greenhouse gases from agricultural soils [3, 4]. In addition, biochar has been used as an adsorbent in the removal of arsenic [5], cadmium [6], and chromium from aqueous solutions.

Pyrolysis process occurs with high heat flux to the biomass with a corresponding high heating rate of the biomass particle and heat transfer to the biomass in a very short time period. Applying heat to biomass at high rate will result into smaller fragments (volatiles) due to the cleavage of cellulose, hemicellulose, protein, and lignin, and other constituents of biomass, unstable above 400°C due to the presence of oxygen in the fragments. Hence, very short residence times are required in the thermal treatment step followed by immediate quenching to impede secondary chemical change of unstable volatiles. Pyrolysis with a short residence time and optimum temperature of exposure with immediate cooling of the vapors can be achieved by taking into account the appropriate size of the reactor and heating rate in terms of heat flux. Heat transfer rate to biomass particles and residence time of vapors in the reactor have a strong influence on the nature and distribution of pyrolysis reaction products. Therefore, the first and foremost consideration for designing biomass pyrolysis reactor (residence time, size, and material of construction) is to draw the clear picture of thermal behavior of the feedstock biomass with the application of heat at different heating rate which corresponds to heat transfer rate to biomass particle in order to estimate the amount of energy to be supplied for pyrolysis.

Various biomass feedstocks show different thermal behavior profiles mainly due to the variation in the composition of biomass constituents. For instance, Parthasarathy and Sheeba [7] have investigated the thermal behavior of bagasse, coir pith, groundnut shell, and casuarinas leaves, which exhibited different degradation temperatures of individual biomass components (hemicellulose, cellulose, and lignin).

Vhathvarothai et al. [8] have studied the copyrolysis of cypress wood chips, macadamia nut shells biomass, and coal using thermogravimetric technique and Kissinger's corrected kinetic equation for kinetic parameters determination and concluding that the activation energy of coal is greater than that of both types of biomass.

Pyrolysis kinetics of raw and hydrothermally carbonized loblolly pine lignocellulosic biomass, performed in a thermogravimetric analyzer, was carried out by Yan et al. [9]. The study revealed that pyrolysis of hydrothermally carbonized biomass progressed less aggressively than that of raw biomass which also showed that after an initial significant decomposition, the pyrolysis reactions continued at much slower rates with temperature.

The current research interest is to create a platform for deoiled karanja seed cake biomass in biorefinery by producing energy and value-added products. The study is aimed at thermal behavior of the biomass under pyrolysis conditions which can be used in the design and development of an effective pyrolysis process. Recent research [10] has indicated the potential of bio-oil and biochar obtained from pyrolysis of the seed cake biomass in areas such as energy production and the production of activated carbon.

Pongamia pinnata or Pongamia glabra known as the karanja tree in India is an evergreen tree of south and Southeast Asia. The Indian Government has initiated karanja tree plantation project in the year 2003 since then 20 million trees have been planted resulting into an annual production of 800 metric tons of karanja seeds in the year 2008 and was increased to 2,00,000 metric tons by 2011 [11]. National oilseeds and vegetable oils Development Board, Ministry of Agriculture, Government of India (2008) has showed the statistics of karanja seed kernals production per one hectare in the range 1958–4405.5 kg kernals/hectare. Nonedible seed oils (30% of entire seed weight) have been successfully processed for biodiesel production (transesterification), the remaining 70% of seed material and deoiled seed cake can be processed by a variety of treatment methods (hydrolysis and fermentation, pyrolysis, and gasification) for energy and chemicals production, which gives an added economic value to the seed cake biorefinery [12].

The aim of the present work was to study the thermal behavior and pyrolysis kinetics of the nonedible karanja seed cake with different temperature programs using thermogravimetric technique. Characteristic properties including proximate–ultimate analyses, surface structure, surface active functional groups, and higher heating value (HHV) of the raw and thermally treated biomass in a fixed bed pyrolysis reactor are evaluated. The present work is the first report on the use of the nonisothermal methods including model-fitting and model-free methods to calculate the kinetic parameters for the pyrolysis of karanja seed cake.

Material and Methods

Materials

Deoiled karanja seed cake used in the present work was obtained from a local producer in Andhra Pradesh, India. The biomass was processed to an average particle size of <1 mm and used as such for thermogravimetric analysis (TGA).

Thermogravimetric analysis

TGA of karanja seed cake was performed in a Mettler Toledo TGA/SDTA 851^e analyzer (Mettler Toledo, Switzerland). The analysis of the samples was performed at three different heating rates, 5, 10, and 20°C/min, using nitrogen with a flow rate of 30 mL/min to a final temperature of 800°C with a sample mass of ~11 ± 0.1 mg. The loss in the weight of the sample was recorded as a function of temperature.

Kinetic studies

Karanja pyrolysis kinetics was studied using the nonisothermal data obtained from different temperature programs at constant heating rate. International Confederation for Thermal Analysis and Calorimetry (ICTAC) kinetics committee recommendations for performing kinetic computations on thermal analysis data were followed to evaluate all the reliable kinetic parameters. Mathematical analysis to determine the kinetic triplet, activation energy (E), frequency factor (A), and reaction model $f(\alpha)$, was performed by the method of Coats and Redfern (model-fitting approach), and model-free isoconversional Kissinger–Akahira–Sunose (KAS) and Ozawa–Flynn–Wall (OFW) methods [13].

The conversion of biomass in the pyrolysis process is represented by the following reaction scheme:

$$\text{Biomass} \rightarrow \text{Biochar} + \text{Volatile}$$

and the kinetic expression with temperature dependence of the kinetic rate constant, given by Arrhenius equation as:

$$\frac{d\alpha}{dt} = k(T)f(\alpha) \tag{1}$$

$$k(T) = A\exp\left(\frac{-E}{RT}\right) \tag{2}$$

where $f(\alpha)$ is the reaction model, and it depends on the reaction mechanism, $k(T)$ is the rate constant, A is frequency factor, E the activation energy, T the absolute temperature, t the time, α the degree of conversion, and R the universal gas constant (8.314 J mol^{-1} K^{-1}). The constant heating rate is defined as:

$$\text{Heating rate, } \beta = \frac{dT}{dt} = \text{Constant} \tag{3}$$

$$dt = \frac{dT}{\beta} \tag{4}$$

Substituting $k(T)$ from equation (2) and dt from equation (4) in equation (1) and after rearrangement, the integration of equation (1) is given as:

$$\int_0^{\alpha} \frac{d\alpha}{f(\alpha)} \int_0^{T} A\exp\left(\frac{-E}{RT}\right)\frac{dT}{\beta} = g(\alpha) \tag{5}$$

Model-fitting approach

The solution of equation (5) based on the approximation given in Coats and Redfern [14] gives rise to the equation form expressed as Coats–Redfern equation, the most frequently used expression to evaluate nonisothermal data to compute kinetic parameters is:

$$\ln\left(\frac{g(\alpha)}{T^2}\right) = \ln\left(\frac{AR}{\beta E}\right)\left(1 - \frac{2RT}{E}\right) - \frac{E}{RT} \tag{6}$$

Conversion (α) in the pyrolysis of karanja seed cake is calculated from TGA data using the equation given below:

$$\text{Conversion, } \alpha = \frac{m_i - m_\alpha}{m_i - m_f} \tag{7}$$

where m_i, m_α, and m_f are initial, instantaneous, and final mass of the samples, respectively.

The data obtained from TGA were analyzed by different kinetic models representing chemical reaction, nucleation and nuclei growth, surface reaction between both the phases, and diffusion models [13, 15]. The TGA data in the pyrolysis temperature range were analyzed by applying 14 kinetic models given in Table 1 using Coats–Redfern equation. The kinetic model which resulted in a

Table 1. Kinetic models used in the solid state reactions.

Model	Mechanism	$f(\alpha)$	$g(\alpha)$
Chemical reaction			
1	First order	$1 - \alpha$	$[-\ln(1-\alpha)]$
2	Second order	$(1-\alpha)^2$	$(1-\alpha)^{-1} - 1$
3	Third order	$(1-\alpha)^3$	$[(1-\alpha)^{-2} - 1]/2$
4	*n*th order	$(1-\alpha)^n$	$[(1-\alpha)^{1-n} - 1]/n - 1$
Random nucleation and nuclei growth			
5	Two-dimensional	$2(1-\alpha)[-\ln(1-\alpha)]^{1/2}$	$[-\ln(1-\alpha)]^{1/2}$
6	Three-dimensional	$3(1-\alpha)[-\ln(1-\alpha)]^{2/3}$	$[-\ln(1-\alpha)]^{1/3}$
Limiting surface reaction between both phases			
7	One dimension	1	α
8	Two dimensions	$2(1-\alpha)^{1/2}$	$1-(1-\alpha)^{1/2}$
9	Three dimensions	$3(1-\alpha)^{2/3}$	$1-(1-\alpha)^{1/3}$
Diffusion			
10	One-way transport	$1/2\alpha$	α^2
11	Two-way transport	$[-\ln(1-\alpha)]^{-1}$	$\alpha+(1-\alpha)\ln(1-\alpha)$
12	Three-way transport	$(2/3)(1-\alpha)^{2/3}/[1-(1-\alpha)^{1/3}]$	$[1-(1-\alpha)^{1/3}]^2$
13	Ginstling–Brounshtein equation	$(2/3)(1-\alpha)^{1/3}/[1-(1-\alpha)^{1/3}]$	$1-2\alpha/3-(1-\alpha)^{2/3}$
14	Zhuravlev equation	$(2/3)(1-\alpha)^{5/3}/[1-(1-\alpha)^{1/3}]$	$[(1-\alpha)^{-1/3} - 1]^2$

straight line fit with high regression coefficient for ln $[g(\alpha)/T^2]$ versus $1/T$ data was concluded to be the best reaction model representing the experimental data.

Model-free approach (isoconversional methods)

A number of isoconversional methods are available for evaluating the biomass pyrolysis kinetics. These methods mainly differ in the approximations of the temperature integral in equation (5) and many of them give rise to linear equations. In equation (5) after introducing the following terms:

$$\text{Let } \frac{E}{RT} = p$$

$$dp = \frac{-E}{R} T^{-2} dT$$

$$\text{As } T \rightarrow 0;\ p \rightarrow \infty \text{ and } T \rightarrow T;\ p \rightarrow p$$

and by substituting $E/RT = p$ on rearrangement the equation is transformed to:

$$g(\alpha) = \int_0^{\alpha} \frac{d\alpha}{f(\alpha)} = \frac{A}{\beta}\frac{E}{R} \int_p^{\infty} p^{-2} \exp(-p) dp \quad (8)$$

KAS method takes the following approximation for the integral in right-side term of equation (8):

$$\int_p^{\infty} p^{-2} \exp(-p) dp = p^{-2} \exp(-p) \quad (9)$$

The solution of equation (9) based on the approximation and suitable modification given as equation (10) was used to determine the activation energy and frequency factor.

$$\ln \frac{\beta}{T^2} = \ln\left(\frac{AE}{Rg(\alpha)}\right) - \frac{E}{RT} \quad (10)$$

OFW method applies Doyle's approximation [16] for the integral term on the right side of equation (8), which becomes

$$\log\left[\int_p^{\infty} p^{-2} \exp(-p) dp\right] = -2.315 + 0.457p \quad (11)$$

By substituting equation (11) in equation (8) and on rearrangement equation (12) was obtained and was used for calculating the Arrhenius parameters:

$$\log(\beta) = \log\left[\frac{AE}{Rg(\alpha)}\right] - 2.315 - 0.457\frac{E}{RT} \quad (12)$$

From KAS method, the plot of $\ln(\beta/T^2)$ versus $1/T$ gives the slope $-E/R$. From OFW method, activation energy was calculated from the slope $(-0.457E/RT)$ of the straight line obtained by plotting $\log\beta$ and $1/T$ data. Frequency factor from the model-free isoconversional methods was determined from the intercept of the plots by making use of the best kinetic model representing the experimental data, $f(\alpha)$, evaluated from Coats–Redfern method.

Results and Discussion

TGA of karanja seed cake

The thermal behavior of karanja seed cake in the temperature range 26 ± 2°C to 800°C was studied at three different heating rates. Reduction in the mass of the sample (11 ± 0.1 mg) with temperature at a heating rate 20°C/min is shown in Figure 1. The complete pyrolysis reaction proceeds with an initial dehydration step followed by decomposition of protein, hemicellulose, cellulose, and lignin [17]. In the initial dehydration step, free/bound moisture and volatile extractives of biomass are carried away by the inert gas flow. The dehydration step occurs to a temperature of 200°C. Most of cellulose and hemicellulose decomposes above 200°C [18] and a vigorous mass loss was observed in the temperature range 250–450°C. Lignin decomposition takes place through a wide temperature range (160–900°C) [19] because of its recalcitrant structure, and in the present study a very slow mass loss after a temperature of 450°C was attributed to the lignin mass loss.

In the initial dehydration step, a 1.35% weight loss was observed as moisture was removed. In the decomposition of biomass constituents, considered as an active pyrolysis zone, a 73.8% weight loss was observed with the remaining 24.85% representing residual matter. During the pyrolysis, the intermolecular associations and weaker chemical bonds are destroyed. The side aliphatic chains may be broken and some small gaseous molecules are produced at lower temperature. At higher temperatures of decomposition, chemical bonds in lignin are broken and the parent molecular skeletons are destroyed [20, 21].

Pyrolysis kinetics

Biomass pyrolysis includes complex heat and mass transfer phenomena, therefore only about 11 mg of karanja seed cake biomass samples were used for TGA to minimize/overcome the endothermic and exothermic effects on the sample due to the furnace temperature. The thermogravimetric data were evaluated using 14 different reaction models given in Table 1. The best correlation was obtained with the third-order reaction model $[f(\alpha) = (1 - \alpha)^3]$ at three different heating rates and the regression plots according to Coats–Redfern method are shown in Figure 2A.

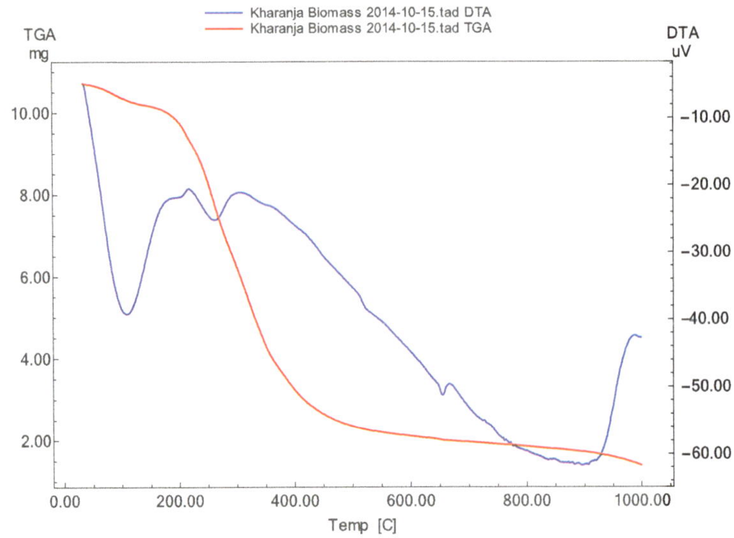

Figure 1. Thermogram of karanja seed cake at 20°C/min heating rate.

The calculation results including the values for E and A by Coats–Redfern method are listed in Table 2. The activation energy data thus obtained ranged from 82.95 to 98.21 kJ/mol. An increase in reaction kinetic parameters with the heating rate employed was observed in the present study. The obtained kinetic triplet was used to simulate the karanja seed cake pyrolysis process. A good agreement was observed between the model and experimental data as shown in Figure 2B.

The data obtained from TGA were also analyzed for isoconversional kinetics according to the KAS and the OFW methods, and standard methods for isoconversional kinetics evaluation. The regression lines obtained using KAS and OFW methods are depicted in Figure 3A and B, respectively.

Activation energies, calculated from the KAS and OFW methods, are presented in Table 3. It is evident from the table that the values obtained employing both methods are similar. Regression coefficients of the plots increased with progressive conversions. Activation energies from KAS method ranged from 122.8 kJ/mol to 106 kJ/mol for conversions ranging from $\alpha = 0.2$ to $\alpha = 0.9$. The activation energies from the OFW method were in the range 126–110 kJ/mol for the same conversions.

Activation energies decreased with an increase in conversion, as high conversions were observed at high

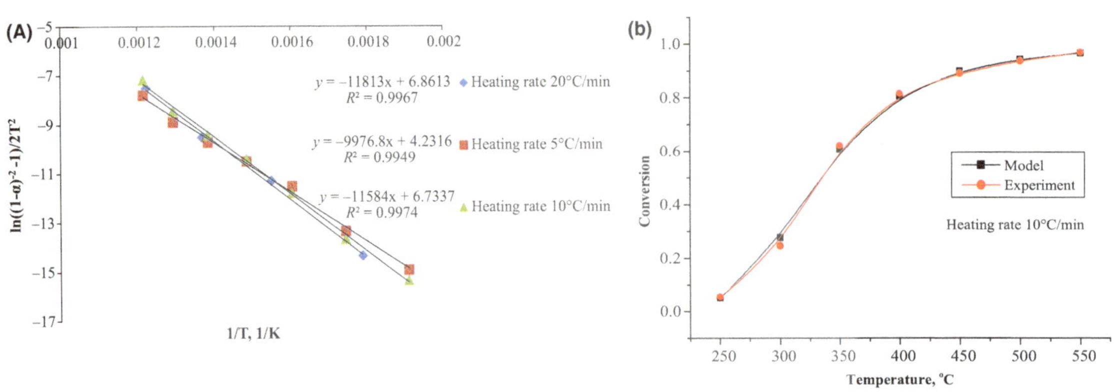

Figure 2. Coats–Redfern method. (A) Third-order reaction model fitted to thermogravimetric (TG) data at different heating rates. (B) Comparison of data from experiment (10°C/min heating rate) and model.

Table 2. Kinetic triplet for karanja seed cake pyrolysis reaction from Coats–Redfern method.

Heating rate, °C/min	Reaction model, $f(\alpha)$	Activation energy (E), kJ mol^{-1}	Frequency factor (A), min^{-1}	Correlation coefficient, R^2
5	$(1-\alpha)^3$	82.95	3.43×10^6	0.9949
10	$(1-\alpha)^3$	96.31	9.73×10^7	0.9974
20	$(1-\alpha)^3$	98.21	2.25×10^8	0.9967

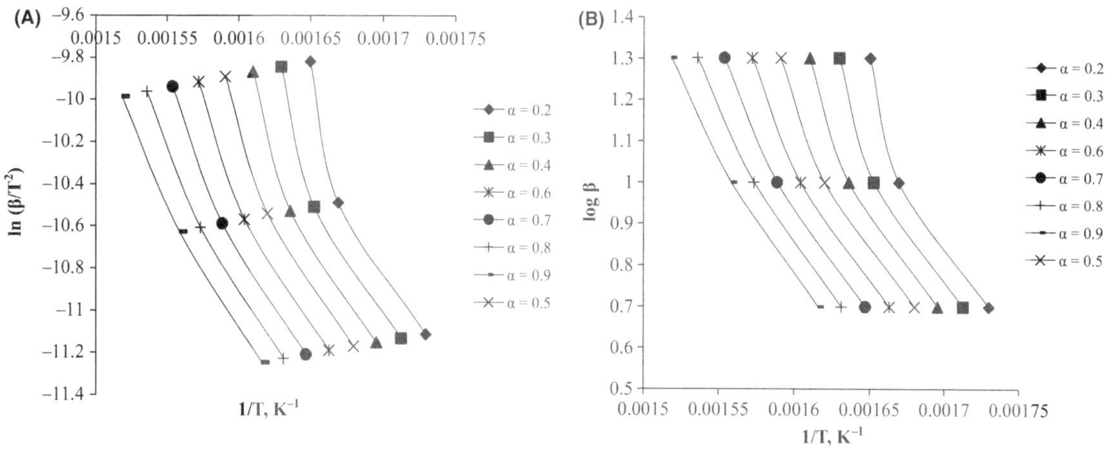

Figure 3. Regression lines for the determination of activation energy by (A) Kissinger–Akahira–Sunose (KAS) and (B) Ozawa–Flynn–Wall (OFW) methods.

Table 3. Arrhenius parameters and regression factors at different conversions from KAS and OFW methods.

	KAS method		OFW method		
Conversion, α	E_a, kJ/mol	R^2	E_a, kJ/mol	R^2	A, min^{-1}
0.2	122.81	0.907	126.03	0.919	1.75×10^{10}
0.3	120.41	0.927	123.87	0.937	1.54×10^{10}
0.4	117.92	0.943	121.61	0.951	1.28×10^{10}
0.5	114.26	0.958	117.95	0.962	8.11×10^{9}
0.6	112.97	0.967	117.10	0.972	9.23×10^{9}
0.7	108.24	0.975	114.92	0.979	9.08×10^{9}
0.8	108.24	0.982	112.83	0.985	1.15×10^{10}
0.9	106.00	0.987	110.81	0.989	2.54×10^{10}

KAS, Kissinger–Akahira–Sunose method; OFW, Ozawa–Flynn–Wall method.

temperatures, due to the requirement of comparatively lower activation energies at elevated temperatures. More reactions were triggered at higher temperatures, leading to a sharp rise in reaction rates with more unstable intermediates and lower activation energies. The reported values of pyrolysis reaction activation energies for hazelnut husk are in the range 128–131 kJ/mol [22], 129 kJ/mol for corn cob [23], and 62–206 kJ/mol for pine wood waste [24]; the activation energy of pyrolysis of karanja seed cake are in the range of the hazelnut husk and corn cob, but there was considerable deviation from the reported activation energy of pine wood biomass. The activation energy of pyrolysis is highly dependent on the extent of conversion and the type and composition of biomass feedstock as seen in the activation energy for pine wood waste which is a woody biomass, rich in lignin, and very different from that of seed cake biomass and nut husk and other biomass types. KAS and OFW methods gave reliable activation energies that were used for the calculation of the pre-exponential factor in order to get a complete picture of the kinetic parameters with progressive conversions, the pre-exponential factors were determined by making use of the reaction model obtained from Coats-Redfern (CRF) method. Calculated pre-exponential factor values from OFW method are shown in Table 3. Therefore, the kinetic expression obtained for pyrolysis of karanja seed cake was:

$$\frac{d\alpha}{dt} = Ae^{-\frac{E}{RT}}(1-\alpha)^3 \quad (13)$$

where the Arrhenius parameters obtained from model-free isoconversional methods were in good agreement with the values determined using model-fitting method.

Obtaining of the pyrolysis kinetic parameters based on TGA experimental data is extremely useful in the design and control of pyrolysis process especially large scale operation with desired composition and yield of products. Moreover, pyrolysis kinetic data link technological parameters, importantly biomass hold up time at preset temperature program that determines pyrolysis products profile.

Characterization of thermally treated biomass

Thermal treatment of karanja seed cake was performed in a fixed bed reactor of 1 cm inner diameter and 35 cm length under slow pyrolysis conditions with a continuous supply of nitrogen gas at 0.1 liter per minute (LPM) flow at different temperatures including 200, 250, 300, 350, 400, 450, and 500°C. The pyrolysis transformed the karanja seed cake to biochar (thermally treated biomass retained in the reactor), bio-oil (condensed liquid product), and the gases. The solid and the liquid obtained were weighed and the yields of the same were expressed gravimetrically. Biochar was characterized by proximate analysis carried out according to ASTM standard methods [25] using Mettler Toledo TGA/SDTA 851^e analyzer (Switzerland), ultimate/elemental analysis using a CHNS Analyzer–ELEMENTARVario microcube model, surface structure was obtained by scanning electron microscopy (SEM) using Hitachi S-3000N scanning electron microscope, Japan and FTIR (Fourier transform infrared spectroscopy) using a Perkin Elmer System 100 Fourier transform infrared spectrometer (Perkin Elmer Ltd., Seer Green, Beaconsfield, Bucks HP9 2FX, United Kingdom) [26].

Pyrolysis temperature showed a profound influence on the product distribution in the reaction. Figure 4 shows the pyrolysis products, biochar, and total volatiles and yields obtained in relation to temperature. Pyrolysis at higher temperature (≥500°C) resulted in increased carbon conversion into volatiles [27, 28] and reduced biochar.

Karanja seed cake, processed in the present work, composed of 25.2% carbohydrates, 17.4% lignin, 13.4% protein, and 4% ash with 40% water and ethanol soluble matter, details of which were presented in our earlier work [29]. Biochar obtained from the thermal treatment of karanja seed cake was analyzed for its composition in terms of proximate and ultimate analysis and the data were presented in Table 4 along with evaluated HHV, carbon carryover, H/C and O/C ratios. From the elemental analysis of feed karanja seed cake and biochar at different temperatures it was observed that the carbon content increased, whereas the hydrogen and oxygen contents decreased with increase in temperature. H/C and O/C values give a picture of the degree of carbonization, aromaticity, and maturity of organic materials. A decrease in oxygen (O/C) with increase in temperature was due to the increased decomposition of oxygen containing functional groups. The percent biochar obtained from the seed cake by experimentation was 56.82%, 53.29%, 52.55%, and 51.52% at 350, 400, 450, and 500°C, respectively, indicating a decrease in the carbon carryover from the seed cake to biochar with temperature. Therefore, the value addition of deoiled karanja seed cake by thermal treatment is a promising pretreatment method to obtain biochar for multiple applications.

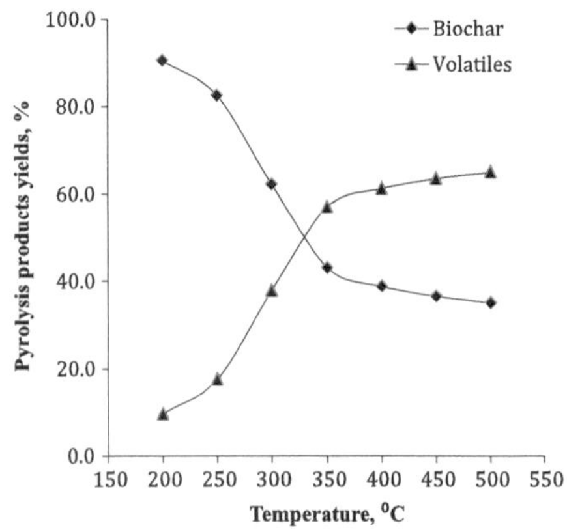

Figure 4. Effect of temperature on biochar and total volatiles yield from pyrolysis.

Scanning electron microscopy

As porosity play a vital role in determining the efficiency of biosorbents in the adsorption of contaminants in effluent treatment, with maximum adsorption occurring when pores are large enough to admit contaminant molecules, SEM of karanja cake and biochar was carried out to assess the surface structure of the samples. The SEM pictures obtained are presented in Figure 5A and B. As is evident from the pictures, the char formed (Fig. 5B) has a highly porous structure compared to the original seed cake biomass (Fig. 5A). This can be attributed to the fact that thermal treatment improves the porous structure of biochar due to the loss of hydrogen, oxygen (major quantity), and carbon (minor quantity) atoms in the form of volatiles from the parent biomass material leaving the skeletal structure.

Table 4. Composition and energy value of biochar obtained from pyrolysis of karanja seed cake at different temperatures.

		Biochar from pyrolysis at			
Element, %wt.	Karanja cake	350°C	400°C	450°C	500°C
Proximate analysis					
Moisture	06.7				03.20
Ash	04.00				15.42
Volatile matter	71.44				14.13
Fixed carbon	17.86				67.25
Ultimate analysis					
N	04.30	05.84	05.91	06.07	05.95
C	44.20	58.39	60.80	63.64	65.06
H	06.50	05.21	04.45	04.61	04.30
S	–	–	–	–	–
O	45.00	30.56	28.84	25.68	24.69
HHV, MJ/kg	18.11	22.9	23.14	24.23	24.5
H/C	1.76	1.07	0.91	0.83	0.80
O/C	0.68	0.35	0.31	0.27	0.26
Carbon carryover[1], %	–	56.82	53.29	52.55	51.52

[1](Carbon content in biochar × Biochar yield)/(Carbon content in karanja seed cake).

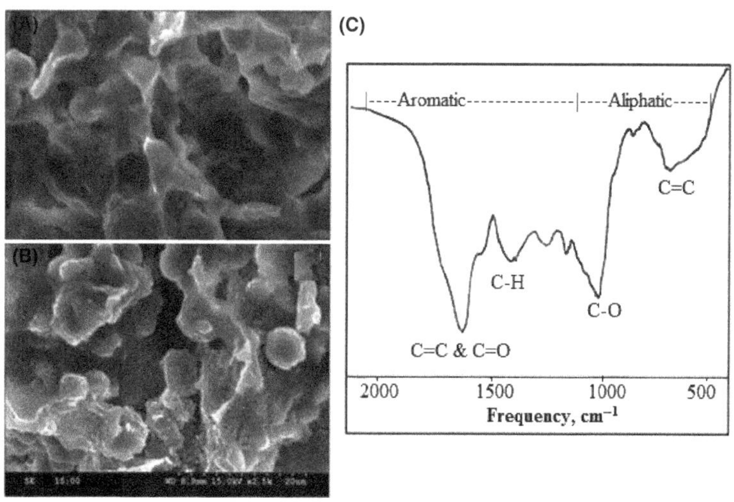

Figure 5. Surface structure (from scanning electron microscopy [SEM]) of (A) karanja seed cake biomass and (B) biochar from karanja seed cake, and (C) Fourier transform infrared spectra of biochar obtained at 500°C.

Fourier transform infrared spectroscopy

Fourier transform infrared spectra of biochar was collected using the KBr pelletization method on a Perkin Elmer System 100 Fourier transform infrared spectrometer (FTIR) with a cesium iodide (CsI) micro-focus accessory and triglycine sulfate (TGS) detector by scanning from 4000 to 400 cm^{-1}, with a resolution of 4 cm^{-1}. The biochar spectra as shown in Figure 5C depicts the vibrations in the frequency range 700–1200 cm^{-1} and 1200–2000 cm^{-1} confirming the aliphatic and aromatic nature of the sample. The peaks at frequencies of 698 (aromatic ring C-H), 1020 (C-O stretching vibration), 1420 (C-H stretching vibration), and 1626 cm^{-1} (C=C and C=O stretching vibrations) show the carbon structure of biochar. This spectrum is comparable to reference spectra obtained by Rutherford et al. [26] who studied changes in composition and porosity occurring during the thermal degradation of wood and wood components. The result further emphasizes the fact that thermal treatment of biomass converts aliphatic carbon to aromatic on prolonged treatment [30].

From the characterization of the seed cake biomass and biochar based on the elemental composition, surface structure, and FTIR, it is evident that the biochar is rich in carbon and has porous structure and can be effectively used for energy generation and as biosorbent for industrial effluent treatment and electrode material in activated carbon electrodes manufacture.

Conclusions

This study is the first report on nonedible seed cake biomass pyrolysis kinetics in particular karanja seed cake biomass, and the work presented is extremely helpful to understand the usefulness of value addition to karanja seed cake through thermal treatment. The TGA pyrolysis reactions at constant heating rate revealed that a temperature of 450–500°C was sufficient for carbonization of the biomass. The KAS, OFW, and CRF methods gave reliable activation energies and pre-exponential factors with third-order reaction model for the seed cake pyrolysis reaction. Characterization results showed that the thermally treated biomass has high porous structure and rich in carbon, suitable for various applications as biosorbent and precursor for carbon materials. The kinetic data thus obtained are extremely useful in design of pyrolysis system for different seed cake biomass.

Conflict of Interest

None declared.

References

1. Song, X. D., X. Y. Xue, D. Z. Chen, P. J. He, and X. H. Dai. 2014. Application of biochar from sewage sludge to plant cultivation: influence of pyrolysis temperature and biochar to soil ratio on yield and heavy metal accumulation. Chemosphere 109:213–220.
2. Suddick, E. C., and J. Six. 2013. An estimation of annual nitrous oxide emissions and soil quality following the amendment of high temperature walnut shell biochar and compost to a small scale vegetable crop rotation. Sci. Total Environ. 465:298–307.
3. Watanabe, A., I. KosukeIkeya, N. Kanazaki, S. Makabe, Y. Sugiura, and A. Shibata. 2014. Five crop season's records of greenhouse gas fluxes from upland fields with repetitive applications of biochar and cattle manure. J. Environ. Manage. 144:168–175.
4. Creamer, E. A., B. Gao, and M. Zhang. 2014. Carbon dioxide capture using biochar produced from sugarcane bagasse and hickory wood. Chem. Eng. J. 249:174–179.
5. Jin, H., S. Carapeda, Z. Chang, J. Gao, Y. Xu, and J. Zhang. 2014. Biocharpyrolytically produced from municipal solid wastes for aqueous As(V) removal: adsorption property and its improvement with KOH activation. Bioresour. Technol. 169:622–629.
6. Xu, D., Y. Zhao, K. Sun, B. Gao, Z. Wang, J. Jin, et al. 2014. Cadmium adsorption on plant- and manure-derived biochar and biochar amended sandy soils: impact of bulk and surface properties. Chemosphere 111:320–326.
7. Parthasarathy, P., and K. N. Sheeba. 2014. Determination of kinetic parameters of biomass samples using thermogravimetric analysis. Environ. Prog. Sustain. Energy 33:256–266.
8. Vhathvarothai, N., J. Ness, and Q. J. Yu. 2014. An investigation of thermal behaviour of biomass and coal during copyrolysis using thermogravimetric analysis. Int. J. Energy Res. 38:1145–1154.
9. Yan, W., S. Islam, C. J. Coronella, and V. R. Vásquez. 2012. Pyrolysis kinetics of raw/hydrothermally carbonized lignocellulosic biomass. Environ. Prog. Sustain. Energy 31:200–204.
10. Rebitanim, N. Z., W. A. W. Ab Karim Ghani, N. A. Rebitanim, and M. A. Mohd Salleh. 2013. Potential applications of wastes from energy generation particularly biochar in Malaysia. Renew. Sustain. Energy Rev. 21:694–702.
11. Gaurav, D., J. Siddarth, and P. S. Mahendra. 2011. Pongamia as a source of biodiesel in India. Smart Grid Renew. Energy 2:184–189.
12. Ramachandran, S., K. S. Sudheer, L. Christian, C. R. Soccol, and A. Pandey. 2007. Oil cakes and their biotechnological applications – a review. Bioresour. Technol. 98:2000–2009.
13. Vyazovkin, S., and C. A. Wight. 1999. Model-free and model-fitting approaches to kinetic analysis of isothermal and nonisothermal data. Thermochim. Acta 340–341:53–68.
14. Coats, A. W., and J. P. Redfern. 1964. Kinetic parameters from thermogravimetric data. Nature 201:68.
15. Vlaev, L. T., I. G. Markovska, L. A. Lyubchev. 2003. Non-isothermal kinetics of pyrolysis of rice husk. Thermochim. Acta 406:1–7.
16. Doyle, C. D. 1965. Series approximations to the equations of thermogravimetric data. Nature 207:290–291.
17. Kok, M. V., and E. Ozgur. 2013. Thermal analysis and kinetics of biomass samples. Fuel Process. Technol. 106:739–743.
18. Carrier, M., A. L. Serani, D. Denux, J. M. Lasnier, F. H. Pichavant, F. Cansell, et al. 2011. Thermogravimetric analysis as a new method to determine the lignocellulosic composition of biomass. Biomass Bioenergy 35:298–307.
19. Burhenne, L., J. Messmer, T. Aicher, and M. P. Laborie. 2013. The effect of the biomass components lignin, cellulose and hemicellulose on TGA and fixed bed pyrolysis. J. Anal. Appl. Pyrol. 101:177–184.

20. Raman, P., W. P. Walawender, L. T. Fan, and J. A. Howell. 1981. Thermogravimetric analysis of biomass, devolatilization studies on feedlot manure. Ind. Eng. Chem. Process Des. Dev. 20:630–636.
21. Singh, R. K., and K. P. Shadangi. 2011. Liquid fuel from castor seeds by pyrolysis. Fuel 90:2538–2544.
22. Ceylan, S., and Y. Topcu. 2014. Pyrolysis kinetics of hazelnut husk using thermogravimetric analysis. Bioresour. Technol. 156:182–188.
23. Gai, C., Y. Dong, and T. Zhang. 2013. The kinetic analysis of the pyrolysis of agricultural residue under non-isothermal conditions. Bioresour. Technol. 127:298–305.
24. Amutio, M., and G. Lopez. 2012. Kinetic study of lignocellulosic biomass oxidative pyrolysis. Fuel 95:305–311.
25. Cantrell, K. B., Martin, J. H., and Ro, K. S. 2010. Application of thermogravimetric analysis for the proximate analysis of livestock wastes. J. ASTM Int. 7: 1–13
26. Rutherford, D. W., L. Robert Wershaw, and L. G. Cox. 2004. Changes in composition and porosity occurring during the thermal degradation of wood and wood components. U.S. Geol. Sci. Invest. Rep. 5292:79.
27. Antony, R. S., D. S. Smart Robin Son, B. C. Pillai, and C. Lee Robert Lindon. 2011. Parametric studies on pyrolysis of pungam oil cake in electrically heated fluidized bed research reactor. Res. J. Chem. Sci. 1:70–80.
28. Singh, R. C., R. Kataki, and T. Bhaskar. 2014. Characterization of liquid and solid product from pyrolysis of pongamia glabra deoiled cake. Bioresour. Technol. 165:336–342.
29. Radhakumari, M., A. S. Ball, K. Suresh Bhargava, and B. Satyavathi. 2014. Optimization of glucose formation in karanja biomass hydrolysis using Taguchi robust method. Bioresour. Technol. 166:534–540.
30. Rutherford, D. W., L. RobertWershaw, E. ColleenRostad, and N. CharleneKelly. 2012. Effect of formation conditions on biochars: compositional and structural properties of cellulose, lignin, and pine biochars. Biomass and Bioenergy 46:693–701.

17

Fuzzy ARTMAP and GARCH-based hybrid model aided with wavelet transform for short-term electricity load forecasting

Swati Takiyar[1], K. G. Upadhyay[1] & Vivek Singh[2]

[1]M.M.M. University of Technology, Gorakhpur, Uttar Pradesh, India
[2]National Hydroelectric Power Corporation Ltd., Faridabad, Haryana, India

Keywords
Artificial neural network, electricity load forecasting, fuzzy ARTMAP, GARCH, wavelet transform

Correspondence
Swati Takiyar, M.M.M. University of Technology, Gorakhpur, Uttar Pradesh, India.

E-mail:takiyar.s@gmail.com

Funding Information
Accepted under 100% fee waiver anniversary edition.

Abstract

With the evolution of the electricity market into a restructured smart version, load forecasting has emerged as an eminent research domain. Many forecasting models have been proposed by researchers for electricity price and load forecasting. This state of art introduces a load time series modeled with a hybrid technique culminating from the logical amalgamation of GARCH, a conventional hard computing method, Fuzzy ARTMAP, an artificial intelligence-based soft computing technique, and wavelet transform, for treating the load time series. The study investigates into the ability of the proposed hybrid model in tackling the electricity load time series forecasting problems. The work under this study also includes comparisons drawn among models which use either one or two of the mentioned techniques and the model proposed. Results certify the efficacy and effectiveness of the model over others.

Introduction

Forecasting [1, 2] knowingly or unknowingly holds an integral stature in every person's life. We often predict stock market, earthquake and weather in our day to day life. A business manager forecasts product sales [3]. People make future plans based on these forecasts. Here it should be understood that it is an impossible task to make exact forecasts, we can only work incessantly towards attaining higher accuracies. Generally forecasting presumes that future occurrences depend upon past or present observable events; it assumes that some aspects of the past pattern will continue in the future. Through observing and studying past data relationships can be established between the event and the parameters persisting at the then moment [4].

Forecasting with load time series is a challenging application because load time series are inherently nonstationary [5], deterministically chaotic, and highly noisy by nature. Above this by no technique the past information can solely determine the futuristic behavior of the electricity market. Therefore, to maintain global competitiveness, market dependency on advance computer technologies is increasing day by day.

Forecasting can either be spatial or temporal in nature. While spatial forecasts are based on area covered, temporal are based on time horizon utilized. Temporal forecasting can further be divided into three subcategories: Short Term, Medium Term and Long Term [6]. Out of these, short-term forecasting is of most importance because of its utility. They help plan capacity building, estimate load flows and prevent overloading to name a few. Forecasting techniques primarily use either statistical tools [7, 8] or artificial intelligence based algorithms [9].

In this study, we present a new method of forecasting electricity loads using Fuzzy ARTMAP (FA) and GARCH along with wavelet transform (WT) for day-ahead forecasting. The approach implemented aims to develop a sturdy, precise and efficient day-ahead load forecasting tool utilizing data filtering technique, employing WT, fused with a computational model implementing conjointly FA (soft computing model), and GARCH (hard computing model). Comparison results of the proposed models' performance with that utilizing only FA and utilizing both FA and GARCH show a convincing reduction in mean absolute percentage error (MAPE). A simple artificial neural network (ANN) model utilizing the same data has also been implemented simply to draw a wider range comparison and highlight the supremacy of the hybrid model over ANN.

The procedure of forecasting proposed can be summed in three steps:

1. Decomposition of past load series using WT.
2. Using FA model fitted to one approximated and two decomposed series of WT and using GARCH on one decomposed series of WT.
3. Then using inverse WT to reconstruct the forecasted load series.

The literature of the study is arranged as follows:

Section II details WT, FA, and GARCH techniques in brief to enhance reader understanding.
Section III elaborates on the implemented hybrid methodology.
Section IV comprises numerical and graphical results and Section V concluding the study.

Any load data series embodies numerous spikes, nonlinearities, and fluctuations. The 2010 hourly New South Wales electricity load series is no exception to it. As illustrated in Figure 1, it is also characterized by chaotic and random changes. This series of 2148 h has been utilized in this study. Where the hourly data of 92 days or 3 months has been used to train the network and the data of next 24 h has been reserved as validation set data and is also the predicted subset.

Wavelet Transform

Wavelet transform is a mathematical model which transforms the original load series (in time domain) into constituent subseries over time domain of a different scale for processing and analysis. WT is most suitable for the nonstationary data (mean and autocorrelation of series are not constant). It is also well known that most of the load data series are nonstationary, hence the utility of WT [10]. The WT is used to decompose the original load series into several other series with resolution of different levels, which is called multiresolution decomposition [11, 12].

Fourier transform (FT) decomposes the original load series into linear combinations as sine and cosine functions whereas by WT the series is decomposed into a sum of more flexible functions which are localized in both time and frequency [13].

Wavelet transform can be classified into two: continuous wavelet transform (CWT) and discrete wavelet transform (DWT).

The CWT of a continuous time signal $x(t)$ is defined as [4, 6]:

$$\psi_{a,b}(t) = \int_{-\infty}^{\infty} x(t)\psi^*_{a,b}(t)dt, \tag{1}$$

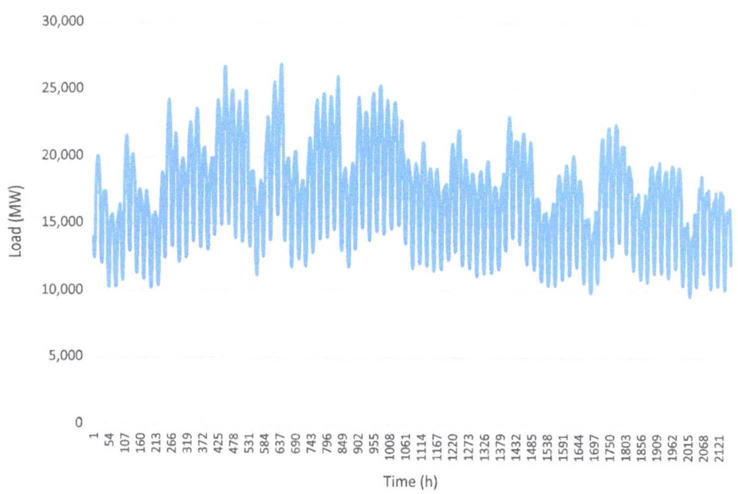

Figure 1. Load data of New South Wales.

where $\psi(t)$ is the mother wavelet, given by eq. (2) where a acts as a scaling parameter and b as a translating parameter.

$$\psi^*_{a,b}(t) = \frac{1}{\sqrt{a}}\psi^*\left(\frac{t-b}{a}\right). \quad (2)$$

Each wavelet is formulated by scaling and translating the mother wavelet. The mother wavelet is an oscillatory function characterized by zero average and finite energy.

$$\psi_{c,d} = \sum_n x(n)\psi^*_{c,d}(n), \quad (3)$$

where

$$\psi^*_{c,d}(n) = \frac{1}{\sqrt{a^c_o}}\psi^*\left(\frac{n-db_o a^c_o}{a^c_o}\right). \quad (4)$$

Here, c acts as a scaling coefficient while d acts as a sampling one.

To implement DWT as a filter, Mallat propounded an algorithm called Mallat multiresolution analysis or the Mallat algorithm [12]. It is a two-staged algorithm where decomposition occurs in the first stage followed by reconstruction in the second one. This study implements a three-level decomposition on the original load series yielding three detailed series (D) and one approximated series (A) as illustrated in Figure 2. Decomposing and reconstructing processes both involve filtering for which both high-pass (HPF) and low-pass filters (LPF) are utilized. While down-sampling occurs during wavelet decomposition, up-sampling and filtering is used in wavelet reconstruction. A Daubenchies wavelet function of order 5 (db5) has been utilized in this study as a mother wavelet.

Fuzzy ARTMAP

The FA network is a supervised learning method based on adaptive resonance theory (ART) [14]. FA network carries out learning without forgetting previously learned information [15]. FA is flexible and adaptive to changes in the environment and is self-organizing by nature [16]. FA network is a recent technique that has been utilized in forecasting applications including load forecasting.

Neural network is another popular artificial intelligence technique utilized in forecasting applications. Most neural networks struggle with the plasticity–stability dilemma which probes into ways by which a network can endure adaptive-ness or plasticity toward new inputs while staying aloof of the noisy data inputs, hence stability [17, 18]. A general neural network encounters hindrance in preserving previously learned knowledge while learning newer concepts. The FA confronts this dilemma with a feedback mechanism laid between the competitive and input layers to allow fresh concepts to be absorbed without losing the knowledge attained previously. This results in a firmer learning environment endowed with faster convergence capability compared to traditional soft computing techniques [19]. This is also confirmed by the results of this study. These properties of FA can improve load forecasting performance as load series data are highly stochastic by nature (Figure 1).

The functional layout of FA network is shown in Figure 3. An ARTMAP system embodies twin art modules (ART_a and ART_b) to fabricate stable recognition categories corresponding to the arbitrary input patterns. ART_a uses ART-1, a type of ART network which accepts only binary input, while ART_b uses FUZZY ART. This setup enables to switch the binary module notations into a corresponding feature in the fuzzy ART module. For example, the

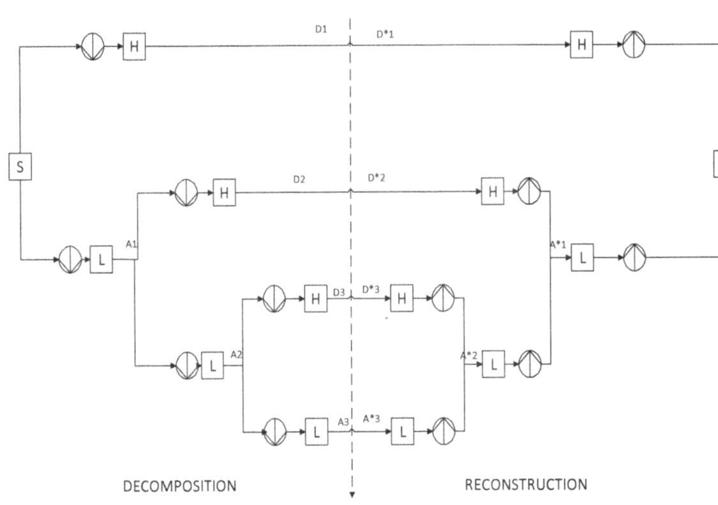

Figure 2. Wavelet transformation (order 3) [23, 28].

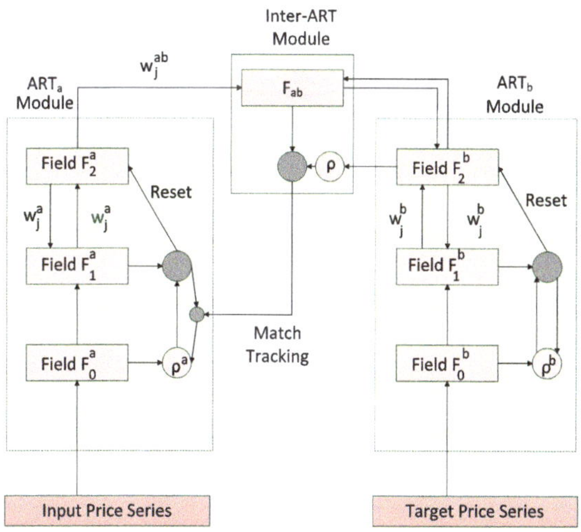

Figure 3. Functional layout of Fuzzy ARTMAP [15, 20, 28].

intersection operator (^) of ART_1 is replaced by the operator (^) in FUZZY ART. The architecture called FA is achieved by the synthesis of fuzzy logic and ART neural network, employing a close formal similarity between two computations of fuzzy subsets and ART category. Also, FA actualizes a new min–max learning rule that collectively minimizes predictive error and maximizes generalization, or code compression. This is achieved by a match tracking process that increases the ART vigilance parameter by the minimum amount needed to correct a predictive error. As a result, the system automatically learns a minimal number of recognition categories, or "hidden units," to meet the criteria of accuracy. Category proliferation is prevented by normalizing input vectors at a preprocessing stage. A normalization procedure called complement coding [15] leads to a symmetric theory in which the AND operator (^) and the OR operator (v) of fuzzy logic plays complementary roles. In training, the best matching category is [19]:

$$J = \arg\max_{0 \le j \le N} T_j(I_{tr}), \tag{5}$$

where

$$T_j(I_{tr}) = \begin{cases} \frac{|I_{tr} \wedge W_j|}{\alpha + |W_j|}, & \text{If } \frac{|I_{tr} \wedge W_j|}{|I_{tr}|} \ge \rho \\ 0, & \text{Otherwise} \end{cases} \tag{6}$$

where T_j = choice function, α = choice parameter, ^ = Fuzzy MIN operator, ρ = vigilance parameter, and $\frac{|I_{tr} \wedge W_j|}{|I_{tr}|} \ge \rho$ is the vigilance criteria. If vigilance criteria satisfy, then resonance occurs. During training, the vigilance criteria vary from baseline vigilance which is the initial value. If vigilance criteria qualify, then category J becomes representative membership function for time series, and the weighing vector of the winning category W_j is updated as per the following equation:

$$W_j^{(\text{new})} = \beta(I_{tr} \wedge W_j^{(\text{old})}) + (1-\beta)W_j^{(\text{old})}. \tag{7}$$

Here β represents the learning rate. If vigilance criteria fail, then category J is deactivated for the present load series by equating choice function equals to zero. If ART_b does not predict the correct output for ART_a, then the vigilance parameter is increased. This is called match tracking, in which the value of the vigilance parameter is slightly increased to a new value [17]:

$$\rho = \frac{|I_{tr} \wedge W_j|}{|I_{tr}|} + \varepsilon, \tag{8}$$

where ε denotes the learning precision.

The scheme resizes a category on predictive success by amplifying the vigilance parameter ρ by a minimal amount essential to verify the predictive error in ART_b. The parameter ρ holds an inverse relationship with the category size. A lower value leads to a broadly generalized category with higher compressed code. This parameter rates the minimum faith that ART_a should have while accepting a category during hypothesis testing which focuses ART_a on a new cluster. The failures at ART_a increase ρ to that threshold value which in turn triggers ART_a under a process called match tracking. This technique reduces generalization essential to correct a predictive error. The combination of these techniques, i.e. ARTMAP

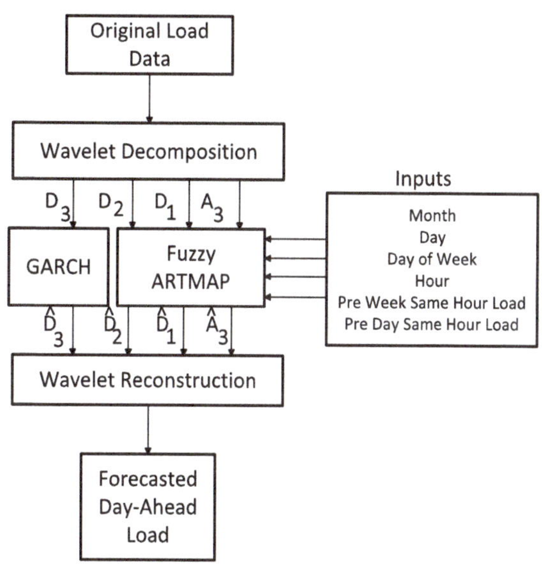

Figure 4. Schematic diagram of the proposed hybrid model.

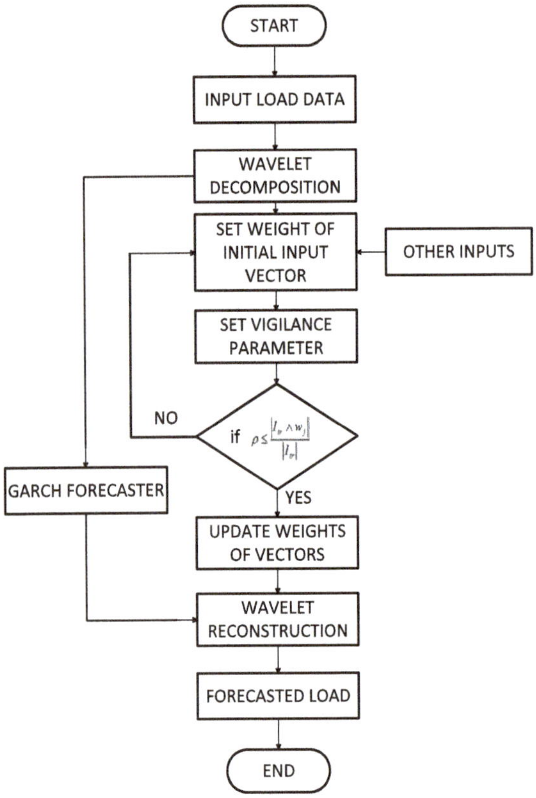

Figure 5. Flowchart for the proposed method for day-ahead load forecasting.

and match tracking leads to a faster learning and erudition from a rare event. The fuzzy ART reduces to ART_1 for a binary input and works as self for a binary input and works as self for an analog vector. Thus the crisp logics of ART_1 with their fuzzy counterparts form a potent module.

Once the training stage is completed, the FA network is used as a classifier of the input load series which is given to ART_a. ART_b is not used during classification process and the learning capability of the network is deactivated during classifying process (i.e. $\beta = 0$). In this stage we get predicted classified labels in the output of ARTMAP. These classified labels are later de-fuzzified to get the forecasted loads.

GARCH

GARCH stands for Generalized Autoregressive Conditional Hetero-skedasticity which is used to model observed time series. GARCH is effectively implemented to highly volatile time series caused by unexpected random effects [20, 21]. The model GARCH (p, q) is defined as:

$$x_t = \mu + \varepsilon_t, \tag{9}$$

where μ is offset and $\varepsilon_t = \sigma_t z_t$.

Considering a time series x_t with a constant mean offset [4]:

$$\sigma_t^2 = c + \sum_{i=1}^{q} \phi_i \varepsilon_{t-i}^2 + \sum_{i=1}^{p} \psi_i \sigma_{t-i}^2, \tag{10}$$

where p is the order of GARCH terms σ^2 and q is the order of ARCH terms ε^2.

As can be seen in eq. (10), in GARCH (p, q) model is $p = 0$, i.e. a GARCH (0, q) model becomes an ARCH (q) model.

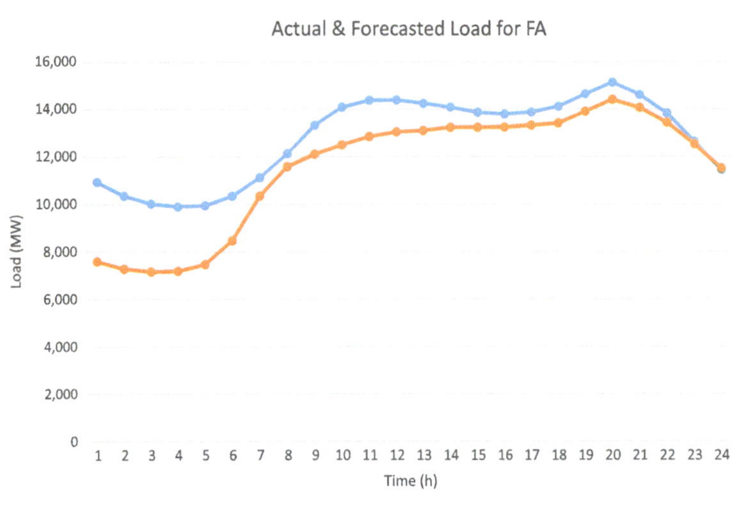

Figure 6. Actual and forecasted load for FA.
FA, Fuzzy ARTMAP.

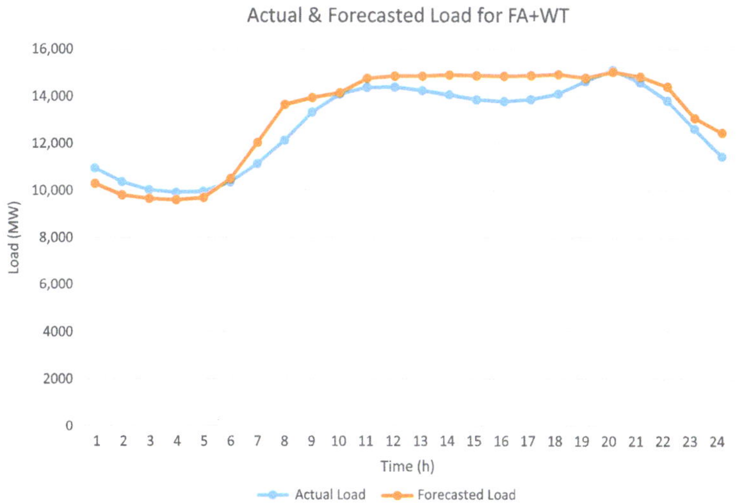

Figure 7. Actual and forecasted load for FA + WT. FA, Fuzzy ARTMAP; WT, wavelet transform.

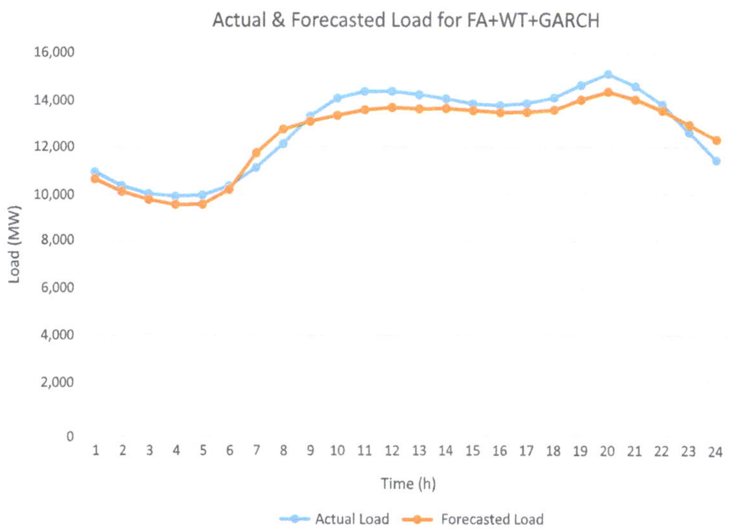

Figure 8. Actual and forecasted load for FA + WT + GARCH. FA, Fuzzy ARTMAP; WT, wavelet transform.

A limitation of the GARCH model is that it can only be specified for stationary time series, hence the below equation must be satisfied for stationary time series only:

$$\sum_{i=1}^{q} \phi_i + \sum_{i=1}^{p} \psi_i < 1. \tag{11}$$

Steps for GARCH modeling [21, 22]:

1. Identify the class of models.
2. Identify the subsets of GARCH models.
3. Parameter estimation.
4. Model validation.
5. Apply forecast.

Proposed Methodology

Figure 4 presents the schematic diagram of the prospective hybrid model for day-ahead electricity load forecasting built on the FA technique combined with GARCH and WT. The procedure for forecasting is as follows:

1. In the proposed hybrid model, input variables are hourly data of electricity load, month, day, day of week, hour, previous week same hour load and previous day same hour load. Only the load series is passed through WT. The load series is decomposed through WT into four components. The decomposed detailed coefficients,

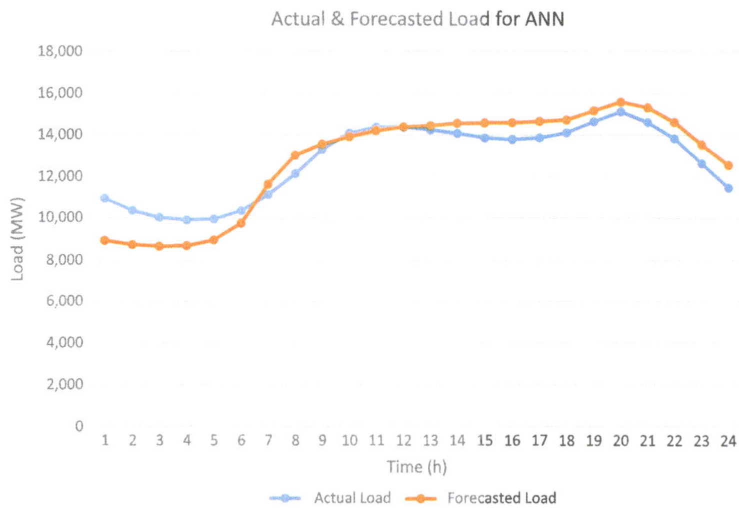

Figure 9. Actual and forecasted load for artificial neural network.

Table 1. Comparative hourly forecasts for respective techniques.

Hours	Actual load	FA		FA + WT		FA + WT + GARCH		ANN	
		Forecasted load	MAPE	Forecasted load	MAPE	Forecasted load	MAPE	Forecasted load	MAPE
1	10940	7599.86	30.53	10304.45	5.81	10654.79	2.61	8945.86	18.23
2	10357	7285.57	29.66	9810.04	5.28	10124.10	2.25	8740.28	15.61
3	10020	7171.58	28.43	9668.12	3.51	9777.74	2.42	8660.86	13.56
4	9906	7190.78	27.41	9607.85	3.01	9566.39	3.43	8691.42	12.26
5	9950	7476.25	24.86	9700.96	2.50	9575.59	3.76	8961.52	9.93
6	10347	8463.64	18.20	10508.24	−1.56	10209.69	1.33	9762.46	5.65
7	11119	10358.68	6.84	12054.61	−8.41	11765.89	−5.82	11636.54	−4.65
8	12130	11593.56	4.42	13668.34	−12.68	12787.44	−5.42	13048.84	−7.57
9	13322	12118.27	9.04	13958.09	−4.77	13122.40	1.50	13583.11	−1.96
10	14081	12509.26	11.16	14160.53	−0.56	13374.91	5.01	13937.94	1.02
11	14369	12851.78	10.56	14778.85	−2.85	13611.55	5.27	14231.49	0.96
12	14376	13049.30	9.23	14868.05	−3.42	13708.11	4.65	14402.53	−0.18
13	14241	13104.27	7.98	14868.00	−4.40	13651.71	4.14	14458.80	−1.53
14	14069	13240.01	5.89	14914.21	−6.01	13668.93	2.84	14580.21	−3.63
15	13852	13243.24	4.39	14887.22	−7.47	13578.73	1.97	14589.42	−5.32
16	13783	13252.21	3.85	14864.27	−7.84	13503.08	2.03	14596.21	−5.90
17	13864	13331.54	3.84	14892.18	−7.42	13519.43	2.49	14663.60	−5.77
18	14107	13415.88	4.90	14943.05	−5.93	13601.57	3.58	14734.08	−4.45
19	14633	13908.10	4.95	14796.38	−1.12	14039.43	4.06	15180.97	−3.74
20	15117	14411.75	4.67	15048.74	0.45	14379.41	4.88	15598.05	−3.18
21	14595	14069.20	3.60	14845.71	−1.72	14042.48	3.79	15325.88	−5.01
22	13813	13441.37	2.69	14420.63	−4.40	13572.74	1.74	14628.50	−5.90
23	12623	12529.19	0.74	13097.21	−3.76	12974.20	−2.78	13556.04	−7.39
24	11446	11518.88	−0.64	12474.47	−8.99	12333.07	−7.75	12557.34	−9.71
Overall MAPE			10.77		4.74	3.56			6.38

ANN, artificial neural network; FA, Fuzzy ARTMAP; MAPE, mean absolute percentage error; WT, wavelet transform.

D1, D2, D3 (high frequency components) and approximated series, A3 (low frequency components) are then obtained by down sampling with HPF and LPF, respectively.

2. The characteristic decomposed load series (D1, D2, and A3) along with other input data are furnished into the FA network and decomposed load series D3 is fed into GARCH.

Table 2. Mean absolute percentage error comparison of various models for load.

MAPE of various models			
ANN	FA	FA + WT	FA + WT + GARCH
6.381%	10.772%	4.745%	3.563%

ANN, artificial neural network; FA, Fuzzy ARTMAP; MAPE, mean absolute percentage error; WT, wavelet transform.

3. The output components of GARCH, FA network, and the decomposed detailed and approximate series are then processed by wavelet reconstruction to produce the day-ahead electricity loads.

This step-by-step summary is shown in Figure 5.

Numerical and Graphical Results

The paper introduces a new hybrid algorithm based on WT, FA, and GARCH, which accounts for the interactions of month, day, day of week, hour, previous week same hour load, and previous day same hour load. The proposed method has worked on the electricity load data of New South Wales. To rank the performance of the proposed model, the results have been compared to other models such as FA, FA + WT, and the most employed artificial intelligence technique, ANN. The summary table of this is illustrated in the Conclusion section.

Above this the outputs of the mentioned models have been tabulated below followed by the respective graphs comparing the forecasted and actual data.

Before one takes a look at these, we have briefed how they have been tabulated and why they serve as an appropriate measure of the efficiency of any forecasting model. Error is defined as the difference between the actual value and the forecasted value for the corresponding period [10, 23–26].

$$\varepsilon_t = A_t - F_t, \tag{12}$$

where ε_t is the error for the period t, A_t is the actual value for the period t, and F_t is the forecasted value for the period t. MAPE or mean average percentage error is the most widely accepted parameter of forecasting error, which mathematically means:

$$\text{MAPE} = \frac{\sum_{t=1}^{N} \left| \frac{\varepsilon_t}{A_t} \right|}{N}. \tag{13}$$

In this study, N has been valued 24 for daily electricity load forecasts. N should be valued 168 when we attempt weekly electricity forecasts. The graph shown below uses $N = 24$ as it predicts the day-ahead forecasts. Figures 6–9 represent the actual versus forecasted data for FA, FA + WT, FA + WT + GARCH, and ANN, respectively. Table 1 presents the actual versus forecasted data for all the techniques, namely FA, FA + Wavelet, ANN, and the proposed model (FA + WT + GARCH).

Conclusion

The hybrid model proposed in this paper is for short-term electricity load forecasting. The model is the aftermath of befitting coalition of FA, GARCH, and WT. While WT looks after the ill-behaved load series, FA captures the nonlinear fluctuations by virtue of stability–plasticity dilemma [27]. The attributes of FA renders the proposed hybrid method robustness and higher efficiency enabling forecasting meeting higher accuracy.

The model has also been compared with FA, FA + WT, and ANN. The results certify the efficacy of the proposed load forecasting hybrid model, as can be seen from Table 2.

Acknowledgments

The authors thank the New South Wales System Operator for providing hourly data of electricity load (http://www.asx.com.au/).

Conflict of Interest

None declared.

References

1. X., Zhang, Q. Wu, and J. Zhang. 2010. Crude oil price forecasting using fuzzy time series, 3rd ISKAM, 978-1-4244-8005-0, Pp. 213–216.
2. Georgilakis, P. S. 2006. Market clearing price forecasting in deregulated electricity market using adaptively trained neural networks, SETN 2006. Lecture notes in artificial intelligence LANI 3955. Springer, Berlin, Heidelberg. Pp. 56–66.
3. Gountis, V. P., and A. G. Bakirtzis. 2004. Bidding strategies for electricity producers in a competitive electricity marketplace. IEEE Trans. Power Syst. 19:356–365.
4. Gross, G., and F. D. Galiana. 1987. Short-term load forecasting. Proc. IEEE 75:1558–1573.
5. Binh, P. T. T., N. T. Hung, P. Q. Dung, and L.-H. Hee. 2012. Load forecasting based on wavelet transform and fuzzy logic. Power System Technology (POWERCON), 2012 IEEE International Conference on, pp. 1–6. doi: 10.1109/PowerCon.2012.6401281.

6. Alfares, H. K., and M. Nazeeruddin. 2002. Electric load forecasting: literature survey and classification of methods. Int. J. Syst. Sci. 33:23–34.
7. Hahn, H., S. Meyer-Nieberg, and S. Pickl. 2009. Electric load forecasting methods: tools for decision making. Eur. J. Operat. Res. 199:902–907.
8. Marin, F. J., and F. Sandoval. 1997. Short-term peak load forecasting: statistical methods versus artificial neural networks. Springer Lect. Notes Comput. Sci. 1240:1334–1343.
9. Sureban, M. S., and S. G. Ankaliki. 2015. Applications of artificial neural networks in various areas of power system: a review. J. Power Sys. Eng. 2:35–44.
10. Al Wadi, M. T. I. S., and M. Tahir Ismaul. 2011. Selecting wavelet transforms model in forecasting financial time series data based on ARIMA Model. Appl. Math. Sci. 5:315–326.
11. Chui, C. K. 1992. An introduction to wavelets. Academic, New York. Pp. 6–18.
12. Mallet, S. 1989. A theory for multiresolution signal decomposition – The wavelet representation. IEEE Trans. Pattern Anal. Mach. Intell. 11:674–693.
13. http://fourier.eng.hmc.edu/e161/lectures/wavelets/node7.html
14. Carpenter, G. A., and S. Grossberg. 1988. The ART of adaptive pattern recognition by a self-organizing neural network. Computer 21:77–88. doi:10.1109/2.33.
15. Carpenter, G., S. Grossberg, N. Markuzon, J. Reynolds, and D. Rosen. 1992. Fuzzy ARTMAP: aneural network architecture for incremental supervised learning of analog multidimensional maps. IEEE Trans. Neural Networks 3:698–713.
16. Lopes, M. L. M., C. R. Minussi, and A. D. P. Lotufo. 2005. Electric load forecasting using a Fuzzy ART and ARTMAP neural network. Appl. Soft. Comput. 5:235–244.
17. Christodoulou, C., and M. Georgiopoulos. 2000. Applications of neural networks in electromagnetics, 1st ed. Artech House, Norwood, MA.
18. Dagher, I., M. Georgiopoulos, G. Heileman, and G. Bebis. 1999. An ordering algorithm for pattern presentation in fuzzy ARTMAP that tends to improve generalization performance. IEEE Trans. Neural Netw. 10:768–778.
19. Zornetzer, S. F., J. L. Davis, and C. Lau, eds. 1990. An introduction to neural and electronic networks. Academic Press Professional Inc, San Diego, CA.
20. Antunes, J. F., de Souza Araujo N. V., and C. R. Miussi. 2013. Multinodal load forecasting using an ART-ARTMAP-Fuzzy neural network and PSO strategy. in PowerTech (POWERTECH), IEEE Grenoble.
21. Garcia, R. C., J. Contreras, M. van Akkeren, and J. B. C. Garcia. 2005. A GARCH forecasting model to predict day-ahead electricity prices. IEEE Trans. Power Syst. 20:867–874.
22. Mathworks 2005. GARCH Toolbox for use with MATLAB. Mathworks User's Guide Version 2.0. Pp. 1–272. Available at http://www.mathworks.com/access/helpdesk/help/pdf_doc/garch/garch.pdf.
23. Mandal, P., A. U. Haque, J. Meng, R. Martinez, and A. K. Srivastava. 2012. A hybrid intelligent algorithm for short-term energy price forecasting in the Ontario market. IEEE 978-1-4673-2729-9.
24. Tan, Z., J. Zhang, J. Wang, and J. Xu. 2010. Day-ahead electricity price forecasting using wavelet transform combined with ARIMA and GARCH models. Appl. Energy 87:3606–3610.
25. Mishra, S., A. Sharma, and G. Panda. 2011. Wind power forecasting model using complex wavelet theory, 978-1-4673-0136-7, IEEE.
26. Mandal, P., A. U. Haque, J. Meng, A. K. Srivastava, and R. Martinez. 2013. A novel hybrid approach using wavelet, firefly algorithm and fuzzy ARTMAP for day-ahead electricity price forecasting. IEEE Trans. Power Syst. 28:1041–1051.
27. Carpenter, G. A., S. Grossberg and D. B. Rosen, 1991. Fuzzy ART: fast stable learning and categorization of analog patterns by an adaptive resonance system. Orig. Res. Art. Neural Netw. 4:759–771.
28. Haque, A. U., P. Mandal, J. Meng, A. K. SrivastavaT.-L. Tseng, and T. Senjyu. 2012. A novel hybrid approach based on wavelet transform and fuzzy ARTMAP network for predicting wind farm power production. Ind. Appl. Soc. Ann. Meet. 1–8.

PERMISSIONS

All chapters in this book were first published in ESE, by Wiley Online Library; hereby published with permission under the Creative Commons Attribution License or equivalent. Every chapter published in this book has been scrutinized by our experts. Their significance has been extensively debated. The topics covered herein carry significant findings which will fuel the growth of the discipline. They may even be implemented as practical applications or may be referred to as a beginning point for another development.

The contributors of this book come from diverse backgrounds, making this book a truly international effort. This book will bring forth new frontiers with its revolutionizing research information and detailed analysis of the nascent developments around the world.

We would like to thank all the contributing authors for lending their expertise to make the book truly unique. They have played a crucial role in the development of this book. Without their invaluable contributions this book wouldn't have been possible. They have made vital efforts to compile up to date information on the varied aspects of this subject to make this book a valuable addition to the collection of many professionals and students.

This book was conceptualized with the vision of imparting up-to-date information and advanced data in this field. To ensure the same, a matchless editorial board was set up. Every individual on the board went through rigorous rounds of assessment to prove their worth. After which they invested a large part of their time researching and compiling the most relevant data for our readers.

The editorial board has been involved in producing this book since its inception. They have spent rigorous hours researching and exploring the diverse topics which have resulted in the successful publishing of this book. They have passed on their knowledge of decades through this book. To expedite this challenging task, the publisher supported the team at every step. A small team of assistant editors was also appointed to further simplify the editing procedure and attain best results for the readers.

Apart from the editorial board, the designing team has also invested a significant amount of their time in understanding the subject and creating the most relevant covers. They scrutinized every image to scout for the most suitable representation of the subject and create an appropriate cover for the book.

The publishing team has been an ardent support to the editorial, designing and production team. Their endless efforts to recruit the best for this project, has resulted in the accomplishment of this book. They are a veteran in the field of academics and their pool of knowledge is as vast as their experience in printing. Their expertise and guidance has proved useful at every step. Their uncompromising quality standards have made this book an exceptional effort. Their encouragement from time to time has been an inspiration for everyone.

The publisher and the editorial board hope that this book will prove to be a valuable piece of knowledge for researchers, students, practitioners and scholars across the globe.

LIST OF CONTRIBUTORS

Jinto M. Anthonykutty, Juha Linnekoski and Ali Harlin
VTT Technical Research Centre of Finland, Biologinkuja 7, Espoo, FI-02044 VTT, Finland

Juha Lehtonen
Department of Biotechnology and Chemical Technology, School of Chemical Technology, Aalto University, PO Box 16100, Aalto FI-00076, Finland

Viktorija Tomkute and Espen Olsen
Department of Mathematical Sciences and Technology, Norwegian University of Life Sciences (NMBU), PO Box 5003, Drøbakveien 31, NO-1432 Ås, Norway

Asbjørn Solheim
SINTEF Materials and Chemistry, Post Office Box 4760, Sluppen, NO-7465 Trondheim, Norway

Simas Sakirzanovas
Institute of Chemistry, Center for Physical Sciences and Technology, A. Gostauto 9, LT-01108 Vilnius, Lithuania

Sonil Nanda and Janusz A. Kozinski
Department of Earth and Space Science and Engineering, Lassonde School of Engineering, York University, Toronto, Ontario, Canada

Sivamohan N. Reddy
Department of Chemical Engineering, Indian Institute of Technology Roorkee, Roorkee, Uttarakhand, India

Sushanta K. Mitra
Department of Mechanical Engineering, Lassonde School of Engineering, York University, Toronto, Ontario, Canada

Thomas Parker
WA3RM AB, Lund, Sweden

Anders Kiessling
Swedish University of Agricultural Sciences, Uppsala, Sweden

Bruno Soria, Eric Gerritsen and Paul Lefillastre
Photovoltaic Modules Laboratory, CEA-INES, 50 avenue du Lac Léman, Le Bourget-du-Lac F-73375, France

Jean-Emmanuel Broquin
CNRS, IMEP-LACH, Grenoble Alpes University, Grenoble F-38000, France

Hai Yu and Paul Feron
CSIRO Energy, 10 Murray Dwyer Circuit, Mayfield West, New South Wales 2304, Australia

Kangkang Li
CSIRO Energy, 10 Murray Dwyer Circuit, Mayfield West, New South Wales 2304, Australia
Department of Chemical Engineering, Curtin University of Technology Australia, GPO Box U1987, Perth, Western Australia 6845, Australia

Moses Tade
Department of Chemical Engineering, Curtin University of Technology Australia, GPO Box U1987, Perth, Western Australia 6845, Australia

Ashleigh Cousins
CSIRO Energy, P.O. Box 883, Kenmore, Queensland 4069, Australia

Weiliang Luo and Jian Chen
Department of Chemical Engineering, Tsinghua University, Beijing 100084, China

Jaya Shankar Tumuluru
Biofuels and Renewable Energy Technology Department, Idaho National Laboratory, 750 University Blvd, Idaho Falls, Idaho 83415-3570

Nies Reininghaus, Clemens Feser, Benedikt Hanke, Martin Vehse and Carsten Agert
NEXT ENERGY EWE Research Centre for Energy Technology, University of Oldenburg, Carl-von-Ossietzky-Str. 15, 26129 Oldenburg, Germany

Francesco Meneguzzo and Lorenzo Albanese
Institute of Biometeorology, CNR, via Caproni 8, Firenze 50145, Italy

Mario Pagliaro and Rosaria Ciriminna
Institute for the Study of Nanostructured Materials, CNR, via U. La Malfa 153, Palermo 90146, Italy

Yoshihiro Nishikawa, Kiyoshi Imai and Keiji Miyao
KONICA MINOLTA, INC., 3-91 Daisen Nishi-machi, Sakai-ku, Sakai City, Osaka 590-8551, Japan

Satoshi Uchida
The University of Tokyo, 4-6-1 Komaba, Meguro-ku, Tokyo 153-8904, Japan

Daisuke Aoki, Hidenori Saito, Shinichi Magaino and Katsuhiko Takagi
Kanagawa Academy of Science and Technology, 3-2-1 Sakado, Takatsu-ku, Kawasaki-shi, Kanagawa 213-0012, Japan

Ahmadreza Ghahremani and Aly E. Fathy
Electrical Engineering Department, University of Tennessee Knoxville, Knoxville, Tennessee

Sunday O. Oyedepo
Mechanical Engineering Department, Covenant University, Ota, Nigeria

Richard O. Fagbenle
Mechanical Engineering Department, Obafemi Awolowo University, Ile Ife, Nigeria

Samuel S. Adefila
Chemical Engineering Department, Covenant University, Ota, Nigeria

Md. Mahbub Alam
Institute for Turbulence-Noise-Vibration Interaction and Control, Shenzhen Graduate School, Harbin Institute of Technology, Shenzhen, China

Lanre Olatomiwa
Power Electronics and Renewable Energy Research Laboratory (PEARL), University of Malaya, 50603 Kuala Lumpur, Malaysia
Department of Electrical Engineering, Faculty of Engineering, University of Malaya, 50603 Kuala Lumpur, Malaysia
Department of Electrical & Electronic Engineering, Federal University of Technology, PMB 65, Minna, Nigeria

Saad Mekhilef and A. S. N. Huda
Power Electronics and Renewable Energy Research Laboratory (PEARL), University of Malaya, 50603 Kuala Lumpur, Malaysia
Department of Electrical Engineering, Faculty of Engineering, University of Malaya, 50603 Kuala Lumpur, Malaysia

Kamilu Sanusi
Department of Electrical Engineering, Waziri Umaru Federal Polytechnics, Birnin-Kabbi, Nigeria

Francesco Meneguzzo and Lorenzo Albanese
Institute of Biometeorology, CNR, via Caproni 8, 50145 Firenze, Italy

Rosaria Ciriminna and Mario Pagliaro
Institute for the Study of Nanostructured Materials, CNR, via U. La Malfa 153, 90146 Palermo, Italy

Michael H. Stephenson
British Geological Survey, Keyworth, Nottingham NG12 5GG, United Kingdom

Satyavathi Bankupalli
Chemical Engineering Division, CSIR - Indian Institute of Chemical Technology, Hyderabad 500007, India

Radhakumari Muktham
Chemical Engineering Division, CSIR - Indian Institute of Chemical Technology, Hyderabad 500007, India
School of Applied Sciences, Royal Melbourne Institute of Technology, Melbourne 3083, Australia

Andrew S. Ball and Suresh K. Bhargava
School of Applied Sciences, Royal Melbourne Institute of Technology, Melbourne 3083, Australia

Swati Takiyar and K. G. Upadhyay
M.M.M. University of Technology, Gorakhpur, Uttar Pradesh, India

Vivek Singh
National Hydroelectric Power Corporation Ltd., Faridabad, Haryana, India

Index